T0135311

Advances in Science, Technology & Innovation

IEREK Interdisciplinary Series for Sustainable Development

Advances in Science, Technology & Innovation (ASTI) is a series of peer-reviewed books based on important emerging research that redefines the current disciplinary boundaries in science, technology and innovation (STI) in order to develop integrated concepts for sustainable development. It not only discusses the progress made towards securing more resources, allocating smarter solutions, and rebalancing the relationship between nature and people, but also provides in-depth insights from comprehensive research that addresses the **17 sustainable development goals (SDGs)** as set out by the UN for 2030.

The series draws on the best research papers from various IEREK and other international conferences to promote the creation and development of viable solutions for a **sustainable future and a positive societal** transformation with the help of integrated and innovative science-based approaches. Including interdisciplinary contributions, it presents innovative approaches and highlights how they can best support both economic and sustainable development, through better use of data, more effective institutions, and global, local and individual action, for the welfare of all societies.

The series particularly features conceptual and empirical contributions from various interrelated fields of science, technology and innovation, with an emphasis on digital transformation, that focus on providing practical solutions to **ensure food, water and energy security to achieve the SDGs.** It also presents new case studies offering concrete examples of how to resolve sustainable urbanization and environmental issues in different regions of the world.

The series is intended for professionals in research and teaching, consultancies and industry, and government and international organizations. Published in collaboration with IEREK, the Springer ASTI series will acquaint readers with essential new studies in STI for sustainable development.

ASTI series has now been accepted for Scopus (September 2020). All content published in this series will start appearing on the Scopus site in early 2021.

More information about this series at http://www.springer.com/series/15883

Zhien Zhang • Wenxiang Zhang •
Mohamed Mehdi Chehimi

Editors

Membrane Technology Enhancement for Environmental Protection and Sustainable Industrial Growth

 Springer

Editors
Zhien Zhang
William G. Lowrie Department of Chemical
and Biomolecular Engineering
The Ohio State University
Columbus, OH, USA

Wenxiang Zhang
Department of Civil and Environmental
Engineering, Faculty of Science and Technology
University of Macau
Macau SAR, China

Mohamed Mehdi Chehimi
ICMPE, CNRS
Université Paris-Est, Marne-la-Vallee
Créteil, France

ISSN 2522-8714   ISSN 2522-8722   (electronic)
Advances in Science, Technology & Innovation
IEREK Interdisciplinary Series for Sustainable Development
ISBN 978-3-030-41297-5   ISBN 978-3-030-41295-1   (eBook)
https://doi.org/10.1007/978-3-030-41295-1

This Springer imprint is published by the registered company Springer Nature Switzerland AG
The registered company address is: Gewerbestrasse 11, 6330 Cham, Switzerland

Contents

Forward Osmosis for Sustainable Industrial Growth

Mónica Rodríguez-Galán, Francisco M. Baena-Moreno,
Fátima Arroyo-Torralvo, and Luis F. Vilches-Arenas

Abstract

In this work, a comprehensive discussion of forward osmosis membrane technology is presented. Forward osmosis is an interesting and promising system to concentrate multiple kind of solutions in different industrial areas as an alternative solution to classical water evaporation. Therefore, the number of publications and works related to this topic has considerably increased in the last years. Several aspects of forward osmosis have been discussed such as membrane fouling, concentration polarization phenomena, the different available draw solutions and the industrial applications in which forward osmosis has been applied. Cellulose triacetate membranes and thin-film composite membranes are the most employed nowadays. Chemical industry, desalination of drinking water, food industry and pharmaceutical industry are analyzed in deep since these are the most studied areas for forward osmosis application. Herein, the potential of forward osmosis for a sustainable industrial growth is widely proved in every sense.

Keywords

Forward osmosis • Membrane • Review • Draw solutions • Feed solutions • Renewable energy • Fouling phenomena • Industrial applications

M. Rodríguez-Galán (✉) · F. M. Baena-Moreno (✉) ·
F. Arroyo-Torralvo · L. F. Vilches-Arenas
Chemical and Environmental Engineering Department,
Technical School of Engineering, University of Seville,
C/Camino de los Descubrimientos s/n, 41092 Sevilla, Spain
e-mail: mrgmonica@us.es

F. M. Baena-Moreno
e-mail: fbaena2@us.es

1 Introduction

The continuous innovation of the different separation processes is the key to design more efficient and robust systems, as well as to improve the environment which is characterized for higher and higher greenhouse gas levels. Together with carbon capture and utilization technologies, the invest on new renewable technologies which imply less pollution is taking an important part of the research sources (Baena-moreno et al. 2018). Among the novel technologies for separation of gas/liquid substances, membranes have achieved a high level of recognition by the experts in this area (Zhang et al. 2018; Baena-Moreno et al. 2019a, b). Membrane separation techniques are technologies frequently employed for industrial and environmental applications. Indeed, high-level activity from researchers can be demonstrated by the elevated number of research papers in this field (Baena-Moreno et al. 2019a, b; Zhang 2016; Zhang et al. 2014; Scholes et al. 2013; Hajilary et al. 2018; Brunetti et al. 2010; Bell et al. 2017).

In addition to industrial and environmental applications, membrane separation processes are widely applied in wastewater treatment, desalination and petrochemical (Brunetti et al. 2016; Farahani et al. 2018; Venzke et al. 2018). For liquids involved-applications membrane filtration, reverse osmosis, forward osmosis and membrane distillation have stood out (Chabanon et al. 2016; Li et al. 2016; Eykens et al. 2017). Among these technologies, forward osmosis (FO) has been intensively studied due to its numeral advantages such as high water recovery, low energy requirements, low fouling phenomena and ease of cleaning (Wang et al. 2018; Cath et al. 2006). Moreover, FO has the advantages of needing less chemicals than other techniques and no applied pressure is needed. Furthermore, its good combination with membrane distillation for draw solution regeneration after FO stage makes this technology suitable for many applications in which low energy requirements are required (Chekli et al. 2016). Nevertheless, there are some aspects yet to be solved such as concentration polarization and reverse salt flux. These aspects will be further discussed in the following sections of this chapter.

FO application in chemical industry usually deals with the originated wastes by manufacturing process. FO has been also proposed for many food industry areas such as dairy industry, juice processing, tea extracts and olive mill wastewater. Furthermore, pharmaceutical industry has also pointed out for FO utilization to treat pharmaceutical liquids and to recover the organic solvents from pharmaceutical active ingredients (Haupt and Lerch 2018).

As a consequence of above-mentioned advantages, an increase of publications has been observed during the last years as represented in Fig. 1. This increase reveals not only the range of applications in which FO can be used, but also the high interest showed by the scientific community in this technology.

Osmosis is defined as the net flux of water from a high concentrated solution to another low concentrated solution through a selective membrane. The semipermeable membrane located between both solutions with different osmotic pressures allows the flux of water and retains the solute, molecules or ions. In this way, water passes from the solution of less osmotic pressure to the solution with higher osmotic pressure, known as draw solution (DS). DS attracts water thanks to the potential difference between both solutions, which causes the flow of water through the membrane (Cath et al. 2006; Zhao et al. 2012).

As mentioned before, a standout characteristic of FO is its good combination with other membrane-based processes. In this sense, hybrid FO processes have turned out in a topic of great interest and discussion for researches in this field. The main reason is that water recovery produces higher dilution of DS which should be regenerated to keep the process economically affordable. This regeneration process cannot be done by stand-alone FO process (Wang et al. 2018). Nevertheless, there is a possibility of coupling FO processes with another separation process such as membrane distillation to regenerate the diluted DS and recover high-purity water. Although many studies have been carried out by experts in this hybrid-membrane-based processes, further research is needed to optimize the overall process for an overall balance (Cath et al. 2006; Zhao et al. 2012).

As it has been explained in the introduction section, the potential of FO processes is noticeable. Thus, a high-quality review is needed for those focusing their efforts in this area of knowledge. For this reason, in this book readers can find an overall dense information about FO processes. To organize the information into appropriately differentiated sections, the following points will be covered: FO technology principles, membrane typically employed in FO and FO applications in industrial areas. To summarize and finalize this document, a discussion section will be added in which not only the main points of this manuscript will be highlighted but also future perspectives of this topic will be exposed.

2 Forward Osmosis Technology Principles

2.1 Osmotic Phenomena

As previously defined, osmotic phenomena consist of the transport of water across a selectively permeable membrane (Wang et al. 2018). This water flux is due to the difference of osmotic pressure between the solutions situated at both sides of the membrane, identifying as draw solution (DS) the one with a high concentration, and "feed solution" the one diluted. This causes the concentrated stream to dilute and hence to decrease its osmotic pressure, while the initially dilute feed solution is concentrated due to the water lost, which increases its osmotic pressure. Therefore, FO takes advantage of this difference and uses it as driving force for water transport through the membrane. A representative scheme of FO process is shown in Fig. 2.

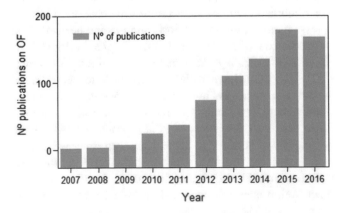

Fig. 1 Number of publications which include FO processes from 2010 to 2016. Modified after Wang et al. (2018)

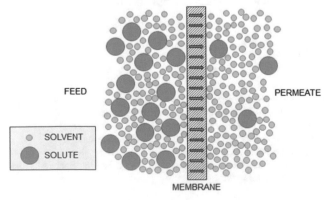

Fig. 2 Representative scheme of FO process

Once both osmotic pressures are equalized, equilibrium is reached in which the water flux is zero. The water flux is defined by the general equation (Eq. 1) which describes water transport in osmosis processes (Cath et al. 2006):

$$J_{\mathrm{w}} = A(\sigma \cdot \Delta\pi - \Delta P) \qquad (1)$$

J_{w} is the water flux; A is a constant related to the pure water permeability of the membrane; σ is the reflection coefficient; $\Delta\pi$ is the osmotic pressure differential; and ΔP is the applied pressure in those applications needed (reverse osmosis and pressure-retarded osmosis).

Compared to conventional water separation technologies, FO includes the following advantages (Zhao et al. 2012; Cath et al. 2006; Chung et al. 2012):

- Low energy consumption
- Possibility of treating two problematic effluents in the same equipment
- High-purity water recovery
- Treatment of liquids which are highly difficult to treat using other membrane processes (due to their fouling tendency).

2.2 Draw Solutions

The DS selected for each application is the key for obtaining a valuable result. Indeed in FO processes, draw solutions are considered a fundamental choice for a successful performance. Thus, the selection of an appropriate DS has been the purpose of some studies (Chekli et al. 2012; Achilli et al. 2010; Ge et al. 2013). The main criterion for selecting an adequate draw solution is the osmotic pressure. This last one has to be higher than the feed solution to produce the highest water flux.

The second more important parameter to take into account for selecting a suitable FO draw solution is the availability for re-concentrating after FO stage. Re-utilizing the DS is necessary to keep the overall economic performance of the process (Ge et al. 2013; Chekli et al. 2012). Furthermore, the reverse flux of the draw solution solute through the membrane to the feed solution should be considered (Shaffer et al. 2015; Lutchmiah et al. 2014).

For these reasons, several draw solutions have been tested by experts in this area (Shaffer et al. 2015; Chekli et al. 2012). Many researchers have put their efforts on finding novel draw solutions, and as a result multitude of compounds have been proposed in the literature. These draw solutions present some advantages as the same time that there are aspects which need to overcome (Ge et al. 2013). Table 1 presents the most typically employed draw solutions which can serve as a guide for those looking for the correct draw solution.

As can be seen in Table 1, $MgCl_2$ and $CaCl_2$ present the highest osmotic pressures. Nevertheless, their flux is not the best ones showed in the literature by some researches (Ge et al. 2013; Achilli et al. 2010).

Nevertheless, many studies have employed NaCl in a wide range of applications since saline water is abundant on earth and easy to obtain. Moreover, NaCl can be easily re-concentrated with RO processes or MD with few energy consumptions. Additionally, NaCl presents high water solubility which is a clear advantage since there is no need of using organic solvents. Therefore, food production and wastewater treatment have applied NaCl as draw solution at industrial scale (Akther et al. 2015).

Not as typical as the employment of inorganic salts as previously exposed but during the last years, some studies have proposed ethanol, sucrose, glucose and fructose as organic draw solutions with acceptable results. Experimental water flux obtained range from 0.24 to 7.5 LMH for these organic solvents and their solubility is higher due to hydrogen bonding.

2.3 Concentration Polarization Phenomenon

An essential phenomenon in pressure osmotic-driven processes is the concentration polarization phenomena, which highly influence in water flux across the membrane area. This phenomenon is carried out by the increase of the feed solution concentration over the membrane surface (Kim et al. 2010; McCutcheon and Elimelech 2006). Figure 3 represents a schematic explanation of this process.

Although this phenomenon takes place in many membrane processes such as nanofiltration, membrane distillation and reverse osmosis, the effect is harder in FO processes (Chen et al. 2004). Due to the concentration polarization, the osmotic pressure gradient decreases compared to the normal operation, which negatively affects water flux (Kim et al. 2010). Therefore, permeated water is lower than expected and bigger membrane areas are needed to keep constant the desired flux (van den Berg and Smolders 1992).

Typically, concentration polarization has been split into two types: external concentration polarization (ECP) and internal concentration polarization (ICP). The first one occurs very frequently in both FO and RO processes, and it takes place into the surface of the active layer due to the difference in the concentration of the solution respecting to the bulk solution. ECP phenomenon has decreasing effects on the osmotic gradient, and hence it caused the inhibition of the water flux across the membrane. Nevertheless, the impact of ICP during FO operation is much higher than

Table 1 Main draw solutions usually employed in FO processes and its characteristics as draw solution

Draw solute	Typical concentration range (M)	Osmotic pressure range (atm)	Molecular weight (g/mol)	Water flux range (LMH)	Approximate unitary cost ($/kg)	References
NaCl	0.5–5	25–250	58.5	5–45	10–15	Phillip et al. (2010), Chou et al. (2010)
KCl	0.5–5	20–230	74.6	3–40	35–40	Achilli et al. (2010), Tan and Ng (2010)
MgCl$_2$	1–5	100–1150	95.2	8–30	25–30	Tan and Ng (2010), Cornelissen et al. (2011)
CaCl$_2$	1–5	100–1100	111	8–30	35–40	Shu et al. (2016), Tang et al. (2014)
NH$_4$HCO$_3$	0.5–5	20–100	79.1	5–25	45–50	Bevacqua et al. (2017), McCutcheon et al. (2005)

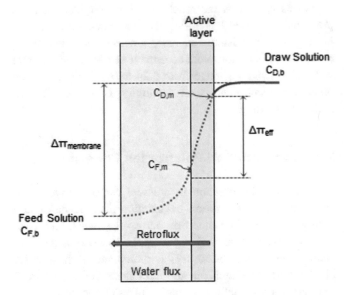

Fig. 3 External concentration polarization process

ECP. Experimental studies have verified that the diminution of water flux in FO is mainly caused by ICP. It occurs within the porous later of the membrane. Even thought, in FO this phenomenon is reduced comparing with other membrane-based processes (Akther et al. 2015).

2.4 Fouling in Forward Osmosis

FO has lower irreversible fouling properties than pressure-driven membrane processes. This statement is due to the lack of applied hydraulic pressure in comparison with reverse osmosis, for example. Many types of fouling can be distinguished in membranes: organic fouling, colloidal particle fouling and cake-enhanced osmotic pressure (Lee et al. 2010).

Organic fouling is typically produced by alginate and humic acids. The compactness and thickness of the fouling layer are the main factors which influence in the behavior showed by the membrane during organic fouling. This structure is affected by chemical and physical conditions of the overall system (Mi and Elimelech 2010).

Colloidal particle fouling was studied in deep by Lee et al. (2010). In their study, they verified the relationship between particle size and flux decline mechanisms by conduction fouling runs with silica particles of about 300 nm. They concluded that there is a greater salt buildup near the membrane surface due to salt intrusion from the draw solution.

Regarding the cake-enhanced osmotic pressure, in the same study than before, Lee et al. (2010) compared the flux decline curves for RO and for FO and concluded that the flux decline in FO is much severer than in RO (Lee et al. 2010).

Even though fouling is a significant problem for RO and FO membranes, the last ones can be cleaned by some techniques. The first and most traditional one is the osmotic backwashing for organic and inorganic fouling and decreases the amount of chemical for cleaning. Moreover, electricity can be used for fouling removal. The application of electrical current for cleaning FO membranes resulted in completely restoring the water flux capacity of the membrane. Nevertheless, membranes have to be clean diary by this method. Other methods such as physical cleaning have proved not to be as effective as others (Akther et al. 2015).

3 Forward Osmosis Membranes

Traditionally, asymmetric porous membranes have been employed for FO applications. Indeed nowadays there is still not better layout for FO processes, which makes asymmetric porous membranes the best candidate for mostly applications. In this kind of membranes, both the structure and transport properties are subjected to changes across the membrane thickness (Zhao et al. 2012; Wang et al. 2018; Cath et al. 2006).

Typically, asymmetric membranes are composed by a dense layer and a support layer. Dense layer thickness is about 0.1–1 μm, whereas the support layer is highly porous and 100–200 μm thick. The separation effectiveness is influenced by the chemical structure of the dense layer, the size of the pores (usually between 0.4 and 1 nm), the pore structure and the thickness of the support layer (Haupt and Lerch 2018).

The desired characteristics of membranes for forward osmosis applications are high density of the active layer for high solute rejection and minimum porosity of the support layer which supposes low concentration polarization (Cath et al. 2006). A typical membrane structure can be seen in Fig. 4.

The first commercial FO membrane was made of cellulose triacetate (CTA), and it was developed by hydration technologies (Wang et al. 2018). From this moment, a high number of CTA-based forward osmosis studies have been carried out by several researchers (Jin et al. 2012; Bell et al. 2017; Zuo et al. 2017; Wang et al. 2016; Bensaadi et al. 2016). CTA membranes are obtained by phase inversion through immersing casted or spun polymer dope into coagulants. Thus, asymmetric membrane structure featured with a skin layer integrally supported by a porous substrate is obtained as product.

CTA membranes show a low fouling propensity since their hydrophilic nature allows the water flux to pass without keeping big molecules trapped. Nevertheless, this characteristic also provokes a higher salt flux from the draw solution to the feed solution, which is equal to lower rejection to salt than other membranes. For this reason, many efforts have been put into reducing the salt flux in CTA membranes. Some of the improvements proposed are exploring new materials and additives, tuning fabrication conditions and introducing new additives. However, not remarkable improvements were achieved and the efforts were focalized to thin-film composite (TFC) membranes (Haupt and Lerch 2018).

TFC membranes are the most extended commercial type of forward osmosis membranes because the characteristics previously explained are achieved by these membranes. In this kind of membranes, the porous layer is an integrally skinned membrane which resulted from a non-solvent-induced phase separation process. TFC membranes are promising in terms of both manufacturing cost and water flux. The polymer needed for CTA membrane production is among 15–18%, whereas for TFC membranes this range is about 8–11% (Haupt and Lerch 2018; Wang et al. 2016).

TFC membranes are fabricated mainly by HTI, Porifera and Oasys Water and typically are embedded in hollow fiber configurations. TFC membrane drawback is the moderate water flux in comparison with CTA membranes, which could be overcome by increasing the draw solution concentration (Ismail et al. 2015). For example, in case of using NaCl as draw solution, water flux differences between TFC and CTA are corrected by increasing NaCl concentration from 0.5 to 1 M (Alsvik and Hägg 2013; Lau et al. 2012).

Nowadays, studies for TFC FO membrane improvements focused on making the support later more interconnected and porous as well as thinner. Moreover, reducing the tortuosity is another factor which needs to be optimized. Furthermore, fouling problems are under strict evaluation continuously in new applications for enhancing the overall performance (Tarboush et al. 2008; Alsvik and Hägg 2013).

4 Forward Osmosis Applications in Industrial Areas

Traditionally, forward osmosis has been employed for water desalination. Nevertheless, its use can be engineered and adapted for many other industrial processes. Nowadays, the benefits previously exposed presented by forward osmosis application have benefited its implementation in many sectors such as the following ones (Mi and Elimelech 2010; Haupt and Lerch 2018):

- Chemical industry
- Desalination of drinking water
- Food industry
- Pharmaceutical industry
- Textile industry
- Coal processing industry
- Electronic industry
- Heavy metal industry.

Therefore, many studies have been published in different branches and this number has been increased considerably (Haupt and Lerch 2018). To further knowledge of the different applications, in this section we have included four points to give the reader a wide comprehension of the

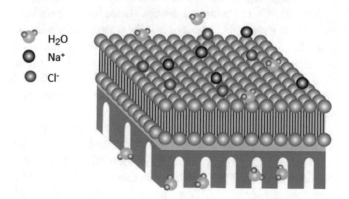

H₂O
Na⁺
Cl⁻

Fig. 4 Membrane structure

application of this topic to the different industries. These four points embrace the four main industries in which forward osmosis applications have a major impact—chemical industry, desalination of drinking water, food industry and pharmaceutical industry.

4.1 Forward Osmosis in Chemical Industry

Forward osmosis in chemical industry refers usually to the wastewater treatment originated from different industries. Many references have reported the implementation of FO with different effluents, and some of them are explained in this section (Soler-Cabezas et al. 2018; Law and Mohammad 2018; Cho et al. 2012; Kalafatakis et al. 2017; Shibuya et al. 2017).

Soler-Cabezas et al. (2018) corroborated the recovery of nitrogen and phosphorus coming from the anaerobically digested sludge concentrate through FO membranes. They tested two draw solutions: a residual effluent from ammonia elimination stage and brine from seawater. Moreover, they tested two different FO membranes: CTA and Aquaporin-based membranes. They obtained a high nitrogen concentration with both membranes and draw solutions. They concluded that the employment of ammonia absorption effluent enhanced the nitrogen concentration.

Law and Mohammad (2018) investigated the role of the FS pH of succinic acid concentration by FO as shown in Fig. 5. They employed CTA membranes and examined both the water flux and the reverse solute flux, using seawater as draw solution. The results revealed that the pH affects the FO performances. A strong effect was pointed out on the succinate rejection for which nearly 100% rejections were achieved at pH above its pKa = 2 value. With real seawater as the draw solution, moderate water fluxes (<4 L m^{-2} h^{-1}) were observed. They concluded that is possible to obtain considerable water fluxes by swinging pH value.

Cho et al. (2012) focused on organic acid separation and dewatering processes using nanofiltration and FO processes.

The organic acids are obtained during biomass decomposition under an anaerobic fermentation process, and their accumulation hinders the microbial metabolism in the fermentation broths. Therefore, the authors of this investigation proposed a novel research to remove organic acids through a combination of nanofiltration and forward osmosis membranes. Using nanofiltration membranes, aqueous organic acids can be selectively separated from pretreated fermentation feed solutions while other organics and many salts can be rejected using these processes by varying pH conditions in the feed. Finally, a low-energy-consuming forward osmosis process was applied for dewatering in the aqueous organic acid solutions to concentrate organic acid.

Kalafatakis et al. (2017) proposed crude glycerol and enzymatically pretreated wheat straw as draw solution in FO application against RO. This DS is generated as second-generation product of bio-refineries. They applied Aquaporin Inside™ TFC membranes, crude glycerol and wheat straw hydrolysate and demonstrated water fluxes up to 10.5 L/m^2/h and 5.37 L/m^2/h, respectively. Furthermore, they concluded with an economic analysis of FO coupled with bioprocessing and proved the reduction cost in the final product.

Shibuya et al. (2017) worked in a membrane process by combining nanofiltration and FO to concentrate sugar with the aim of bio-ethanol product from the liquid fraction of rice straw. They found that the commercial NF membrane known as "ESNA3" was more adequate for removal of fermentation inhibitors (such as acetic acid) than the FO membrane, whereas the commercial FO membrane "TFC-ES" was more adequate for concentration of the sugars than the NF membrane. As they found this, they proposed the next process for the liquid fraction: (NF (+H$_2$O)) → enzymatic hydrolysis → FO concentration as shown in Fig. 6.

4.2 Desalination of Drinking Water

Even though big efforts have been done to carry this application of FO to industrial levels, there are researchers focused on developing new systems for water desalination in order to optimize the global performance. Herein, some of the most relevant works are briefly summarized (Blandin et al. 2016; Valladares Linares et al. 2016; Fan et al. 2016; Zhao et al. 2016). Figure 7 shows a general scheme for desalination of drinking water.

First, Blandin et al. (2016) dedicated a very interesting review to describe the state of the art of different systems for combining water reuse and desalination through FO-RO hybrid systems. FO-RO combination can be an alternative to new desalination facilities or to implementation of stand-alone water reuse schemes. They exposed that FO-RO

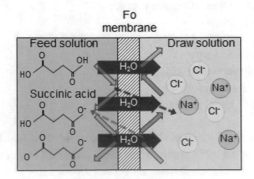

Fig. 5 Succinic acid concentration by forward osmosis. Modified after Law and Mohammad (2018)

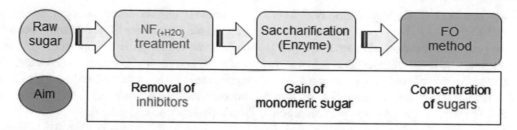

Fig. 6 Flowchart for the nanofiltration forward osmosis system showed by Shibuya et al. (2017). Modified after Shibuya et al. (2017)

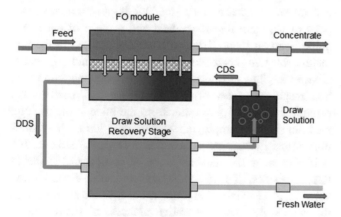

Fig. 7 General scheme for drinking water desalination

system has to overcome the technical limitation such as low FO permeation flux to become economically attractive. They found as recent developments (i.e., high-performance FO membranes and pressure-assisted osmosis) improve water flux. However, water flux improvement is associated with drawbacks, such as increased fouling behavior, lower rejection of trace organic compounds and limitation in FO membrane mechanical resistance, which need to be better considered. They concluded that further work is required regarding upscaling to apprehend full-scale challenges in terms of mass transfer limitation, pressure drop, fouling and cleaning strategies on a module scale.

To point out the importance of the economic analyses of forward osmosis such as cheap product like water, Valladares Linares et al. (2016) presented a detailed economic analysis on capital and operational expenses (CAPEX and OPEX) for many systems: (i) a hybrid forward osmosis–low-pressure reverse osmosis (FO-LPRO) process, (ii) a conventional seawater reverse osmosis (SWRO) desalination process and (iii) a membrane bioreactor–reverse osmosis–advanced oxidation process (MBR-RO-AOP) for wastewater treatment and reuse. They took as the main parameters for the life cycle costs the water quality, production capacity, energy consumption, materials, maintenance, operation, RO and FO module costs and chemicals. As main results, they obtained that compared to SWRO, the FO-LPRO systems have a 21% higher CAPEX and a 56% lower OPEX due to

savings in energy consumption and fouling control. In terms of the total water cost per cubic meter of water produced, the hybrid FO-LPRO desalination system has a 16% cost reduction compared to the benchmark for desalination, mainly SWRO. Compared to the MBR-RO-AOP, the FO-LPRO systems have a 7% lower CAPEX and 9% higher OPEX, resulting in no significant cost reduction per m^3 produced by FO-LPRO.

Fan et al. (2016) develop a novel simple method to prepare thermosensitive poly(ionic liquid) (PIL) hydrogels as smart draw agents for FO desalination. In their study, they ensure that these polyelectrolyte hydrogels produce a high osmotic pressure that can draw a large amount of desalinated water from brackish water into the hydrogel through a semipermeable membrane. Due to the thermosensitive nature of the PIL hydrogels, the liquid water can be easily recovered and the hydrogel can be reused by temperature cycling. More advantages were described such as the non-toxicity, and negligible leakage of the draw agents makes recovered liquid water suitable for drinking.

More draw solutes applicable to seawater desalination in FO processes with high level of success were reviewed by Zhao et al. (2016). In their work, they identified the main characteristics for draw solutions such as high osmotic pressure, low reverse flux and easy regeneration mechanism, as explained before in this manuscript. They reviewed a special kind of draw solutes which were created to satisfy the characteristics pointed out. The draw solutes were developed by them and known as "multi-functional FO draw solutes." Mainly, these solutions include Na^+-functionalized carbon quantum dots, thermoresponsive copolymers, hydrophilic magnetic nanoparticles and thermoresponsive magnetic nanoparticles.

4.3 Food Industry

Food industry is subjected to changes which makes the process more viable economically but with the drawback of having high-quality specifications regarding human health. In this sense, the use of FO as clean technology has been demonstrated. In this point, some novel papers can be found

(Rastogi 2016; Aydiner et al. 2016; Raghavarao et al. 2005; Petrotos et al. 2010).

Rastogi (2016) presented a high-quality review which is possible to obtain high quantity of relevance information. They exposed the necessity of their paper because of its novel features, which include the concentration of liquid foods at ambient temperature and pressure without significant fouling of membrane. Therefore, they exposed that the characteristics explained made the technology commercially attractive. They explain that the asymmetric membrane used for FO poses newer challenges to account both external and internal concentration polarizations leading to significant reduction in water flux. They concluded that FO is an emerging technology for water recovery from liquids due to its novel features, which include no use of solvent for extraction, low energy consumption, higher retention of thermolabile components and attainment of higher concentration. Furthermore, recovery and regeneration of draw solution is a future challenge to achieve by the scientific community to scale up forward osmosis in food industry.

Dairy wastewater management is a potential area in which many improvements can be gotten by implementing membrane processes. In this sense, Aydiner et al. (2016) presented an hierarchical prioritization of innovative treatment systems which consist of multiple criteria decision making. In their study, four innovative systems—which include membrane distillation, FO, nanofiltration and RO—are compared to a traditional system in terms of hierarchically prioritized for environmentally benign treatment. Despite the existence of both greatest importance of energy consumption and higher energy requirements of novel solutions, major prominent features to be provided by real-scale applications of innovative systems were determined as more valuable multiple outcomes in environmental protection and economic profit. They concluded that sustainable dairy wastewater management in a more desirable manner than now could be accomplished by means of technically highly efficient, economically cost-effective and environmentally eco-innovative achievements by membrane technology employment.

With liquid foods, we refer to fruit juices and natural colors. These compounds have gained importance during the last years in food industry due to the high number of beneficial components for human health. Fruit juices and natural colors are extracted with a low solid content which inevitably suppose a high water consumption. Therefore a concentration stage previous its commercialization is obligatory. For this reason, Raghavarao et al. (2005) proposed an interesting review of the different available techniques for concentrating liquid foods. Concentrating liquid foods provides a reduction in transport, packaging and storage costs. Both fruit and natural juices are sensitive to temperature changes hence process which work at ambient temperature

are needed. In this sense, in their review these authors exposed the importance of membrane processes since one of its best characteristics is the operation at around 25–30 °C. They explain how important is concentration by using membranes such as ultrafiltration, nanofiltration, reverse osmosis and forward osmosis for the concentration of liquid foods (fruit juices and colors).

In the case of Petrotos et al. (2010), an investigation to concentrate tomato juice was developed by FO application. They designed a novel membrane module which consists of a stainless steel flat module in which it can distinguish two parts of a square flange screwed and a piece of flat membrane between them. The configuration resulted in the formation of two chambers of special morphology which allowed the flow of tomato juice and osmotic medium in the two respective sides of the membrane and enabled the osmotic transfer of pure water from the juice to the osmotic medium side. By employing sodium chloride as draw solution, tomato juice was concentrated from 5.5°Brix to 16° Brix, achieving the standard levels of commercial available tomato sauces. To re-concentrate the post-diluted draw solution, electrodialysis is proposed as a viable alternative to the commonly used evaporative process. As main conclusion, these authors affirmed that the use of NaCl brines as osmotic media in a combined FO–electrodialysis process operated at ambient temperature and low pressure allows to produce tomato concentrates of commercial interest (Fig. 8).

4.4 Pharmaceutical Industry

A high value-added industry with high economic benefits when applying novel low-energy technologies is pharmaceutical industry. In this sense, FO along with other membrane processes is suitable technologies to apply in this industry and many researchers have developed astounding applications (Yang et al. 2009; Wang et al. 2011; Xie et al. 2012; Jin et al. 2012; Cui and Chung 2018).

First, Yang et al. (2009) started an important way in the field of pharmaceutical by demonstrating the prospect of dual-layer polybenzimidazole–polyethersulfone/polyvinylpyrrolidone hollow fiber nanofiltration membranes. They built the dual-layer hollow fiber membrane via extrusion technology that has an ultra-thin selective skin around 10 μm, fully open-cell water channels underneath and a microporous sponge-like support structure. Afterward, experimental results show that the newly developed dual-layer hollow fiber nanofiltration membrane can achieve a high throughput for lysozyme enrichment and less protein fouling when using it as a FO membrane. In addition, the high divalent salt rejection toward Mg^{2+} at around 90% of this dual-layer membrane ensures the enriched lysozyme product with high purity and without change and denaturing.

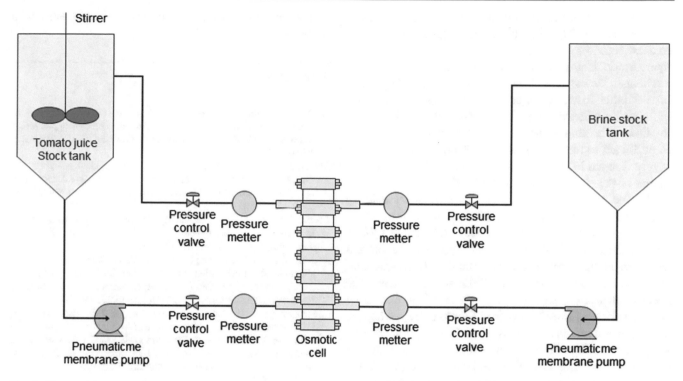

Fig. 8 Tomato juice concentration by forward osmosis membrane scheme. Modified after Petrotos et al. (2010)

Concentrating protein solution in a more energy-efficient way is an interesting path in which forward osmosis can take place. Wang et al. (2011) proposed for the first time an integrated forward osmosis–membrane distillation combination to concentrate protein solutions, specifically a bovine serum albumin (BSA) solution. A hydrophilic polybenzimidazole (PBI) nanofiltration hollow fiber membrane and a hydrophobic polyvinylidene fluoride–polytetrafluoroethylene hollow fiber membrane were fabricated and employed in the FO and MD processes, respectively. As draw solution, NaCl was chosen to dehydrate proteins in FO, while distillate water is a by-product during the re-concentration of diluted NaCl draw solution in MD. To determine suitable operating conditions for the hybrid system, independent characterizations were carried out for both FO and MD processes using different NaCl concentrations as draw solutes in FO and different feed temperatures in MD. They found that the integrated system is stable in continuous operation when the dehydration rate across the FO membrane is the same as the water vapor rate across the MD membrane.

Another point to highlight is the importance of membranes in pharmaceutical industry is the capability to reject pharmaceuticals and to obtain high-quality water. Jin et al. (2012) proposed a study in which the rejection of four pharmaceutical compounds, carbamazepine, diclofenac, ibuprofen and naproxen, by FO was investigated. For the first time, the rejection efficiency of the pharmaceutical compounds was compared between commercial CTA-based membranes and

TFC polyamide-based membranes. The rejection behavior was related to membrane interfacial properties, physico-chemical characteristics of the pharmaceutical molecules and feed solution pH. TFC polyamide membranes exhibited excellent overall performance, with high water flux, excellent pH stability and great rejection of all pharmaceuticals investigated (>94%). For commercial CTA-based FO membranes, hydrophobic interaction between the compounds and membranes exhibited strong influence on their rejection under acidic conditions. The pharmaceutical rejection was well correlated to their hydrophobicity. Under alkaline conditions, both electrostatic repulsion and size exclusion contributed to the removal of deprotonated molecules. The pharmaceutical rejection by CTA-HW membrane at pH = 8 followed the order: diclofenac (99%) > carbamazepine (95%) > ibuprofen (93%) $\approx$ naproxen (93%). The main conclusion of their paper was the meaning of the results for FO membrane synthesis, modification and their application in water purification.

Further efforts in rejection of pharmaceuticals were done by Xie et al. (2012). They study the effects of feed solution pH and membrane orientation on water flux, and the rejection of carbamazepine and sulfamethoxazole was investigated using a bench-scale FO system. They found that water flux was pH-dependent in both membrane orientations. In addition, water flux increased while the specific reverse salt flux and hydrogen ion flux decreased with increasing feed solution pH. The rejection of neutral carbamazepine was generally pH-independent in both membrane orientations.

The rejection of carbamazepine in the PRO mode was lower than that in the FO mode due to the higher concentration gradient caused by concentrative ICP in porous supporting layer. Steric hindrance was probably the main separation mechanism for the neutral carbamazepine in the FO process. On the other hand, the rejection of sulfamethoxazole was significantly affected by the FS pH in both membrane orientations. Variation in the rejection sulfamethoxazole could be attributed to the electrostatic repulsion between the negatively charged FO membrane surface and varying effective charge of the sulfamethoxazole molecule.

Organic solvents employed in pharmaceutical industry along with the potential concentration of pharmaceutical ingredients were investigated by Cui and Chung (2018). They demonstrate and evaluate FO process for solvent recovery. In this demonstration, organic solvent was conducted in different solvents with different draw solutes. The overall process shows rejections >98% when recovering organic solvents from different feed solutions, even when the feed concentration is as high as 20 wt%. More importantly, all systems exhibit relatively low ratios of reverse solute flux to solvent flux, indicating that the adverse effects of using hazardous draw solutions could be minimized. Nevertheless, the use of non-hazardous draw solutes such as citric acid is highly recommended to remove any potential risk.

5 Conclusions

This study confirms that a range of applications is available to apply FO technology on an industrial scale. This novel low-energy technology has demonstrated to be technically developed enough. The high volume of different draw solutions employed by experts in this area makes the process adaptable for a number of particular cases, while further efforts should be focused on reducing fouling aspects as well as on diminishing the concentration polarization. Among the different types of membranes for FO processes, CTA membranes were the first to be developed whereas TFC membranes have proved to be the best. In this sense, the next challenge is finding new kind of membranes which enhances water flux and/or avoids reverse flux.

Several industrial areas have been chosen for implementing FO processes with remarkable results. Some examples described in this paper are chemical industry, desalination of drinking water, food industry, pharmaceutical industry, textile industry, coal processing industry, electronic industry and heavy metal industry. Among them, the first four industries are necessarily to be pointed out since a lot of studies have been done to show its potential during the last years. Nevertheless further efforts should be put into developing alternative systems to re-concentrate the spent DS and thus implementing FO in many industrial areas with economic success.

References

Achilli, A., Cath, T. Y., & Childress, A. E. (2010). Selection of inorganic-based draw solutions for forward osmosis applications. *Journal of Membrane Science.* https://doi.org/10.1016/j.memsci.2010.08.010.

Akther, N., Sodiq, A., Giwa, A., Daer, S., Arafat, H. A., & Hasan, S. W. (2015). Recent advancements in forward osmosis desalination: A review. *Chemical Engineering Journal.* https://doi.org/10.1016/j.cej.2015.05.080.

Alsvik, I. L., & Hägg, M. B. (2013). Pressure retarded osmosis and forward osmosis membranes: Materials and methods. *Polymers.* https://doi.org/10.3390/polym5010303.

Aydiner, C., Sen, U., Koseoglu-Imer, D. Y., & Dogan, E. C. (2016). Hierarchical prioritization of innovative treatment systems for sustainable dairy wastewater management. *Journal of Cleaner Production.* https://doi.org/10.1016/j.jclepro.2015.08.107.

Baena-moreno, F. M., Rodríguez-galán, M., Vega, F., Alonso-fariñas, B., Arenas, L. F. V., & Navarrete, B. (2018). Carbon capture and utilization technologies: A literature review and recent advances. *Energy Sources, Part A: Recovery, Utilization, and Environmental Effects, 41*(12), 1403–1433. https://doi.org/10.1080/15567036.2018.1548518.

Baena-Moreno, F. M., Rodríguez-Galán, M., Vega, F., Vilches, L. F., & Navarrete, B. (2019). Review: Recent advances in biogas purifying technologies. *International Journal of Green Energy, 16*(5), 401–412. https://doi.org/10.1080/15435075.2019.1572610.

Baena-Moreno, F. M., Rodríguez-Galán, M., Vega, F., Vilches, L. F., Navarrete, B., & Zhang, Z. (2019). Biogas upgrading by cryogenic techniques. *Environmental Chemistry Letters, 17*, 1251–1261. https://doi.org/10.1007/s10311-019-00872-2.

Bell, E. A., Poynor, T. E., Newhart, K. B., Regnery, J., Coday, B. D., & Cath, T. Y. (2017). Produced water treatment using forward osmosis membranes: Evaluation of extended-time performance and fouling. *Journal of Membrane Science.* https://doi.org/10.1016/j.memsci.2016.10.032.

Bensaadi, S., Arous, O., Kerdjoudj, H., & Amara, M. (2016). Evaluating molecular weight of PVP on characteristics of CTA membrane dialysis. *Journal of Environmental Chemical Engineering.* https://doi.org/10.1016/j.jece.2016.02.003.

Bevacqua, M., Tamburini, A., Papapetrou, M., Cipollina, A., Micale, G., & Piacentino, A. (2017). Reverse electrodialysis with NH_4HCO_3-water systems for heat-to-power conversion. *Energy.* https://doi.org/10.1016/j.energy.2017.07.012.

Blandin, G., Verliefde, A. R. D., Comas, J., Rodriguez-Roda, I., & Le-Clech, P. (2016). Efficiently combining water reuse and desalination through forward osmosis-reverse osmosis (FO-RO) hybrids: A critical review. *Membranes.* https://doi.org/10.3390/membranes6030037.

Brunetti, A., Bernardo, P., Drioli, E., & Barbieri, G. (2010). Membrane engineering: Progress and potentialities in gas separations. *Membrane Gas Separation.* https://doi.org/10.1002/9780470665626.ch14.

Brunetti, A., Sellaro, M., Drioli, E., & Barbieri, G. (2016). Membrane engineering and its role in oil refining and petrochemical industry. https://doi.org/10.4018/978-1-4666-9975-5.ch005.

Cath, T. Y., Childress, A. E., & Elimelech, M. (2006). Forward osmosis: Principles, applications, and recent developments. *Journal of Membrane Science.* https://doi.org/10.1016/j.memsci.2006.05.048.

Chabanon, E., Mangin, D., & Charcosset, C. (2016). Membranes and crystallization processes: State of the art and prospects. *Journal of Membrane Science*. https://doi.org/10.1016/j.memsci.2016.02.051.

Chekli, L., Phuntsho, S., Kim, J. E., Kim, J., Choi, J. Y., Choi, J. S., et al. (2016). A comprehensive review of hybrid forward osmosis systems: Performance, applications and future prospects. *Journal of Membrane Science*. https://doi.org/10.1016/j.memsci.2015.09.041.

Chekli, L., Phuntsho, S., Shon, H. K., Vigneswaran, S., Kandasamy, J., & Chanan, A. (2012). A review of draw solutes in forward osmosis process and their use in modern applications. *Desalination and Water Treatment*. https://doi.org/10.1080/19443994.2012.672168.

Chen, J. C., Li, Q., & Elimelech, M. (2004). In situ monitoring techniques for concentration polarization and fouling phenomena in membrane filtration. *Advances in Colloid and Interface Science*. https://doi.org/10.1016/j.cis.2003.10.018.

Cho, Y. H., Lee, H. D., & Park, H. B. (2012). Integrated membrane processes for separation and purification of organic acid from a biomass fermentation process. *Industrial and Engineering Chemistry Research*. https://doi.org/10.1021/ie301023r.

Chou, S., Shi, L., Wang, R., Tang, C. Y., Qiu, C., & Fane, A. G. (2010). Characteristics and potential applications of a novel forward osmosis hollow fiber membrane. *Desalination*. https://doi.org/10.1016/j.desal.2010.06.027.

Chung, T. S., Zhang, S., Wang, K. Y., Jincai, Su., & Ling, M. M. (2012). Forward osmosis processes: Yesterday, today and tomorrow. *Desalination*. https://doi.org/10.1016/j.desal.2010.12.019.

Cornelissen, E. R., Harmsen, D. J. H., Beerendonk, E. F., Qin, J. J., & Kappelhof, J. W. M. N. (2011). Effect of draw solution type and operational mode of forward osmosis with laboratory-scale membranes and a spiral wound membrane module. *Journal of Water Reuse and Desalination*. https://doi.org/10.2166/wrd.2011.042.

Cui, Y., & Chung, T. S. (2018). Pharmaceutical concentration using organic solvent forward osmosis for solvent recovery. *Nature Communications*. https://doi.org/10.1038/s41467-018-03612-2.

Eykens, L., De Sitter, K., Dotremont, C., Pinoy, L., & Van der Bruggen, B. (2017). Membrane synthesis for membrane distillation: A review. *Separation and Purification Technology*. https://doi.org/10.1016/j.seppur.2017.03.035.

Fan, X., Liu, H., Gao, Y., Zou, Z., Craig, V. S. J., Zhang, G., & Liu, G. (2016). Forward-osmosis desalination with poly(ionic liquid) hydrogels as smart draw agents. *Advanced Materials*. https://doi.org/10.1002/adma.201600205.

Farahani, D. A., Hossein, M., & Chung, T. S. (2018). Solvent resistant hollow fiber membranes comprising P84 polyimide and amine-functionalized carbon nanotubes with potential applications in pharmaceutical, food, and petrochemical industries. *Chemical Engineering Journal*. https://doi.org/10.1016/j.cej.2018.03.153.

Ge, Q., Ling, M., & Chung, T. S. (2013). Draw solutions for forward osmosis processes: Developments, challenges, and prospects for the future. *Journal of Membrane Science*. https://doi.org/10.1016/j.memsci.2013.03.046.

Hajilary, N., Rezakazemi, M., & Shirazian, S. (2018). Biofuel types and membrane separation. *Environmental Chemistry Letters*. https://doi.org/10.1007/s10311-018-0777-9.

Haupt, A., & Lerch, A. (2018). Forward osmosis application in manufacturing industries: A short review. *Membranes*. https://doi.org/10.3390/membranes8030047.

Ismail, A. F., Padaki, M., Hilal, N., Matsuura, T., & Lau, W. J. (2015). Thin film composite membrane—Recent development and future potential. *Desalination*. https://doi.org/10.1016/j.desal.2014.10.042.

Jin, X., Shan, J., Wang, C., Wei, J., & Tang, C. Y. (2012). Rejection of pharmaceuticals by forward osmosis membranes. *Journal of Hazardous Materials*. https://doi.org/10.1016/j.jhazmat.2012.04.077.

Kalafatakis, S., Braekevelt, S., Carlsen, V., Lange, L., Skiadas, I. V., & Gavala, H. N. (2017). On a novel strategy for water recovery and recirculation in biorefineries through application of forward osmosis membranes. *Chemical Engineering Journal*. https://doi.org/10.1016/j.cej.2016.11.092.

Kim, S. J., Ko, S. H., Kang, K. H., & Han, J. (2010). Direct seawater desalination by ion concentration polarization. *Nature Nanotechnology*. https://doi.org/10.1038/nnano.2010.34.

Lau, W. J., Ismail, A. F., Misdan, N., & Kassim, M. A. (2012). A recent progress in thin film composite membrane: A review. *Desalination*. https://doi.org/10.1016/j.desal.2011.04.004.

Law, J. Y., & Mohammad, A. W. (2018). Osmotic concentration of succinic acid by forward osmosis: Influence of feed solution pH and evaluation of seawater as draw solution. *Chinese Journal of Chemical Engineering*. https://doi.org/10.1016/j.cjche.2017.10.003.

Lee, S., Boo, C., Elimelech, M., & Hong, S. (2010). Comparison of fouling behavior in forward osmosis (FO) and reverse osmosis (RO). *Journal of Membrane Science*. https://doi.org/10.1016/j.memsci.2010.08.036.

Li, D., Yan, Y., & Wang, H. (2016). Recent advances in polymer and polymer composite membranes for reverse and forward osmosis processes. *Progress in Polymer Science*. https://doi.org/10.1016/j.progpolymsci.2016.03.003.

Lutchmiah, K., Verliefde, A. R. D., Roest, K., Rietveld, L. C., & Cornelissen, E. R. (2014). Forward osmosis for application in wastewater treatment: A review. *Water Research*. https://doi.org/10.1016/j.watres.2014.03.045.

McCutcheon, J. R., & Elimelech, M. (2006). Influence of concentrative and dilutive internal concentration polarization on flux behavior in forward osmosis. *Journal of Membrane Science*. https://doi.org/10.1016/j.memsci.2006.07.049.

McCutcheon, J. R., McGinnis, R. L., & Elimelech, M. (2005). A novel ammonia-carbon dioxide forward (direct) osmosis desalination process. *Desalination*. https://doi.org/10.1016/j.desal.2004.11.002.

Mi, B., & Elimelech, M. (2010). Organic fouling of forward osmosis membranes: Fouling reversibility and cleaning without chemical reagents. *Journal of Membrane Science*. https://doi.org/10.1016/j.memsci.2009.11.021.

Petrotos, K. B., Tsiadi, A. V., Poirazis, E., Papadopoulos, D., Petropakis, H., & Gkoutsidis, P. (2010). A description of a flat geometry direct osmotic concentrator to concentrate tomato juice at ambient temperature and low pressure. *Journal of Food Engineering*. https://doi.org/10.1016/j.jfoodeng.2009.10.015.

Phillip, W. A., Yong, J. S., & Elimelech, M. (2010). Reverse draw solute permeation in forward osmosis: Modeling and experiments. *Environmental Science and Technology*. https://doi.org/10.1021/es100901n.

Raghavarao, K. S. M. S. S. M. S., Nagaraj, N., Ganapathi Patil, B., Babu, R., & Niranjan, K. (2005). Athermal membrane processes for the concentration of liquid foods and natural colours. *Emerging Technologies for Food Processing*. https://doi.org/10.1016/B978-012676757-5/50012-8.

Rastogi, N. K. (2016). Opportunities and challenges in application of forward osmosis in food processing. *Critical Reviews in Food Science and Nutrition*. https://doi.org/10.1080/10408398.2012.724734.

Scholes, C. A., Ho, M. T., Wiley, D. E., Stevens, G. W., & Kentish, S. E. (2013). Cost competitive membrane-cryogenic post-combustion carbon capture. *International Journal of Greenhouse Gas Control*. https://doi.org/10.1016/j.ijggc.2013.05.017.

Shaffer, D. L., Werber, J. R., Jaramillo, H., Lin, S., & Elimelech, M. (2015). Forward osmosis: Where are we now? *Desalination*. https://doi.org/10.1016/j.desal.2014.10.031.

Shibuya, M., Sasaki, K., Tanaka, Y., Yasukawa, M., Takahashi, T., Kondo, A., & Matsuyama, H. (2017). Development of combined nanofiltration and forward osmosis process for production of ethanol from pretreated rice straw. *Bioresource Technology*. https://doi.org/10.1016/j.biortech.2017.03.158.

Shu, Li., Obagbemi, I. J., Liyanaarachchi, S., Navaratna, D., Parthasarathy, R., Aim, R. B., & Jegatheesan, V. (2016). Why does pH increase with CaCl₂ as draw solution during forward osmosis filtration. *Process Safety and Environmental Protection*. https://doi.org/10.1016/j.psep.2016.06.007.

Soler-Cabezas, J. L., Mendoza-Roca, J. A., Vincent-Vela, M. C., Luján-Facundo, M. J., & Pastor-Alcañiz, L. (2018). Simultaneous concentration of nutrients from anaerobically digested sludge centrate and pre-treatment of industrial effluents by forward osmosis. *Separation and Purification Technology*. https://doi.org/10.1016/j.seppur.2017.10.058.

Tan, C. H., & Ng, H. Y. (2010). A novel hybrid forward osmosis—nanofiltration (FO-NF) process for seawater desalination: Draw solution selection and system configuration. *Desalination and Water Treatment*. https://doi.org/10.5004/dwt.2010.1733.

Tang, Y., Kai, M., & Ng, H. Y. (2014). Impacts of different draw solutions on a novel anaerobic forward osmosis membrane bioreactor (AnFOMBR). *Water Science and Technology*. https://doi.org/10.2166/wst.2014.116.

Tarboush, A., Belal, J., Rana, D., Matsuura, T., Arafat, H. A., & Narbaitz, R. M. (2008). Preparation of thin-film-composite polyamide membranes for desalination using novel hydrophilic surface modifying macromolecules. *Journal of Membrane Science*. https://doi.org/10.1016/j.memsci.2008.07.037.

Valladares Linares, R., Li, Z., Yangali-Quintanilla, V., Ghaffour, N., Amy, G., Leiknes, T., & Vrouwenvelder, J. S. (2016). Life cycle cost of a hybrid forward osmosis—Low pressure reverse osmosis system for seawater desalination and wastewater recovery. *Water Research*. https://doi.org/10.1016/j.watres.2015.10.017.

van den Berg, G. B., & Smolders, C. A. (1992). Diffusional phenomena in membrane separation processes. *Journal of Membrane Science*. https://doi.org/10.1016/0376-7388(92)80121-Y.

Venzke, C. D., Giacobbo, A., Ferreira, J. Z., Bernardes, A. M., & Rodrigues, M. A. S. (2018). Increasing water recovery rate of membrane hybrid process on the petrochemical wastewater treatment. *Process Safety and Environmental Protection*. https://doi.org/10.1016/j.psep.2018.04.023.

Wang, K. Y., Teoh, M. M., Nugroho, A., & Chung, T. S. (2011). Integrated forward osmosis-membrane distillation (FO-MD) Hybrid system for the concentration of protein solutions. *Chemical Engineering Science*. https://doi.org/10.1016/j.ces.2011.03.001.

Wang, X., Zhao, Y., Yuan, Bo., Wang, Z., Li, X., & Ren, Y. (2016). Comparison of biofouling mechanisms between cellulose triacetate (CTA) and thin-film composite (TFC) polyamide forward osmosis membranes in osmotic membrane bioreactors. *Bioresource Technology*. https://doi.org/10.1016/j.biortech.2015.11.087.

Wang, Y. N., Goh, K., Li, X., Setiawan, L., & Wang, R. (2018). Membranes and processes for forward osmosis-based desalination: Recent advances and future prospects. *Desalination*. https://doi.org/10.1016/j.desal.2017.10.028.

Xie, M., Price, W. E., & Nghiem, L. D. (2012). Rejection of pharmaceutically active compounds by forward osmosis: Role of solution pH and membrane orientation. *Separation and Purification Technology*. https://doi.org/10.1016/j.seppur.2012.03.030.

Yang, Q., Wang, K. Y., & Chung, T. S. (2009). A novel dual-layer forward osmosis membrane for protein enrichment and concentration. *Separation and Purification Technology*. https://doi.org/10.1016/j.seppur.2009.08.002.

Zhang, Z. (2016). Comparisons of various absorbent effects on carbon dioxide capture in membrane gas absorption (MGA) process. *Journal of Natural Gas Science and Engineering*. https://doi.org/10.1016/j.jngse.2016.03.052.

Zhang, Z., Chen, F., Rezakazemi, M., Zhang, W., Lu, C., Chang, H., & Quan, X. (2018). Modeling of a CO₂-piperazine-membrane absorption system. *Chemical Engineering Research and Design*. https://doi.org/10.1016/j.cherd.2017.11.024.

Zhang, Z., Yan, Y., Zhang, L., Chen, Y., & Ju, S. (2014). CFD investigation of CO₂ capture by methyldiethanolamine and 2-(1-Piperazinyl)-ethylamine in membranes: Part B. Effect of membrane properties. *Journal of Natural Gas Science and Engineering*. https://doi.org/10.1016/j.jngse.2014.05.023.

Zhao, D., Chen, S., Guo, C. X., Zhao, Q., & Lu, X. (2016). Multi-functional forward osmosis draw solutes for seawater desalination. *Chinese Journal of Chemical Engineering*. https://doi.org/10.1016/j.cjche.2015.06.018.

Zhao, S., Zou, L., Tang, C. Y., & Mulcahy, D. (2012). Recent developments in forward osmosis: Opportunities and challenges. *Journal of Membrane Science*. https://doi.org/10.1016/j.memsci.2011.12.023.

Zuo, Y. C., Chi, X. Y., Xu, Z. L., & Guo, X. J. (2017). Morphological controlling of CTA forward osmosis membrane using different solvent-nonsolvent compositions in first coagulation bath. *Journal of Polymer Research*. https://doi.org/10.1007/s10965-017-1311-7.

Current Strategies for the Design of Anti-fouling Ion-Exchange Membranes

Le Han

Abstract

Ion-exchange membranes (IEMs) have been established as a key component in industrial processes including water desalination. Nevertheless, IEMs suffer different degrees of fouling problems from cation-exchange membrane (CEM) to anion-exchange membrane (AEM), which significantly impedes the efficient application of this membrane technology. Preparing anti-fouling membranes is a fundamental strategy to deal with pervasive fouling problems from a variety of foulants. In this review, the category in diverse IEM materials and the fouling types are firstly summarized. Then, based on the current understanding of the fouling mechanisms between the foulant and membranes, two strategies of anti-fouling design for future IEMs are commented, mitigation of fouling by surface modification and by membrane synthesis. In particular, current work regarding the membrane surfacial coating/deposition by specific organic/inorganic materials to tune the surface physicochemical properties and new functional polymer block fabrication/development is discussed. Both the advantages and disadvantages are reviewed on the current strategies.

Keywords

Ion-exchange membrane • Anti-fouling • Surface modification • Membrane synthesis

Abbreviations

A-3	Trianchor
DBS	Dodecylbenzene sulfonate
GO	Graphene oxide
HA	Humic acid
HACC-Ag	Np hydroxypropyltrimethyl ammonium chloride chitosan–nanosilver particles
NB-8	Disodium salt α,ω oligooxipropylene-bis (o-urethane-2.4,2.6 tolueneurylbenzensulfonic acid)
PAH	Poly(allylamine hydrochloride)
PDA	Polydopamine
PDADMAC	Poly(diallyldimethylammonium chloride)
PS-DVB	Styrene-divinylbenzene copolymer
PSS	Poly(sodium 4-styrene sulfonate)
SDA	Sulfonated dopamine
SDBS	Sodium dodecyl benzene sulfonate
SDS	Sodium dodecyl sulfate

L. Han (✉)
Key Laboratory of the Three Gorges Reservoir Region's Eco-environment, Ministry of Education, College of Environment and Ecology, Chongqing University, Chongqing, 400044, People's Republic of China
e-mail: lehan@cqu.edu.cn

1 Introduction

The electromembrane-based separation having the advantages of excellent selectivity, low space and chemical requirement, and operational simplicity has found its wide applications in water treatment, involving the desalination of complex saline solutions in food, beverage, drug and chemical industries as well as in biotechnology and municipal wastewater (Ruan et al. 2018; Suwal et al. 2015; Jaroszek and Dydo 2016; Merle et al. 2011; Zeng et al. 2019; Sarapulova et al. 2018; Lee et al. 2013). These processes are considered as environment-friendly technology, where migration of ions and charged solutes through ion-exchange membranes (IEMs) often occurs. The representative processes include electrodialysis (ED), reverse electrodialysis (RED), bipolar membrane electrodialysis (BMED), electromembrane filtration (EMF), electrodialysis with filtration membrane (EDFM) and so on (Luiz et al. 2017; Tamburini et al. 2017; İpekçi et al. 2018; Ponomarev et al. 1989; Bazinet et al. 2012) (Table 1).

Table 1 Applications of ion-exchange membrane

Types	Principle	Driving mode	Some Applications	References
ED	ED is an electrochemical separation process where selective transport of ions across the IEMs occurs under electrical field via external direct current	External direct current	Brackish water desalination, table salt production, recovery of useful materials	Frioui et al. (2017)
RED	RED is one method to obtain the salinity gradient energy power, which is a chemical potential energy arising from the controlled mixing of a high salinity stream and a low salinity stream	Salinity gradient energy power	Energy storage, pollutants abatement and nanofluidic/microfluidic RED devices	Mei and Tang (2018), Kingsbury et al. (2015), Egmond et al. (2016)
BMED	Bipolar membrane electrodialysis (BMED) was developed based on conventional ED by introducing a bipolar membrane. The water molecules at the interphase of the bipolar membrane are split into hydrogen and hydroxide ions, and then hydrogen/hydroxide ions combine with anions and cations migrating from the feed solution compartment through the AEM/CEM to produce the corresponding acid and base	External direct current	Chemical and biochemical applications, food processing	Reig et al. (2016a, b, 2017)
EDI	The electrodeionization (EDI) is IEM-based desalination process combining ED and ion-exchange resins. The ion-exchange resin material is a conductor which serves as a bridge between the ion-exchange membranes, making the whole resistance of the cell much lower than in a normal ED condition	External direct current	Producing ultrapure water, removal of chromium, cobalt, copper and nickel from wastewater	Alvarado et al. (2009), Bhadja et al. (2015), Yeon et al. (2003), Arar et al. (2011), Dermentzis (2010)
EDR	The electrodialysis reversal (EDR) method is utilized via periodic polarity changes during ED operation	External direct current	Brackish water desalination, water softening and the production of potable water	Lee et al. (2013), Valero and Arbós (2010)

As a key composition to any electromembrane processes, IEMs are typically composed of hydrophobic substrates, immobilized ion-functionalized groups and movable counterions. Since they do not contain distinct pores, IEMs are often very dense membranes. IEMs can be divided into anion-exchange membrane (AEM) and cation-exchange membrane (CEM) according to their charge, where in any case high permselectivity, conductivity, stability (mechanical, chemical and thermal stabilities), low cost and simple fabrication procedures are desirable (Suwal et al. 2015; Ran et al. 2017; Campione et al. 2018). Although IEMs are not prone to fouling comparing to the other membrane processes (e.g., pressure-driven ones), recent reports show that the fouling of IEM is an un-negligible issue in the operation of the IEM-based process, making anti-fouling a new desirable property for IEM (Liu et al. 2019; Mikhaylin and Bazinet 2016).

Fouling is the phenomenon of undesirable attachment of colloidal matter, inorganic compounds or macromolecules to the membrane surface or inside the materials like inner pore or free volume (Yooprasertchuti and Dechadilok 2018; Cifuentes-Araya et al. 2011; Park et al. 2003). This often results in an increased electrical resistance or energy consumption, decreased ion migration or permselectivity, even a decreased membrane lifetime with physical damages (Park et al. 2003), largely hampering its industrial application (Mikhaylin and Bazinet 2016).

The membrane fouling usually originates from the chemical and/or physical interactions between membranes and foulants. Such an interaction refers to many key material properties such as hydrophilicity, charge nature and surface roughness. For example, since most organic substances are negatively charged in natural waters, the AEMs (of positively charged nature) are more prone to fouling than CEM (of negatively charged nature) under the electrostatic interaction (Wang et al. 2017, 2011).

Therefore, the main strategy for mitigating or preventing membrane fouling is to prevent undesirable adhesive interaction between the foulant and the membrane. This review paper thus outlines the mechanisms of IEM fouling and surmises latest studies where the modification of the membrane surface and development of new membrane structure are attempted toward the anti-fouling effect. Hopefully, the review will shed light on advancing the development of future anti-fouling IEMs.

2 Category of IEMs

IEMs are typically thin polymeric films containing fixed charged groups which are ionized in water (Strathmann 2010). Figure 1 summarizes the common categories of IEM regarding the charge nature and morphology, where based on the charge nature (or type of ionic functional groups), IEMs are broadly classified into CEMs and AEMs (Zeng et al. 2019; Ran et al. 2017), and based on their morphology, IEMs can be classified into the homogeneous and heterogeneous ones. The forming category according to the functional charged group is further illustrated since the ion permselectivity is highly related.

IEMs can selectively allow the passage of oppositely charged ions (counterions) while obstruct similarly charged ions (co-ions), with the counterions permselectivity following the theory elucidated by Donnan. CEMs generally contain negatively charged groups (e.g., $-SO_3^-$, $-COO^-$, $-PO_3^{2-}$, $-PO_3H^-$). Diverse polymer materials including poly(ether sulfone) (PES), poly(ether ketone) (PEK), polybenzimidazole (PBI), polyimide (PI), poly(phenylene), polyphosphazene and polyvinylidene fluoride (PVDF) were investigated as the backbones for CEMs (Ran et al. 2017). AEMs are generally prepared from positively charged polyelectrolytes, and quaternized ammonium is the most conventional conducting groups for AEMs (Nie et al. 2015), where imidazole, phosphonium, tertiary sulfonium cations and metal-cation-based polyelectrolytes also serve as promising candidates for AEMs (Yang et al. 2015; Stokes et al. 2011; Zhang et al. 2012; Disabb-Miller et al. 2013). The topological architectures of polymeric ionomers generally contain main chain, side chain, block copolymers and densely functionalized types (Ran et al. 2017) (Table 2).

3 Category of Ion-Exchange Membrane Fouling

Previous section introduces the membrane material background, and this section will focus on the foulant part, in order to link to the possible membrane–foulant interaction.

Indeed, understanding the commonly reported fouling types and their mechanisms is crucial to the development of anti-fouling strategies for IEM-based processes. The foulants intend to adhere to the membrane materials by complicated foulant–membrane and foulant–foulant attractive interactions from the physicochemical point of view. While thermodynamically, this is a process toward the minimization of Gibbs free energy of the system (Zhao et al. 2018a; Zhang et al. 2016).

The fouling behavior of any foulant can be simplified to approaching–adsorption–accumulation (Wang et al. 2017; Zhao et al. 2018a). First, the foulants (or their precursors) approach or contact a membrane surface. Then, these foulants adsorb or attach onto the membrane surface via electrostatic, hydrophobic, van der Waals, hydrogen bonding or other interactions; finally, foulants accumulate or aggregate together with each other and then form cake, gel, biofilm or scaling layers on the membrane surface.

Various membrane–foulant classifications exist. For example, according to intrinsic characteristics of fouling behaviors, membrane–foulants can be divided into non-migratory (e.g., organic colloids, natural organic matters, biomacromolecules), proliferative (e.g., microorganisms, bacterial, living cells) and inorganic foulants (e.g., precipitated salts and sparingly soluble salts) (Zhao et al. 2018a; Zhang et al. 2016; Dydo and Turek 2013; Husson et al. 2013; Luo et al. 2012). For simplicity, organic fouling, biofouling, inorganic scaling and colloidal fouling are used to describe the category of IEMs fouling depicted in Fig. 2 and Table 3.

3.1 Organic Fouling

Common organic foulants to IEMs involve organic substances such as oil, carbohydrates, proteins, humic acid and aromatics (Husson et al. 2013; Banasiak and Schäfer 2009; Guo et al. 2014; Shi et al. 2011; Bukhovets et al. 2010; Sri Mulyati et al. 2012). These organic substances easily stick to the surface of the membrane and/or lodge themselves inside the membrane-free volume, and the induced fouling dynamics is generally relevant to the foulant physical–chemical properties such as the molecular structure, hydrophobicity, charging nature, solubility, mobility and bulk solution concentration (Mikhaylin and Bazinet 2016).

For example, a permeable organic anion which is small enough to penetrate into membrane-free volume possibly fouls the membrane inner phase due to a low mobility in the confined zone. Given a high bulk concentration and its low solubility, organic molecule adsorption at the membrane surface also occurs, causing a drastic increase in the membrane stack resistance (Bukhovets et al. 2010).

Generally, anion-exchange membranes are more susceptible to fouling by organic compounds under electrostatic attraction effect. Researchers have focused on fouling of AEMs by organic foulants of negative charge, such as bovine serum albumin (BSA), humate, carboxylic acids and anionic surfactants (Sri Mulyati et al. 2012). The mixture of proteins formed an intense layer on membrane surfaces, which was ascribed to solute charge rather than particle size and variation in molecular packing (Bukhovets et al. 2010).

Other foulant properties can also be influential. Hydrophobicity of sodium dodecylbenzene sulfonate (SDBS) due to the indissociable organic chain may foul the

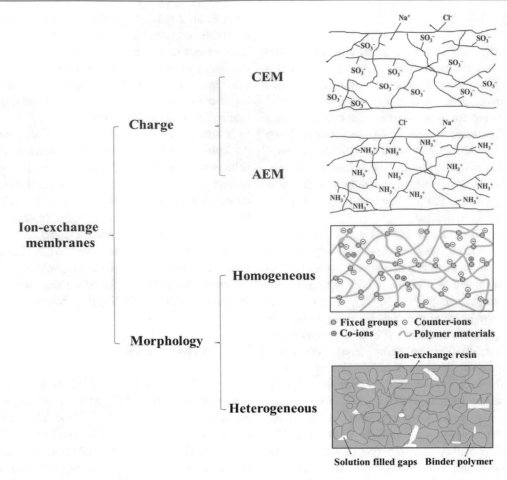

Fig. 1 Classification of IEMs based on charge nature and morphology. Reproduced with permission (Tongwen Xu 2008)

Table 2 Composition of IEMs

Membrane types	Polymer materials	Fixed charged groups	References
CEM	Poly(ether sulfone) (PES), poly(ether ketone) (PEK), polybenzimidazole (PBI), polyimide (PI), poly(phenylene), polyphosphazene, polyvinylidene fluoride (PVDF) etl	$-SO_3^-$, $-COO^-$, $-PO_3^{2-}$, $-PO_3H^-$, $-C_6H_4OH$ etl	Ran et al. (2017), Tongwen Xu (2008)
AEM		$-NH_3^+$, $-RNH_2^+$, $-R_2NH^+$, $-R_3N^+$, $-R_3P^+$, $-R_2S^+$ etl	Nie et al. (2015), Tongwen Xu (2008)

hydrophobic polymer chain of IEMs under hydrophobic–hydrophobic interaction (Sri Mulyati et al. 2012). Structural size of sodium dodecyl sulfate (SDS) also troubles the AEM where this organic foulant may be trapped inside the membrane polymer phase to form a densely structured membrane polymer (Zhao et al. 2018c).

3.2 Biofouling

Biofouling often refers to the accumulation of unwanted bio-organisms on the membrane surface or inside the membrane polymer phase. This is more reported in the pressure-driven membrane processes like microfiltration, ultrafiltration, nanofiltration and reverse osmosis, where the membrane permeability/productivity decreases with increasing transmembrane pressure (Tijing et al. 2015).

Biofouling in the IEM-based technologies is rarely reported, mainly in the microorganism-coped desalination process (e.g., the microbial desalination cells) or energy production processes (e.g., microbial fuel cell) (Luo et al. 2012; Choi et al. 2011). Aquatic biofouling dynamics can be described as the following four stages: the adsorption of a conditioning layer, adhesion of bacteria, growth of a biofilm

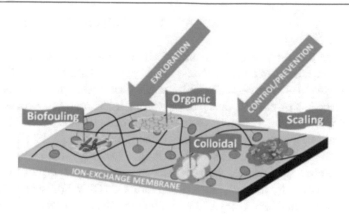

Fig. 2 Category of ion-exchange membrane fouling. Reproduced with permission (Mikhaylin and Bazinet 2016)

Table 3 Fouling of ion-exchange membrane

Foulant type	Foulant	Observation	References
Organic	Oil, carbohydrates, proteins, humic acid and aromatics	These organic substances easily stick to the surface of the membrane and/or lodge themselves inside the membrane-free volume, and the induced fouling dynamics is generally relevant to the foulant physical–chemical properties such as the molecular structure, hydrophobicity, charging nature, solubility, mobility and bulk solution concentration	Husson et al. (2013), Banasiak and Schäfer (2009), Guo et al. (2014), Shi et al. (2011)
Biofouling	Bio-organisms, extracellular polymeric substances	Biofilm formation and development on the membrane surface, and mainly in the microorganism-coped desalination process (e.g., the microbial desalination cells) or energy production processes (e.g., microbial fuel cell)	Luo et al. (2012), Choi et al. (2011), Pontié et al. (2012)
Inorganic scaling	Magnesium, calcium and carbonate	The membrane scalant or salt/ion precipitation occurs when the equilibrium of salt solution shifts toward a decreasing solubility until below the respective salt concentration	Hayes and Severin (2017), Araya-Farias and Bazinet (2006), Cifuentes-Araya et al. (2012)
Colloidal	Clay minerals, colloidal silica, iron oxide, aluminum oxide, manganese oxide, organic colloid and other non-dissolved suspended solids	The colloids treated by ED are mostly negatively charged, which often leads to the interaction with positively charged ion-exchange groups of AEM	Lee et al. (2003), Cohen and Probstein (1986), Mondor et al. (2009)

and macrofouling (Choi et al. 2011; Pontié et al. 2012) as shown in Fig. 3. Normally, biofilm formation and development on the membrane surface can be observed in such processes, where cell number, nutrient status and microbial biochemical properties are correlated (Luo et al. 2012; Choi et al. 2011; Tijing et al. 2015).

Usually, the microorganisms in biofilms live in a self-produced matrix of hydrated extracellular polymeric substances (EPSs), consisting of mainly polysaccharides, proteins, nucleic acids and lipids (Kochkodan et al. 2014). These EPSs form the immediate environment of microbial, providing the mechanical stability of biofilms, mediating their adhesion to surfaces and forming a cohesive, three-dimensional polymer network that interconnects and transiently immobilizes biofilm cells (Komlenic 2010). Hydrophobicity and negatively charged nature of EPS are often highlighted (Tijing et al. 2015; Nguyen et al. 2012; Kochkodan and Hilal 2015).

3.3 Inorganic Scaling

The major scaling ions present in solutions potentially to the IEM-based processes include magnesium, calcium, carbonate and so on (Hayes and Severin 2017; Araya-Farias and Bazinet 2006). The membrane scalant or salt/ion precipitation occurs when the equilibrium of salt solution shifts toward a decreasing solubility until below the respective salt concentration (Cifuentes-Araya et al. 2011; Momose et al. 1991). Given a type of the potential scaling ion, the precipitation equilibrium is largely affected by not only the ion concentration but also the solution pH and temperature. Indeed, the influence of the solution pH on the inorganic scaling is often reported. A basic condition, e.g., excessive presence of OH^-, easily results in divalent precipitation such as $Ca(OH)_2$ and $Mg(OH)_2$ (following Eqs. 1 and 2). Even in a weak alkaline solution, these two divalent cations possibly precipitate with carbonate ions (following Eqs. 3 and 4) (Mikhaylin and Bazinet 2016).

$$Mg^{2+}_{(aqueous)} + 2OH^-_{(aqueous)} \leftrightarrows Mg(OH)_{2(solid)} \qquad (1)$$

$$Ca^{2+}_{(aqueous)} + 2OH^-_{(aqueous)} \leftrightarrows Ca(OH)_{2(solid)} \qquad (2)$$

$$Mg^{2+}_{(aqueous)} + CO^{-3}_{2(aqueous)} \leftrightarrows MgCO_{3(solid)} \qquad (3)$$

$$Ca^{2+}_{(aqueous)} + CO^{-3}_{2(aqueous)} \leftrightarrows CaCO_{3(solid)} \qquad (4)$$

Inorganic scaling observed on IEM surface may be of various ionic species, contributing to the formation of a multilayer-typed scaling, where not only the ion type but the ionic ratio matters regarding the nucleation and crystal growth (Cifuentes-Araya et al. 2011, 2012).

3.4 Colloidal Fouling

Colloidal particles are mainly clay minerals, colloidal silica, iron oxide, aluminum oxide, manganese oxide, organic colloid and other non-dissolved suspended solids, which present in natural and processed waters. The diameter of colloid particles is between 10 Å and 2 μm (Lee et al. 2003; Cohen and Probstein 1986). Colloidal particles have a net charge that plays an important role in colloidal stability (according to the theory of Derjaguin, Landau, Verwey, and Overbeek (DLVO)), which may also cause colloids to adhere to the membrane surface. The colloids treated by ED are mostly negatively charged, which often leads to the interaction with positively charged ion-exchange groups of AEM (Mikhaylin and Bazinet 2016; Lee et al. 2003; Mondor et al. 2009). The fouling potential of the negatively charged silica sol in ED by adsorption on the surface of an AEM was investigated (Lee et al. 2003). The factors affecting colloidal fouling are concentration of fouling particles as well as dissolved salt concentration, pH, temperature, membrane properties, mode of operation and hydrodynamic conditions (Mikhaylin and Bazinet 2016).

4 Mitigation of Fouling by Surface Modification

Previous sections mentioned the common foulant types and the fouling mechanism where the physicochemical properties such as the hydrophilicity and charging nature are highlighted. Considering the IEMs are featured as highly charged materials, often of both hydrophobicity due to the polymer chain and hydrophilicity due to the charged dissociable group, as well as

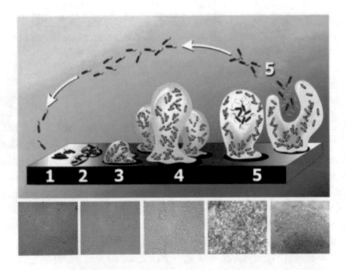

Fig. 3 Biofilm life cycle. Stages in the development and dispersion of biofilm are shown proceeding from right to left. The lower panel shows photomicrographs of bacteria at each of the five stages shown in the schematic above. Reproduced with permission (Flemming et al. 2011)

of varying surface roughness on the dense film, mitigation of the membrane fouling via membrane surface modification on its hydrophilicity, electrostatic charge and roughness are proposed, which may strongly modify the interaction between the membrane and the foulants.

4.1 Surface Hydrophilicity and Charge

It has been widely acknowledged that increasing the hydrophilicity of the membrane surface and imparting a negative surface charge are very effective means to prevent fouling, considering the common foulants are often of hydrophobicity and negative charge (Goh et al. 2018).

The general consensus on the anti-fouling mechanism of a hydrophilic surface emphasizes the critical role of the formation of a hydration layer formed via hydrogen bonding between water and the hydrophilic functional groups on the surface (Pashley 1981a, b). By largely reducing the possible hydrophobic–hydrophobic interactions, this layer can prevent or reduce undesirable adsorption of foulants which is energetically highly unfavorable. Also, this hydration layer contributes a new steric hindrance to the hydrophobic foulant, effectively reducing the possible direct contact. Thus, surface hydrophilization is often recommended to reduce membrane fouling, especially toward the organics, colloids and microorganisms (Wang and Lin 2017).

The surface charge of membranes is also an important consideration in reducing the membrane fouling where foulants are charged. Usually, it is appropriate to use a membrane carrying the same charge as the foulants, giving an electrostatic repulsion contributing to anti-fouling effect (Kochkodan et al. 2014). Thus, there have been a number of attempts to incorporate new ionizable functional groups on the surface of IEM as anti-fouling strategy (Ulbricht 2006; Kato et al. 2003).

High hydrophilicity can decrease the surface charge on the membrane surface, thus reducing the adhesion of pollutants (Vaselbehagh et al. 2014). To sum up, a high hydrophilicity prevents the adsorption of foulants through reducing hydrophobic–hydrophobic interactions, while a negative surface charge prevents the adsorption of negatively charged foulants through enhancing the electrostatic repulsion. Such an anti-fouling strategy has seen many applications, as listed in Table 4 (Ruan et al. 2018; Sri Mulyati et al. 2012; Vaselbehagh et al. 2014; Mulyati et al. 2013; Zhao et al. 2018b; Grebenyuk et al. 1998; Li et al. 2018b; Hao et al. 2018).

4.2 Surface Roughness

Membrane surface roughness as an important characteristic of membrane morphology was correlated to the membrane fouling phenomenon (Kochkodan et al. 2014; Hao et al. 2018; Fernandez-Gonzalez et al. 2017). Membranes with rougher surfaces are observed to be more favorable for foulants' attachment: A greater roughness increased the total surface area to which foulants can approach and attach; i.e., the ridge–valley structure favors accumulation of foulants at the surface (Hao et al. 2018). Similarly, a membrane with a smooth surface is thus not easily fouled, due to a reduced contact area for foulant, less propensity for foulant–membrane interaction and a presumably higher shear rate (Kochkodan et al. 2014; Fernandez-Gonzalez et al. 2017).

Membrane fouling is clearly mitigated via constructing a smoother and denser sulfonated dopamine (SDA) layer by introducing the sulfonated group, compared to the counterpart of dopamine (DA)/AEM (Ruan et al. 2018). AEM images clearly report the membrane morphology change due to mussel adhesive mimetic random copolymer immobilization on the membrane as shown in Fig. 4. DA/AEM (Ra = 36.5 nm) had a rougher surface than pristine AEM (Ra = 11.7 nm) which may be caused by the nonuniform polymerization and aggregation of DA, while the SDA-coated one exhibits very smooth surface (Ra = 5.08 nm).

Decreasing surface roughness by surface modification with polydopamine (PDA) enhances the anti-fouling potential of AEM (Li et al. 2018b). The introduction of PDA coating helps to construct a smoother and denser composite-modified layer which consists of graphene oxide (GO) and PDA (shown in Fig. 5). The electrodeposition of GO on AEM (GO-M) (Ra = 33.9 nm) led to more ridges and valleys than the pristine membrane (PM) (Ra = 23.5 nm), while the surface roughness of GO@PDA-M gave a Ra value to 26.5 nm. The good anti-fouling ability could be maintained in the long-term operation, with nearly negligible increase of AEM potential ($\sim$0.6 V) for the prepared GO@PDA-modified one after 20 h fouling, much lower than that for GO-modified AEM ($\sim$2.0 V).

5 Mitigation of Fouling by New Membrane Synthesis

In addition to the anti-fouling modification of the IEM surface, another promising strategy is to synthesize new functional membranes. The key factor is to incorporate functional groups of less sensitivity to fouling (e.g., aliphatic polymer or nanocomposite ones) and optimize the packing in the newly synthesized membrane phase (Chen et al. 2010).

5.1 Aliphatic-Hydrocarbon-Based Membrane

Almost all AEMs have been prepared from derivatives of styrene–divinylbenzene copolymers. As a consequence,

Table 4 Anti-fouling strategy of AEM

Functional materials	Fabrication method	Feed solution	Performance	References
SDA	Polymerization and deposition	0.05 M NaCl and 174 mg/L SDBS	The transition time of SDA-modified AEM (112 min) showed better anti-fouling performance than the original AEM (76 min)	Ruan et al. (2018)
PSS	Electrodeposition	0.05 M NaCl and SDBS	Surface modification of PSS resulted in better anti-fouling properties at SDBS concentrations that range from 1×10^4 kg/m^3 to 6×10^4 kg/m^3	Sri Mulyati et al. (2012)
PSS-PAH	Electrodeposition	0.05 M NaCl, and 52 mg/L SDBS	The transition time increased more than 150 min by seven layers, compared to 0 min without surface modification	Mulyati et al. (2013)
PSS-PDADMAC	Electrodeposition	0.1 M NaCl containing 75 mg/L SDS	Modification of AEM showed better anti-fouling performance than the pristine AEM with higher SDS (100 mg/L)	Zhao et al. (2018b)
A-3, NB-8	Chemical adsorption	0.05 N Na$_2$SO$_4$ and 30–90 mg/L HA, 0.05 N Na$_2$SO$_4$ and 30–100 mg/L DBS	With 30 mg/L DBS, modification of AEM with A-3 no fouling occurred. Modification of AEM with NB-8 agent permitted electrodialysis of solutions containing 100 and 60 mg/L of DBS and HA, respectively	Grebenyuk et al. (1998)
GO-PDA	Electrodeposition	0.1 M NaCl and 150 mg/L SDBS	After being fouled by 150 mg/L SDBS for 4 h, the desalination rate of GO and PDA surface-modified AEM increased by 39.3% compared to pristine membrane	Li et al. (2018b)
PDA	Chemical adsorption	0.05 M NaCl and SDBS	The transition time for PDA-modified AMX membrane was extremely longer than that for the pristine membrane above the critical micelle concentration of SDBS	Vaselbehagh et al. (2014)
PDA-PSS-HACC-Ag Np	Chemical adsorption and electrodeposited	0.05 M NaCl and 1.0 g/L SDBS	The time elapsed until the occurrence of fouling (transition time) of the modified AEM (125 min) is much longer than that of the original one (60 min)	Hao et al. (2018)

aromatic foulant-induced fouling toward the aromatic-hydrocarbon-based AEMs could be serious. Figure 6 illustrates the structure of a commercially available hydrocarbon-type AEM and a suggested mechanism for fouling by aromatic compounds where two aspects of the fouling mechanism are indicated: affinity between anions and the oppositely charged fixed groups of the AEM, and affinity as π–π interactions between the aromatic membrane matrix and compounds (Tanaka et al. 2011).

According to Allison, organic molecules causing an irreversible membrane fouling usually have fixed charged groups and aromatic rings (Allison 2005). Hydrophobic interactions between benzene rings of organic molecule and the membrane polymer are the main reason for strong irreversible surface adsorption (Bukhovets et al. 2010). Accordingly, a promising anti-fouling strategy is to fabricate an aliphatic polymer backbone matrix (avoiding affinity interaction between the organic foulants and membrane surface) and a hydrophilic membrane surface (Liu et al. 2019).

Tanaka et al. (2011) prepared AEM with aliphatic hydrocarbon matrix and deduced that the permeability coefficient for sodium dodecyl benzene sulfonate (DBS) of AEMs with aliphatic matrix was lower than that of aromatic AEMs in ED system. AEMs with various membrane structures were prepared by introducing various amines: trimethylamine (TMA), triethylamine (TEA), tri-n-propylamine (TPA) and tri-n-butylamine (TBA) into precursor membranes prepared from chloromethylstyrene (CMS)-divinylbenzene (DVB) and glycidyl methacrylate (GMA)-DVB. The voltage change through the AEMs during ED operation using solutions containing DBS as a foulant indicated that aliphatic AEMs show lower fouling than aromatic ones (Higa et al. 2014).

Wang et al. (2017) compared the interaction (fouling) among varying model organics [sodium dodecyl benzene sulfonate (SDBS), bovine serum albumin (BSA), sodium humate and sodium alginate (SA)] and the varying AEMs (TWEDAI and TWEDAII). TWEDAI and TWEDAII are made by aliphatic monomers and aromatic monomers in a

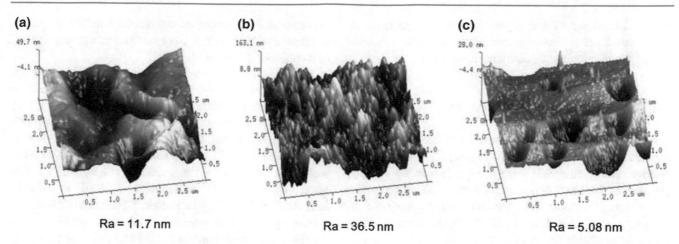

Fig. 4 AFM images and mean surface roughness (Ra) of **a** the pristine AEM, **b** the DA/AEM and **c** the SDA/AEM (deposition time is 48 h). Reproduced with permission (Ruan et al. 2018)

Fig. 5 Characteristic of AFM 3D images of pristine membrane (PM), GO-modified membrane (GO-M), PDA-modified membrane (PDA-M) and GO@PDA-modified membrane (GO@PDA-M). Reproduced with permission (Li et al. 2018a)

nonwoven fabric, respectively. After being fouled by organics, the TWEDAI membrane made of an aliphatic monomer remained of relatively lower electrical resistance, indicating the best anti-fouling performance. This is mainly attributed to its small original resistance and the structure stability. The result is in line with the finding of Tanaka et al. (2011), Audinos (1997) showed that the fouling potential of the aromatic membrane and the aromatic substances was higher than those of the others (aromatic membrane–aliphatic substances, aliphatic membrane–aromatic substance and aliphatic membrane–aliphatic substances).

5.2 Nanocomposite Membranes

Recently, many studies on membrane technology have focused on inorganic and organic nanocomposites. Nanocomposite materials have become increasingly important due to their extraordinary properties, which arise from the synergism between the advantages of both inorganic and organic components (Zuo et al. 2009; Xiao et al. 2011; Kango et al. 2013; Heinz et al. 2017).

Generally, membranes with lower surface roughness, higher surface charge density and larger surface hydrophilicity have better anti-fouling properties (Vatanpour et al. 2011). Considering of the conductivity and strong hydrophilicity of sulfonated Fe_2O_3, based on sulfonated poly (2,6-dimethyl-1,4-phenylene oxide) (sPPO) and sulfonated iron oxide created, researcher (Hong and Chen 2014) once made an organic–inorganic nanocomposite IEM via the blending method. This kind of nanocomposite ion-exchange membrane shows great prospects for anti-fouling and energy generation in reverse electrodialysis.

Carbon nanotubes (CNTs) are one type of inorganic nanomaterial that have gained a lot of attention due to their high flexibility, low mass density, large aspect ratio, excellent mechanical property and good electronic conductivity. It has been reported that ion pathways exist at the interface of nanomaterials and polymer; hence, long-distance ionic pathways could be formed using elongated nanomaterials (nanotubes or nanofibers). They largely improved the inner structure of membrane and facilitate ion transport. In addition, oxidized multi-walled CNTs (O-MWCNTs) were found to effectively improve the anti-fouling properties of

pressure-driven membranes due to their ability to change membrane surface morphology, surface charge density and hydrophilicity (Tong et al. 2016).

CNT-based anti-fouling nanocomposite AEM was made from a commercial polyethylene anion-exchange membrane and a negative thin layer (Fernandez-Gonzalez et al. 2017). This layer is composed of sulfonated poly(2,6-dimethyl-1,4-phenylene oxide) (sPPO) and two nanomaterials of oxidized multi-walled carbon nanotubes (CNTs-COO$^-$) and sulfonated iron oxide nanoparticles (Fe_2O_3-SO_4^{2-}). The novel nanocomposite membranes showed a relevant improvement in fouling resistance. CNT-based anti-fouling nanocomposite CEM was also synthesized using oxidized multi-walled carbon nanotubes (O-MWCNTs) blended with sulfonated poly(2,6-dimethyl-1,4-phenylene oxide) (sPPO) (Tong et al. 2016). The nanocomposite CEM showed simultaneous improvement of membrane anti-fouling performance and energy generation performance in reverse electrodialysis systems.

6 Conclusion

This paper aims to overview the current fouling to ion-exchange membranes and the promising strategies to overcome it. Fouling is a major drawback hampering the industrial application of these processes. However, if the phenomena are well studied and understood, it is possible to find a right solution in order to minimize or completely avoid the fouling. Various foulants may interact and foul the ion-exchange membranes through the hydrophobic–hydrophobic interaction, electrostatic attraction and so on. Thus, the corresponding anti-fouling strategy is proposed based on such foulant–membrane interacting mechanisms.

This paper presents a wide range of techniques allowing fouling investigations and approaches for the following

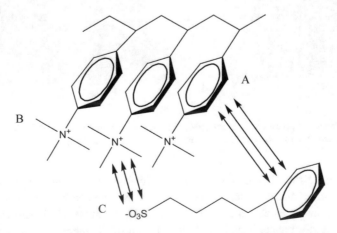

Fig. 6 Suggested mechanism of organic fouling: (A) anion-exchange membrane; (B) anion-exchange group of the membrane; and (C) organic anion (membrane matrix containing sulfonic acid and aromatic groups). Reproduced with permission (Tanaka et al. 2011)

fouling control or/and prevention. The modern tendencies are directed to the creation of membrane materials with anti-fouling properties. Membrane surface modification is one effective strategy. Deposition of novel surface layer by engineering hydrophilicity and charging nature has been a mature solution to prevent foulant adhesion to the hydrophobic and electrostatic attractive membranes. Anti-fouling effect can also be achieved by making the membrane surface smooth.

Moreover, synthesis of new functional membrane is another anti-fouling strategy. By incorporating good functional groups of less sensitivity to fouling (e.g., aliphatic polymer or nanocomposite ones) into novel membrane phase, the developed membrane exhibited extraordinary anti-fouling potential, as well as enhancement to other membrane property due to the new functional materials.

However, the modified membranes usually have anti-fouling properties against certain fouling type, which could be ineffective for complex solutions containing different fouling types. From this point, the perspective direction to investigate the effectiveness of already existed modified membranes against different fouling agents as well as IEM modification to create new membrane resistant to fouling with different nature requires more work.

References

Allison, R. P. (2005). Electrodialysis treatment of surface and waste waters. *GE Water and Process Technologies*, 1–5.

Alvarado, L., Ramirez, A., & Rodriguez-Torres, I. (2009). Cr(VI) removal by continuous electrodeionization: Study of its basic technologies. *Desalination, 249*, 423–428.

Arar, O., Yuksel, U., Kabay, N., & Yuksel, M. (2011). Removal of Cu^{2+} ions by a micro-flow electrodeionization (EDI) system. *Desalination, 277*, 296–300.

Araya-Farias, M., & Bazinet, L. (2006). Electrodialysis of calcium and carbonate high-concentration solutions and impact on membrane fouling. *Desalination, 200*, 624.

Audinos, R. (1997). Ion-exchange membrane processes for clean industrial chemistry. *Chemical Engineering & Technology, 20*, 247–258.

Banasiak, L. J., & Schäfer, A. I. (2009). Removal of boron, fluoride and nitrate by electrodialysis in the presence of organic matter. *Journal of Membrane Science, 334*, 101–109.

Bazinet, L., Brianceau, S., Dubé, P., & Desjardins, Y. (2012). Evolution of cranberry juice physico-chemical parameters during phenolic antioxidant enrichment by electrodialysis with filtration membrane. *Separation and Purification Technology, 87*, 31–39.

Bhadja, V., Makwana, B. S., Maiti, S., Sharma, S., & Chatterjee, U. (2015). Comparative efficacy study of different types of ion exchange membranes for production of ultrapure water via electrodeionization. *Industrial & Engineering Chemistry Research, 54*, 10974–10982.

Ben Sik Ali, M., Hamrouni, B., & Dhahbi, M. (2010). Electrodialytic defluoridation of brackish water: Effect of process parameters and water characteristics. *CLEAN–Soil, Air, Water, 38*(7), 623–629.

Bukhovets, A., Eliseeva, T., & Oren, Y. (2010). Fouling of anion-exchange membranes in electrodialysis of aromatic amino acid solution. *Journal of Membrane Science, 364*, 339–343.

Campione, A., Gurreri, L., Ciofalo, M., Micale, G., Tamburini, A., & Cipollina, A. (2018). Electrodialysis for water desalination: A critical assessment of recent developments on process fundamentals, models and applications. *Desalination, 434*, 121–160.

Chen, S., Li, L., Zhao, C., & Zheng, J. (2010). Surface hydration: Principles and applications toward low-fouling/nonfouling biomaterials. *Polymer, 51*, 5283–5293.

Choi, M. J., Chae, K. J., Ajayi, F. F., Kim, K. Y., Yu, H. W., Kim, C. W., & Kim, I. S. (2011). Effects of biofouling on ion transport through cation exchange membranes and microbial fuel cell performance. *Bioresource Technology, 102*, 298–303.

Cifuentes-Araya, N., Pourcelly, G., & Bazinet, L. (2011). Impact of pulsed electric field on electrodialysis process performance and membrane fouling during consecutive demineralization of a model salt solution containing a high magnesium/calcium ratio. *Journal of Colloid and Interface Science, 361*, 79–89.

Cifuentes-Araya, N., Pourcelly, G., & Bazinet, L. (2012). Multistep mineral fouling growth on a cation-exchange membrane ruled by gradual sieving effects of magnesium and carbonate ions and its delay by pulsed modes of electrodialysis. *Journal of Colloid and Interface Science, 372*, 217–230.

Cohen, R. D., & Probstein, R. F. (1986). Colloidal fouling of reverse-osmosis membranes. *Journal of Colloid and Interface Science, 114*, 194–207.

Dermentzis, K. (2010). Removal of nickel from electroplating rinse waters using electrostatic shielding electrodialysis/ electrodeionization. *Journal of Hazardous Materials, 173*, 647–652.

Disabb-Miller, M. L., Zha, Y., DeCarlo, A. J., Pawar, M., Tew, G. N., & Hickner, M. A. (2013). Water uptake and ion mobility in cross-linked bis(terpyridine)ruthenium-based anion exchange membranes. *Macromolecules, 46*, 9279–9287.

Dong, Z. Q., Wang, B. J., Ma, X. H., Wei, Y. M., & Xu, Z. L. (2015). FAS grafted electrospun poly(vinyl alcohol) nanofiber membranes with robust superhydrophobicity for membrane distillation. *ACS Applied Materials & Interfaces, 7*, 22652–22659.

Dydo, P., & Turek, M. (2013). Boron transport and removal using ion-exchange membranes: A critical review. *Desalination, 310*, 2–8.

Fernandez-Gonzalez, C., Zhang, B., Dominguez-Ramos, A., Ibañez, R., Irabien, A., & Chen, Y. (2017). Enhancing fouling resistance of polyethylene anion exchange membranes using carbon nanotubes and iron oxide nanoparticles. *Desalination, 411*, 19–27.

Flemming, H.-C., Wingender, J., & Szewzyk, U. (2011). *Biofilm highlights*. Berlin: Springer.

Frioui, S., Oumeddour, R., & Lacour, S. (2017). Highly selective extraction of metal ions from dilute solutions by hybrid electrodialysis technology. *Separation and Purification Technology, 174*, 264–274.

Goh, P. S., Lau, W. J., Othman, M. H. D., & Ismail, A. F. (2018). Membrane fouling in desalination and its mitigation strategies. *Desalination, 425*, 130–155.

Grebenyuk, R. D. C. V. D., Peters, S., & Linkov, V. (1998). Surface modification of anion-exchange electrodialysis membranes to enhance anti-fouling characteristics. *Desalination, 115*, 313–329.

Guo, H., Xiao, L., Yu, S., Yang, H., Hu, J., Liu, G., & Tang, Y. (2014). Analysis of anion exchange membrane fouling mechanism caused by anion polyacrylamide in electrodialysis. *Desalination, 346*, 46–53.

Hao, L., Liao, J., Jiang, Y., Zhu, J., Li, J., Zhao, Y., Van der Bruggen, B., Sotto, A., & Shen, J. N. (2018). "Sandwich"-like structure modified anion exchange membrane with enhanced monovalent selectivity and fouling resistant. *Journal of Membrane Science, 556,* 98–106.

Hayes, T. D., & Severin, B. F. (2017). Electrodialysis of highly concentrated brines: Effects of calcium. *Separation and Purification Technology, 175,* 443–453.

Heinz, H., Pramanik, C., Heinz, O., Ding, Y., Mishra, R. K., Marchon, D., Flatt, R. J., Estrela-Lopis, I., Llop, J., Moya, S., & Ziolo, R. F. (2017). Nanoparticle decoration with surfactants: Molecular interactions, assembly, and applications. *Surface Science Reports, 72,* 1–58.

Higa, M., Tanaka, N., Nagase, M., Yutani, K., Kameyama, T., Takamura, K., & Kakihana, Y. (2014). Electrodialytic properties of aromatic and aliphatic type hydrocarbon-based anion-exchange membranes with various anion-exchange groups. *Polymer, 55,* 3951–3960.

Hong, J. G., & Chen, Y. (2014). Nanocomposite reverse electrodialysis (RED) ion-exchange membranes for salinity gradient power generation. *Journal of Membrane Science, 460,* 139–147.

Husson, E., Araya-Farias, M., Desjardins, Y., & Bazinet, L. (2013). Selective anthocyanins enrichment of cranberry juice by electrodialysis with ultrafiltration membranes stacked. *Innovative Food Science & Emerging Technologies, 17,* 153–162.

İpekçi, D., Altıok, E., Bunani, S., Yoshizuka, K., Nishihama, S., Arda, M., & Kabay, N. (2018). Effect of acid-base solutions used in acid-base compartments for simultaneous recovery of lithium and boron from aqueous solution using bipolar membrane electrodialysis (BMED). *Desalination, 448,* 69–75.

Jaroszek, H., & Dydo, P. (2016). Ion-exchange membranes in chemical synthesis—A review. *Open Chemistry, 14,* 1–19.

Kango, S., Kalia, S., Celli, A., Njuguna, J., Habibi, Y., & Kumar, R. (2013). Surface modification of inorganic nanoparticles for development of organic–inorganic nanocomposites—A review. *Progress in Polymer Science, 38,* 1232–1261.

Kato, K., Uchida, E., Kang, E. T., Uyama, Y., & Ikada, Y. (2003). Polymer surface with graft chains. *Progress in Polymer Science, 28,* 209–259.

Kingsbury, R. S., Chu, K., & Coronell, O. (2015). Energy storage by reversible electrodialysis: The concentration battery. *Journal of Membrane Science, 495,* 502–516.

Kochkodan, V., & Hilal, N. (2015). A comprehensive review on surface modified polymer membranes for biofouling mitigation. *Desalination, 356,* 187–207.

Kochkodan, V., Johnson, D. J., & Hilal, N. (2014). Polymeric membranes: Surface modification for minimizing (bio)colloidal fouling. *Advances in Colloid and Interface Science, 206,* 116–140.

Komlenic, R. (2010). Biofouling: Rethinking the causes of membrane biofouling. *Filtration & Separation, 47,* 26–28.

Lee, H. J., Park, J. S., Kang, M. S., & Moon, S. H. (2003). Effects of silica sol on ion exchange membranes: Electrochemical characterization of anion exchange membranes in electrodialysis of silica sol containing-solutions. *Korean Journal of Chemical Engineering, 20,* 889–895.

Lee, H.-J., Song, J.-H., & Moon, S.-H. (2013). Comparison of electrodialysis reversal (EDR) and electrodeionization reversal (EDIR) for water softening. *Desalination, 314,* 43–49.

Li, Y., Shi, S., Cao, H., Zhao, Z., Su, C., & Wen, H. (2018). Improvement of the antifouling performance and stability of an anion exchange membrane by surface modification with graphene oxide (GO) and polydopamine (PDA). *Journal of Membrane Science, 566,* 44–53.

Li, Y., Shi, S., Gao, H., Zhao, Z., Su, C., & Wen, H. (2018). Improvement of the antifouling performance and stability of an anion exchange membrane by surface modification with graphene oxide (GO) and polydopamine (PDA). *Journal of Membrane Science, 566,* 44–53.

Liu, Y., Yang, S., Chen, Y., Liao, J., Pan, J., Sotto, A., & Shen, J. (2019). Preparation of water-based anion-exchange membrane from PVA for anti-fouling in the electrodialysis process. *Journal of Membrane Science, 570–571,* 130–138.

Luiz, A., McClure, D. D., Lim, K., Leslie, G., Coster, H. G. L., Barton, G. W., & Kavanagh, J. M. (2017). Potential upgrading of bio-refinery streams by electrodialysis. *Desalination, 415,* 20–28.

Luo, H., Xu, P., Jenkins, P. E., & Ren, Z. (2012). Ionic composition and transport mechanisms in microbial desalination cells. *Journal of Membrane Science, 409–410,* 16–23.

Mei, Y., & Tang, C. Y. (2018). Recent developments and future perspectives of reverse electrodialysis technology: A review. *Desalination, 425,* 156–174.

Merle, G., Wessling, M., & Nijmeijer, K. (2011). Anion exchange membranes for alkaline fuel cells: A review. *Journal of Membrane Science, 377,* 1–35.

Mikhaylin, S., & Bazinet, L. (2016). Fouling on ion-exchange membranes: Classification, characterization and strategies of prevention and control. *Advances in Colloid and Interface Science, 229,* 34–56.

Momose, N. H. T., Arimoto, O., & Yamaguchi, K. (1991). Effects of low concentration levels of calcium and magnesium in the feed brine on the performance of a membrane chlor-alkali cell. *Journal of the Electrochemical Society, 138,* 735–741.

Mondor, M., Ippersiel, D., Lamarche, F., & Masse, L. (2009). Fouling characterization of electrodialysis membranes used for the recovery and concentration of ammonia from swine manure. *Bioresource Technology, 100,* 566–571.

Mulyati, S., Takagi, R., Fujii, A., Ohmukai, Y., & Matsuyama, H. (2013). Simultaneous improvement of the monovalent anion selectivity and antifouling properties of an anion exchange membrane in an electrodialysis process, using polyelectrolyte multilayer deposition. *Journal of Membrane Science, 431,* 113–120.

Nguyen, T., Roddick, F. A., & Fan, L. (2012). Biofouling of water treatment membranes: A review of the underlying causes, monitoring techniques and control measures. *Membranes, 2,* 804–840.

Nie, G., Li, X., Tao, J., Wu, W., & Liao, S. (2015). Alkali resistant cross-linked poly(arylene ether sulfone)s membranes containing aromatic side-chain quaternary ammonium groups. *Journal of Membrane Science, 474,* 187–195.

Park, J.-S., Lee, H.-J., Choi, S.-J., Geckeler, K. E., Cho, J., & Moon, S.-H. (2003). Fouling mitigation of anion exchange membrane by zeta potential control. *Journal of Colloid and Interface Science, 259,* 293–300.

Park, J. S., Lee, H. J., & Moon, S. H. (2003). Determination of an optimum frequency of square wave power for fouling mitigation in desalting electrodialysis in the presence of humate. *Separation and Purification Technology, 30,* 101–112.

Pashley, R. M. (1981a). Hydration forces between mica surfaces in aqueous electrolyte solutions. *Journal of Colloid and Interface Science, 80,* 153–162.

Pashley, R. M. (1981b). DLVO and hydration forces between mica surfaces in Li+, Na+, K+, and Cs+ electrolyte solutions A correlation of double-layer and hydration forces with surface cation exchange properties. *Journal of Colloid and Interface Science, 83,* 531–546.

Ponomarev, M. I., Verbich, S. V., Grebenyuk, V. D., & Teselkin, V. V. (1989). Removal of microparticles from water by electromembrane filtration. *Colloid Journal of the USSR, 51,* 153–157.

Pontié, M., Ben Rejeb, S., & Legrand, J. (2012). Anti-microbial approach onto cationic-exchange membranes. *Separation and Purification Technology, 101,* 91–97.

Ran, J., Wu, L., He, Y., Yang, Z., Wang, Y., Jiang, C., Ge, L., Bakangura, E., & Xu, T. W. (2017). Ion exchange membranes:

New developments and applications. *Journal of Membrane Science, 522*, 267–291.

Reig, M., Casas, S., Gibert, O., Valderrama, C., & Cortina, J. L. (2016). Integration of nanofiltration and bipolar electrodialysis for valorization of seawater desalination brines: Production of drinking and waste water treatment chemicals. *Desalination, 382*, 13–20.

Reig, M., Casas, S., Valderrama, C., Gibert, O., & Cortina, J. L. (2016). Integration of monopolar and bipolar electrodialysis for valorization of seawater reverse osmosis desalination brines: Production of strong acid and base. *Desalination, 398*, 87–97.

Ruan, H., Zheng, Z., Pan, J., Gao, C., Van der Bruggen, B., & Shen, J. (2018). Mussel-inspired sulfonated polydopamine coating on anion exchange membrane for improving permselectivity and anti-fouling property. *Journal of Membrane Science, 550*, 427–435.

Sarapulova, V., Nevakshenova, E., Nebavskaya, X., Kozmai, A., Aleshkina, D., Pourcelly, G., & Pismenskaya, N. (2018). Characterization of bulk and surface properties of anion-exchange membranes in initial stages of fouling by red wine. *Journal of Membrane Science, 559*, 170–182.

Shi, S., Cho, S.-H., Lee, Y.-H., Yun, S.-H., Woo, J.-J., & Moon, S.-H. (2011). Desalination of fish meat extract by electrodialysis and characterization of membrane fouling. *Korean Journal of Chemical Engineering, 28*, 575–582.

Sri Mulyati, R. T., Fujii, A., Ohmukai, Y., & Tatsuo Maruyama, H. M. (2012). Improvement of the antifouling potential of an anion exchange membrane by surface modification with a polyelectrolyte for an electrodialysis process. *Journal of Membrane Science, 417*, 137–143.

Stokes, K. K., Orlicki, J. A., & Beyer, F. L. (2011). RAFT polymerization and thermal behavior of trimethylphosphonium polystyrenes for anion exchange membranes. *Polym. Chem., 2*, 80–82.

Strathmann, H. (2010). Electrodialysis, a mature technology with a multitude of new applications. *Desalination, 264*, 268–288.

Suwal, S., Doyen, A., & Bazinet, L. (2015). Characterization of protein, peptide and amino acid fouling on ion-exchange and filtration membranes: Review of current and recently developed methods. *Journal of Membrane Science, 496*, 267–283.

Tamburini, A., Tedesco, M., Cipollina, A., Micale, G., Ciofalo, M., Papapetrou, M., Van Baak, W., & Piacentino, A. (2017). Reverse electrodialysis heat engine for sustainable power production. *Applied Energy, 206*, 1334–1353.

Tanaka, N., Nagase, M., & Higa, M. (2011). Preparation of aliphatic-hydrocarbon-based anion-exchange membranes and their anti-organic-fouling properties. *Journal of Membrane Science, 384*, 27–36.

Tijing, L. D., Woo, Y. C., Choi, J.-S., Lee, S., Kim, S.-H., & Shon, H. K. (2015). Fouling and its control in membrane distillation—A review. *Journal of Membrane Science, 475*, 215–244.

Tong, X., Zhang, B., & Chen, Y. (2016). Fouling resistant nanocomposite cation exchange membrane with enhanced power generation for reverse electrodialysis. *Journal of Membrane Science, 516*, 162–171.

Xu, T. W., & Huang, C. (2008). *Preparation and application technology of ion exchange membrane.* Chemical Industry Press.

Ulbricht, M. (2006). Advanced functional polymer membranes. *Polymer, 47*, 2217–2262.

Valero, F., & Arbós, R. (2010). Desalination of brackish river water using electrodialysis reversal (EDR). *Desalination, 253*, 170–174.

van Egmond, W. J., Saakes, M., Porada, S., Meuwissen, T., Buisman, C. J. N., & Hamelers, H. V. M. (2016). The concentration gradient flow battery as electricity storage system: Technology potential and energy dissipation. *Journal of Power Sources, 325*, 129–139.

Vaselbehagh, M., Karkhanechi, H., Mulyati, S., Takagi, R., & Matsuyama, H. (2014). Improved antifouling of anion-exchange membrane by polydopamine coating in electrodialysis process. *Desalination, 332*, 126–133.

Vatanpour, V., Madaeni, S. S., Moradian, R., Zinadini, S., & Astinchap, B. (2011). Fabrication and characterization of novel antifouling nanofiltration membrane prepared from oxidized multiwalled carbon nanotube/polyethersulfone nanocomposite. *Journal of Membrane Science, 375*, 284–294.

Wang, Q., Yang, P., & Cong, W. (2011). Cation-exchange membrane fouling and cleaning in bipolar membrane electrodialysis of industrial glutamate production wastewater. *Separation and Purification Technology, 79*, 103–113.

Wang, W., Fu, R., Liu, Z., & Wang, H. (2017). Low-resistance anti-fouling ion exchange membranes fouled by organic foulants in electrodialysis. *Desalination, 417*, 1–8.

Wang, Z., & Lin, S. (2017). The impact of low-surface-energy functional groups on oil fouling resistance in membrane distillation. *Journal of Membrane Science, 527*, 68–77.

Xiao, X., Wu, C., Cui, P., Luo, J., Wu, Y., & Xu, T. (2011). Cation exchange hybrid membranes from SPPO and multi-alkoxy silicon copolymer: Preparation, properties and diffusion dialysis performances for sodium hydroxide recovery. *Journal of Membrane Science, 379*, 112–120.

Yang, C., Wang, S., Jiang, L., Hu, J., Ma, W., & Sun, G. (2015). 1,2-Dimethylimidazolium-functionalized cross-linked alkaline anion exchange membranes for alkaline direct methanol fuel cells. *International Journal of Hydrogen Energy, 40*, 2363–2370.

Yeon, K. H., Seong, J. H., Rengaraj, S., & Moon, S. H. (2003). Electrochemical characterization of ion-exchange resin beds and removal of cobalt by electrodeionization for high purity water production. *Separation Science and Technology, 38*, 443–462.

Yooprasertchuti, K., & Dechadilok, P. (2018). Relaxation effect on intrapore diffusivities of highly charged colloidal particles confined in porous membranes. *Transport in Porous Media, 123*, 341–366.

Zeng, L., Zhao, T. S., Wei, L., Jiang, H. R., & Wu, M. C. (2019). Anion exchange membranes for aqueous acid-based redox flow batteries: Current status and challenges. *Applied Energy, 233–234*, 622–643.

Zhang, B., Gu, S., Wang, J., Liu, Y., Herring, A. M., & Yan, Y. (2012). Tertiary sulfonium as a cationic functional group for hydroxide exchange membranes. *RSC Advances, 2*, 12683.

Zhang, R., Liu, Y., He, M., Su, Y., Zhao, X., Elimelech, M., & Jiang, Z. (2016). Antifouling membranes for sustainable water purification: Strategies and mechanisms. *Chemical Society Reviews, 45*, 5888–5924.

Zhao, X., Zhang, R., Liu, Y., He, M., Su, Y., Gao, C., & Jiang, Z. (2018). Antifouling membrane surface construction: Chemistry plays a critical role. *Journal of Membrane Science, 551*, 145–171.

Zhao, Z., Shi, S., Cao, H., Li, Y., & Van der Bruggen, B. (2018). Layer-by-layer assembly of anion exchange membrane by electrodeposition of polyelectrolytes for improved antifouling performance. *Journal of Membrane Science, 558*, 1–8.

Zhao, Z., Shi, S., Cao, H., Shan, B., & Sheng, Y. (2018). Property characterization and mechanism analysis on organic fouling of structurally different anion exchange membranes in electrodialysis. *Desalination, 428*, 199–206.

Zuo, X., Yu, S., Xu, X., Bao, R., Xu, J., & Qu, W. (2009). Preparation of organic–inorganic hybrid cation-exchange membranes via blending method and their electrochemical characterization. *Journal of Membrane Science, 328*, 23–30.

Aging and Degradation of Ion-Exchange Membranes

Le Han

Abstract

Recent years have seen the widespread use of ion-exchange membranes (IEM) in chemical, environmental and energy source, and new functional membrane materials have been continuously proposed. Nevertheless, there are yet few studies on the mechanism of the aging and degradation of IEM materials during long-term use, which severely limits the further application of ion-exchange membranes. This paper reviews the existing research on the aging and degradation of ion-exchange membranes in different cases and summarizes the following three major mechanisms, solute adsorption/penetration/fouling, membrane burning under overlimiting current region, and membrane degradation under different conditions. Finally, prevention and control measures such as antioxidant membrane and polarization inhibition are proposed.

Keywords

Ion-exchange membrane • Membrane aging • Membrane degradation • Fouling

1 Introduction

One major worldwide issue is water shortage, and membrane-based technology is considered one promising solution. The ion-exchange membrane (IEM), one typical charged and dense polymer-based membrane, thanks to its high permselectivity, low electrical resistance, good mechanical, and form stability and high chemical stability, can be used in various fields of water treatment such as chemical industry, wastewater treatment, seawater desalination, and food processing.

Among the application fields of IEM, the desalination of seawater and the treatment of industrial solution are the most popular ones (Sata 2004; Strathmann 2010; Tanaka 2015a; Mulder 1998). There are also many IEM-based technologies available for water issues, such as electrodialysis (ED), bipolar membrane electrodialysis (BMED), electrodeionization (EDI). Taking ED as an example, many successful industrial applications are available since nearly a half century ago: antioxidants can be extracted from winery waste (Sarapulova et al. 2018); amino acids can be recovered from fermentation broth (Sata 2004); and edible salt can be produced from seawater (Hirayama 1993). Indeed, since in 1992, people in Japan concentrated salt from seawater using ED technology, and the yield reached up to 1.4 million tons (Xu and Huang 2008). It is also possible to use BMED to produce acid and base from neutral salts (Raucq et al. 1993) and EDI to produce pure water or achieve desalination (Sata 2004).

The IEM consists of three basic components, polymer backbone, fixed group, and movable ions on the group. According to the types of charge of the fixed group, IEM can be mainly divided into cation-exchange membrane (CEM) and anion-exchange membrane (AEM). These membranes can selectively allow the passage of oppositely charged ions (counter-ions) while obstructing similarly charged ions (co-ions). In any case, the desired IEM should have high selective permeability, high ion-exchange capacity, low swelling degree, and high mechanical strength, etc. (Tanaka 2015a; Xu 2005; Hui et al. 2016).

Despite of long lifetime than other dense membranes, IEM suffers aging and functional polymer degradation during the long-term use. The aging and degradation of the IEM will result into many membrane problems, like membrane damage, increased materials resistance, fouling, and so on, greatly reducing the membrane lifetime and process efficiency. The clear definition for membrane ageing and degradation are not given in IEM field yet. Often membrane

L. Han (✉)
Key Laboratory of the Three Gorges Reservoir Region's Eco-environment, Ministry of Education, College of Environment and Ecology, Chongqing University, Chongqing, 400044, People's Republic of China
e-mail: lehan@cqu.edu.cn

© Springer Nature Switzerland AG 2021
Z. Zhang et al. (eds.), *Membrane Technology Enhancement for Environmental Protection and Sustainable Industrial Growth*, Advances in Science, Technology & Innovation,
https://doi.org/10.1007/978-3-030-41295-1_3

degradation is seen as one main driver for membrane aging. The degradation of membrane refers to the loss of availability of the membrane due to the destruction of the stability and/or functionality of the polymer in the IEM under certain physical or chemical factors, such as high temperature or high pH. For example the oxidizing substances in the water like O_2, Cl_2, etc., generated by the polar chamber reaction, can cause oxidative damage to the membrane, rendering functional group degradation.

Thus, it is necessary to summarize the mechanism of aging and degradation of ion-exchange membrane. At the same time, the review paper also proposed prevention and control measures for the aging and degradation of ion-exchange membranes according to the current research.

2 The Membrane Characteristics and Their Relationship of the Ion-Exchange Membrane

The main membrane characteristics related to the aging and degradation of the IEM include exchange capacity, water content and swelling degree, thickness and contact angle of membrane, mechanical strength, conductivity, transport number, permeability. Because these performance parameters vary upon the degree of membrane aging and degradation, we list some details of them in Table 1.

The ion-exchange capacity and water content are two critical parameters of IEM directly or indirectly affecting IME performance. A simple plot between these two is seen in Fig. 1, with the water content of the membrane plotted as the Y-axis and the exchange capacity as the X-axis, then the reciprocal of its slope referring to the concentration of the fixed group (Xu and Huang 2008). In general, for given mechanical strength of the membrane, the higher the concentration of the fixed group, the better the performance of the membrane, and the direction pointer is marked on Fig. 1.

3 Aging and Degradation in IEM Applications

In the applications of IEM, such as electrodialysis (Ghalloussi et al. 2013), electrodialysis reverse (Zhang 2017), biopolar membrane electrodialysis (Mani 1991; Hwang and Choi 2006), electrodeionization (Ting-qing et al. 2014), fuel cell (Tanaka 2015b), etc., the aging and degradation of the IEM result in changes of the membrane characteristics, e.g., decreased ion-exchange capacity, increased membrane resistance, inactivation of functional groups, changes of the hydrophilicity, etc., which in turn reduces the use efficiency of the membrane.

3.1 Electrodialysis

The electrodialysis (ED) process is a combination of an electrochemical process and a dialysis diffusion process. Driven by an external direct current (DC) electric field, ions migrate under fixed directions and are selectively allowed to pass through the respective IEMs. With assistance of repeatable units consisting of alternatively arranged AEM and CEM, two solutions of different salinity are produced, the diluate and the concentrate. Next to the diluate and the concentrate chambers, there are anode chamber where the oxidation reaction occurs (the anode solution is acidic) and cathode chamber where reduction reaction occurs (the cathode solution is alkaline). Ghalloussi et al. (2013) used the IEM supplied by Eurodia Industry SA to find that the ion-exchange capacity of almost all membranes decreased significantly after two years of dealing with a solution containing a weak organic acid in the ED process, and the change of other membrane characteristics is also shown in Table 2.

All the investigated membranes experienced ion-exchange capacity decrease after use with only exception as CEM2, which is the relatively stable sample in any aspects of membrane characteristics (not further detailed). This decrease varies from ca. 70% for CEM1 to ca. 85% for AEM1. Varying water contact angle before and after use for the IEM is also observed, indicating the membrane hydrophilicity changed. Used CEM1 becomes more hydrophobic, in line with decreased water content, volume fraction of the inter-gel solution. Clearly opposite change was found for AEM1 which is more hydrophilic after use. The above data indicates clear aging phenomenon.

Loss of hydrophilic sulfonic acid functional sites of CEM1 results in a decrease in water content and membrane thickness as well as an increase in contact angle. Thus, the membrane becomes denser and its pores are narrower, which was manifested by a decrease in the volume fraction of the inter-gel solution. Lower ion-exchange capacity and narrower pores result in a decrease in the concentration and mobility of the counter-ions, increased adsorption of common ions, resulting in a loss of specific conductivity and permselectivity, but contributes to higher conductivity.

One can conclude that for CEM1, the great loss of ion-exchange capacity (activated sites of the function groups) and the hydrophilicity as well as associated decreased conductivity and permeability are all the consequence of membrane aging. The aging on AEM1 probably differs from that for CEM1, considering the great probability of organic anion adsorption (likely due to the hydrophilic groups such as carboxyl and hydroxyl groups of the organic acids). That explained the hydrophilization of AEM1, together with increased thickness and water content.

Table 1 Main membrane characteristics related to the aging and degradation of the IEMs

Membrane characteristics	Definition	Determination methods	Commonly reported values	References
Ion-exchange capacity	A chemical property index of the concentration of the reactive group in the reaction membrane and its ability to exchange with the counter-ion	Titration	1–3 meq/g	Sata (2004), Xu and Huang (2008), Tanaka (2015a)
Water content	The intrinsic water bound to the reactive group in the membrane	$\frac{W_{\text{wet}} - W_{\text{dry}}}{W_{\text{dry}}} \times 100\%$, $\frac{W_{\text{wet}} - W_{\text{dry}}}{W_{\text{wet}}} \times 100\%$	25–50%	Xu and Huang (2008), Tong (2016)
Swelling degree	The percentage change of the area or volume of an ionic membrane after immersion in a solution	Refer to that for water content	–	Xu and Huang (2008)
Membrane thickness	The thickness of dry or wet membrane, generally expressed in units of micrometers (µm) or millimeters (mm)	Micrometer	0.4–0.7 mm for the heterogeneous, 0.1–0.2 mm for the homogeneous	Sata (2004), Tong (2016)
Hydrophilicity	Be used to characterize the hydrophilicity of the membrane	Contact Angle measurement	–	
Mechanical strength	Burst strength refers to the pressure of the ion-exchange membrane subjected to vertical direction	Hydraulic blasting	–	Tong (2016)
	Tensile strength, the ion-exchange membrane is subjected to a horizontal tensile force	Expressed as the stress strength per unit area	–	Xu and Huang (2008), Tong (2016)
Conductivity	Generally expressed by electrical conductivity or resistivity, and it is also commonly expressed by membrane surface resistance	Electrical conductivity or resistivity or surface resistance	2–10 Ω cm^2	Sata (2004), Xu and Huang (2008)
Transport number	Refers to the equivalent percentage of ions moving through the membrane, which characterizes the membrane's selective permeability to counter-ion, that is, the repulsion to the common ion	Measuring the membrane potential	–	Sata (2004), Xu and Huang (2008)
Permeability	Reflects the selective permeation ability of ion-exchange membranes for different ions, which is related to many factors such as ion-exchange capacity, membrane thickness	Measuring the ion concentration before and after electrifying to calculate the ion migration number and then obtained the permeability	–	Tanaka (2015a), Xu and Huang (2008)

3.2 Reverse Electrodialysis

Reverse electrodialysis (RED) is a technology based on the energy conversion process of charged ions through ion-exchange membranes to generate electricity. Obviously, IEM is one of the core elements of RED process, and its physical and chemical properties play a decisive role in the performance of RED (Xia et al. 2018; Ngai Yin et al. 2014).

During the research of RED, Zhang (2017) found that the ion-exchange capacity and the contact angle of the AEM (tailor-made and commercial ones) changed after adding bovine serum albumin (BSA) and sodium alginate (SA) aqueous solution (content is 0.5%) as organic pollutants.

The ion-exchange capacity and the contact angle of the AEM before and after fouling are shown in Table 3. The ion-exchange capacity of the contaminated AEM increased, indicating pollutants absorption. Meanwhile, the contact angles of the two AEMs both increased after pollution. These change of the AEM characteristics reflected the membrane aging.

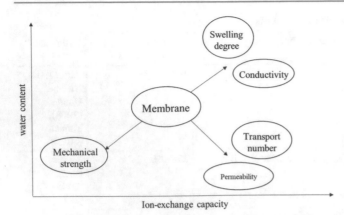

Fig. 1 Complex relationship between ion-exchange membrane characteristics. Reproduced with permission (Xu and Huang 2008). Note the shift directions of one specific parameter can be discussed given fixed influential ones other than ion-exchange capacity (IEC) and water content (WC). (1) To the upper right hand, change of the membrane swelling degree and conductivity is in line with increase of IEC and WC. (2) To the lower right hand, the change of the membrane transport number and permeability is in line with increase of IEC and decrease of WC. (3) To the lower left hand, change of the membrane mechanical strength is in line with decrease of IEC and WC

3.3 Bipolar Membrane Electrodialysis

Bipolar membrane (BPM) is a relatively new type of ion-exchange composite membrane, which is composed of a cation-exchange layer, an interface hydrophilic layer and an anion-exchange layer (Tanaka 2015c). Hydrolysis occurs when the BPM is reversely pressurized. Bipolar membrane electrodialysis (BMED) is based on the principle of hydrolysis separation and electrodialysis to convert the salt solution to the corresponding acid and base without introducing a new component.

When the BPM was first applied to the industry, Mani (1991) observed a decrease in the selective permeability and voltage of the BPM during the conversion of the water-soluble salt to its corresponding acid and base by using BMED. Ui-Son Hwang and Choi (2006) investigated the stability of commercial bipolar membranes immersed in concentrated NaOH solution at different temperatures and different times in the study of BMED and examined the physical and electrochemical changes of the alkali-treated membrane; membrane degradation was reported. FTIR

Table 2 Membrane characteristics of the investigated CEM and AEM (comparisons between the new ones and the used ones): ion-exchange capacity (IEC), water content (%W), thickness (T_m), contact angle ($\theta°$), electric conductivity in 0.1 M NaCl (κ_m), volume fraction of inter-gel solution (f_2) linking to water content, and apparent counter-ion transport number ($t_{counterion}$). More detail information of IEM is not given due to confidential reasons (Ghalloussi et al. 2013)

Membranes	CEM				AEM			
	CEM1N	CEM1U	CEM2N	CEM2U	AEM1N	AEM1U	AEM2N	AEM2U
IEC (meq/g of dry IEM)	2.66	0.77	1.11	1.08	1.45	0.23	1.94	1.40
%W	28.2	23.7	21.1	28.4	28.7	40.3	24.9	23.10
T_m (um)	176	160	170	175	148	344	160	170
$\theta°$	38 ± 2	79 ± 2	43 ± 2	48 ± 2	69 ± 2	52 ± 2	67 ± 2	62 ± 2
κ_m (mS/cm)	6.1	2.3	5.1	4.8	5.6	2.9	7.9	5.0
f_2	0.12	0.10	0.11	0.07	0.11	0.39	0.12	0.30
$t_{counterion}$	0.97	0.60	0.92	0.91	0.99	0.99	0.92	0.95

IEMxy where IEM was CEM or AEM, *x* referred to the membrane (*x* = 1 or 2), and *y* to the membrane state (*y* = N for new samples and *y* = U for used samples)

Table 3 Ion-exchange capacity and contact angle of AEM before and after fouling

Membrane	Status	IEC (meq/g)	Water contact angle
AEM-Type I	Pristine	1.41 ± 0.01	1.41 ± 0.01
	BSA fouling	1.47 ± 0.02	1.47 ± 0.02
	SA fouling	1.50 ± 0.01	1.50 ± 0.01
AEM-Type II	Pristine	1.12 ± 0.07	1.12 ± 0.07
	BSA fouling	1.23 ± 0.01	1.23 ± 0.01
	SA fouling	1.31 ± 0.02	1.31 ± 0.02

Reproduced with permission (Zhang 2017)

spectrum shows the absorption peak of the original BP-1 membrane at a wave number of 1020 cm^{-1} that corresponds to the C–N stretching of the quaternary amine. In the alkali-treated membrane, the intensity of this peak decreased with the immersion time, indicating that the quaternary amine groups in the BP-1 membrane (anion-exchange layer, AEL for short) decomposed.

As shown in Table 4, the water content of the BP-1 membrane immersed in the 2.5 M NaOH solution did not change significantly, yet for those immersed in a 5.0 M NaOH solution, the water content of the membrane increases sharply when the temperature is 40 °C. After soaking in BPMs at a high temperature, the hydrolysis resistance of the membrane and the membrane resistance all increased due to the change of the AEL structure of the BPM, and the resistance of the BPM may also increase due to the decomposition of the ion-exchangeable groups.

3.4 Electrodeionization

Electrodeionization (EDI) is a new separation technique that combines ion-exchange resin and IEM to realize deionization process under the action of DC electric field (Xu and Huang 2008). The most important feature of EDI is the automatic regeneration of the mixed bed ion-exchange resin filled in the electrolysis dialysis chamber by H$^+$ and OH$^-$ produced by water splitting, thereby achieving continuous deep desalting. Compared with ED, EDI has broader application prospects in the fields of electronics, medicine, energy, and other industries and laboratories due to its high degree of advancement and practicability. It is also expected to become the mainstream technology for pure water manufacturing.

During EDI process, crystals are deposited on the IEM after 1800 h, giving granular crystals-deposited membrane surface instead of smooth one in the beginning (Ting-qing et al. 2014). These crystals seriously increased the membrane resistance and decreased membrane ion-exchange capacity and selective permeability by blocking the mass transfer in IEMs (from bulk to membrane surface and in the resin phase) in the diluted compartment, as shown in Table 5.

3.5 Fuel Cell

Fuel cell (FC) is a kind of power generation device that directly converts the chemical energy stored in fuel and oxidant into electricity without pollution with high efficiency (50–80%) (Sata 2004). The electrode provides a place for electron transfer, the anode catalyzes the oxidation process of fuel such as hydrogen, and the cathode catalyzes the reduction process of oxidant such as oxygen. The membrane electrode of the fuel cell is composed of a gas diffusion layer, anode catalytic layer, IEM, cathode catalytic layer, and gas diffusion layer. The anode and cathode of the hydrogen FC are separated by an ion-exchange membrane, which is the core part of the battery and plays a key role in battery performance.

In this process, aging and degradation of the IEM occurred, causing cell voltage declined (with the decreasing rate of a unit cell voltage was 4 mV/1000 h and 2.2 mV/1000 h in a 20-unit cell (Tanaka 2015b). The finding of Yu et al. (Jingrong et al. 2003) showed that a very low sulfur element content near the cathode side is observed after the use of fuel cell as compared with that before using, providing possible evidence for the occurrence of oxidation degradation of the membrane during the process of FC.

Table 4 The water content and electric resistance of alkali-treated BPMs

Temperature (°C)	NaOH concentration (mol/L)	Water content (g H$_2$O/g dry membrane)	Electric resistances with immersion time ($\times 10^3$ Ω m^2)				
			1 day	2 day	3 day	5 day	7 day
20	2.5	0.16	–	–	–	–	1.64
	5.0	0.16	–	–	–	–	1.78
30	2.5	0.16	–	–	–	–	1.63
	5.0	0.17	1.67	1.78	1.76	1.79	1.84
40	2.5	0.17	–	–	–	–	1.64
	5.0	0.19	1.70	1.76	1.87	1.92	2.03
50	2.5	0.18	1.62	1.63	1.71	1.73	1.78
	5.0	0.25	1.85	1.86	1.93	2.30	2.55

Reproduced with permission (Hwang and Choi 2006)

Table 5 Scanning electron microscopy (SEM) picture and membranes resistance of used and new IEM

Ion-exchange membranes		SEM picture	Resistance (Ω)	Membranes resistance ($\Omega\ cm^2$)
AEM	Fresh		12.8	9.46
	After 1800-hour use		20.2	14.93
CEM	Fresh		15.4	11.34
	After 1800-hour use		22.3	16.45

Reproduced with permission (Ting-qing et al. 2014)

4 The Mechanism of Aging and Degradation of IEM

The aging and degradation of the IEM result into the varying of the membrane characteristics like increased resistance or even membrane damage, which greatly reduces the membrane lifetime and process efficiency. The mechanism of aging and degradation of IEM are summarized as the following two parts, fouling and degradation under extreme conditions.

4.1 Membrane Fouling

The phenomenon that foulant deposit on the surface and pores of the membrane often under the concentration polarization condition is called membrane fouling (Sata et al. 1996). Membrane fouling is also a major cause of aging of IEM. The interaction between pollutants and membranes mainly includes electrostatic interaction, van der Waals force, solvation, and spatial stereoscopic effects. According to the nature of pollutants, it can be divided into organic pollution and inorganic pollution. Inorganic pollution is mainly caused by the formation of Ca^{2+}, Mg^{2+} or other multivalent ions on the surface or inside of the ionic membrane, and it is caused by polarization or solution supersaturated. Organic pollution is formed by the deposition or penetration of proteins, humic acids, surfactants or other macromolecular organic substances on the surface of the membrane or inside the membrane free volume or pore, mainly due to the electrostatic interaction or bonding between functional groups even of same charge. For example, Fig. 2 illustrates the structure of a commercially available hydrocarbon-type AEM and a suggested mechanism for fouling by aromatic compounds where two aspects of fouling mechanism are indicated: affinity between anions and the oppositively charged fixed groups of the AEM, and affinity as π–π interactions between the aromatic membrane matrix and compounds (Tanaka et al. 2011). In addition, the molecular size of the organic matter and the network structure of the film also affect the interaction of the organic matter to the membrane base film (Lu et al. 2015).

It is reported that fouling caused a clear loss of ion-exchange sites for both AEM and CEM, as well as other membrane characteristics' variation (Ghalloussi et al. 2013). The adsorption of organic colloidal particles in the membrane free volume occurred and a higher swelling degree was observed due to the hydrophilic nature of foulant. Higher swelling of the IEM will cause breakage of some bonds and polymer chains, resulting in macroscopic defects (defects, voids, cavities) in the membrane that are filled with external solutions. In long-term operation, these defects may form through non-selective channels. Colloidal associations formed by organic fouling (e.g., amino acids) in the nano-pore may contaminate the membrane (Sata 2004),

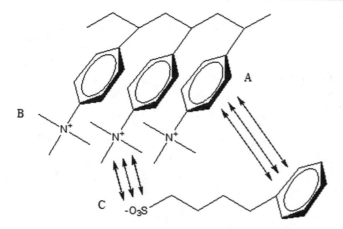

Fig. 2 Suggested mechanism of organic fouling (A) anion-exchange membrane; (B) anion-exchange group of the membrane; and (C) representative organic anion (containing sulfonic acid and aromatic groups). Reproduced with permission (Tanaka et al. 2011)

resulting into a reduction in the effective radius of the ion conducting channels, reducing the mobility of ions in the nano-pore, affecting the ion permeability.

4.2 Membrane Degradation Under Extreme Conditions

Membrane fouling mentioned in Sect. 4.1 frequently happened under common conditions. However, whether membranes keep stable under some extreme conditions such as extreme pH or high temperature should be discussed.

4.2.1 Membrane Degradation Under Extreme pH
Under extreme pH conditions, AEMs are easily affected (Ghigo et al. 2010). Among the different cationic-charged groups, the quaternary ammonium groups are dissociated throughout the pH range, while the other groups are only weakly dissociated. Commercially available AEMs are mostly quaternary ammonium groups (Xu and Huang 2008). However, quaternary ammonium groups are easily prone to degradation by OH⁻ nucleophilic attack (Ghigo et al. 2010). Taking the AEM used in alkaline fuel cells as an example, there are three kinds of degradation mechanism proposed as rearrangement, elimination and nucleophilic substitution (Merle et al. 2011), as shown in Table 6. (1) Different rearrangements mechanisms tend to take place, i.e., Sommelet–Hauser and the Stevens rearrangements. Both the products of the two rearrangements are tertiary amines. The migrating group can be an alkyl or a benzyl moiety, but in the former case, only the Stevens rearrangements will take place (Ghigo et al. 2010). (2) The elimination also includes two mechanism, Hofmann elimination and E1 elimination. Hofmann elimination is a process that the hydroxyl ions attack a beta-hydrogen of the ammonium, resulting into the

formation of an alkene, an amine, and a water molecule (Merle et al. 2011; Norcross 1993). This degradation occurs slowly at modest temperatures (60 °C), but it proceeds considerably faster resulting in faster degradation at higher temperature (100 °C). When the atom carrying the charges is bulky, another mechanism called E1 elimination occurs. This elimination is relatively rare but can occur on the carbon located in the alpha or beta position of the ammonium (Cope and Mehta 1963). In this case, the hydroxyl ions attack the hydrogen of the methyl group belonging to the ammonium. Then a rearrangement occurs, leading to the formation of an alkene and an amine. (3) The degradation mechanism by nucleophilic substitution (SN2) results from the hydroxyl attack of alpha-hydrogen on the ammonium, and then alcohol and amine are formed (Merle et al. 2011).

4.2.2 Membrane Degradation Under High Temperature
In addition to the extreme pH conditions responsible for some membrane degradation, high temperature is another not negligible factor. It is reported that CEM is much easier affected by temperature than AEM, and sulfonate group is the mostly affected fixed group among the different fixed group in CEM such as sulfonate groups, phosphate groups, carboxylic groups, and so on (Xu and Huang 2008). Taking the perfluorosulfonic acid (PFSA) membranes as an example, the decomposition of carbon-based radicals is mainly caused by the change of temperature. Elevated temperature on the membrane causes a reduction in water content, which ultimately leads to irreversible drying. For PFSA membranes, the temperature must reach 150 °C before the chemical structure is significantly affected. This thermal stability is due to the C–F bond strength and the shielding effect of fluorine (Samms et al. 1996). Above 200 °C, the loss of sulfonate groups begins to occur. Samms et al. (1996) and Wilkie (Wilkie et al. 2010) both proposed a degradation mechanism for thermal degradation of perfluorosulfonic acid membranes which includes cleavage of C–S bonds to form SO_2, an OH· radical and carbon-based radicals which further degraded, as shown in Fig. 3.

In addition to the loss of sulfonate groups, the change of temperature will also cause changes in the membrane structure. Collier et al. (2006) reported that after the freeze/thaw cycle, some molecular level rearrangement occurs in the membrane. Chain entanglement and aggregation reduction in the hydrophilic region may lead to an opening up of the molecular structure, ultimately leading to a decrease in strength. Fluctuations in temperature may cause the water in the fuel cell to freeze, thereby expanding its volume and then decreasing again upon melting. These volume changes may have a detrimental effect on the life of the membrane. On the other hand, the increased reactor temperature resulted from some membrane structural damage such as cracks, tears,

Table 6 Possible membrane degradation mechanisms under extreme pH

Degradation mechanism		Reaction mechanism	References
Rearrangement	Sommelet-Hauser rearrangement		Ghigo et al. (2010)
	Stevens rearrangement		Ghigo et al. (2010)
Elimination	Hofmann elimination		Merle et al. (2011), Norcross (1993)
	E1 elimination		Cope and Mehta (1963)
Nucleophilic substitution	SN2		Merle et al. (2011)

Reproduced with permission

perforations, or pinhole blisters (of the CEM) will also lead to membrane degradation (Taniguchi et al. 2004). When perforations or pinholes occur in the membranes, the reactant gases cross and react on the surface of the catalyst. The heat of this reaction may cause the membrane to soften or even melt, thereby increasing gas crossover and destructive cycling, reducing fuel efficiency and thermodynamic efficiency (Laconti et al. 2003; Xie et al. 2005).

4.2.3 Changes of pH and Temperature Caused by Overlimiting Current

The generation of overlimiting current can also cause changes in pH and temperature in the chamber, possibly resulting into aging and degradation of the ion-exchange membranes.

Under the overlimiting current (Region III in Fig. 4, the water splitting often occurs at the membrane interface (Choi and Moon 2003), and such a phenomenon is explained by the proposed mechanism as catalytic theory and electric field theory (Simons 1984). The former mechanism suggests that OH^- and H^+ ions may be produced in charge transfer reactions between charged groups and water, occurring around a thin layer on the surface of the IEM due to the

reversible protonation of weakly basic groups such as tertiary amines (Simons 1984). The electric field theory (Li et al. 1998) suggests that K_d (dissociation constant) can be increased as much as 10^7 times because of a Second Wien effect due to the strong electric field (10^8 V m^{-1}) at the membrane–solution interface under the deep concentration polarization.

Rubinstein et.al. published a series of papers (Rubinstein 1984; Rubinstein and Shtilman 1979; Rubinstein and Maletzki 1991; Rubinstein and Segel 1979; Rubinstein et al. 1988) in which the occurrence of the overlimiting current region was explained by another mechanism called electroconvection. Near the limiting current density the salt concentration at the membrane surface is very low and the basic assumption of electroneutrality does not hold anymore. The electric double layer would be drastically distorted and a weak space charge is build up near the membrane surface. Due to a nonuniform ion conductance through the membrane, the electric field is not uniform. The interaction of space charges with the electric field gives rise to a spatially inhomogeneous bulk force that is bound to set the fluid in the depletion diffusion layer in motion, which is called electroconvection. It was thought that the significant

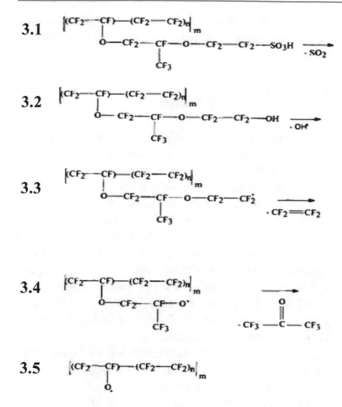

Fig. 3 Possible radical decomposition mechanism for PFSA. Reproduced with permission (Samms et al. 1996)

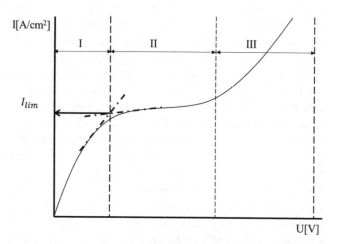

Fig. 4 Current–voltage curve. Reproduced with permission (Tanaka 2015d). It shows a typical example of a measured current–voltage curve for a cation-exchange membrane. The curve has the S-type shape including region I, exhibiting ohmic behavior, region II, in which the current varies slowly with voltage and exhibits a plateau presenting the limiting current density; and region III, corresponding to the overlimiting current sphere in which the current increases gradually

concentration polarization formed in the diffusion boundary layer was disturbed by turbulent convection when the current exceeded the limiting current density, leading to hydrodynamic mixing. As a consequence, the concentration

at the membrane surface increased to a sufficient level for generation of the overlimiting current (Krol et al. 1999). It has be noted that much attention in this has been focused on cation-exchange membranes as it often was assumed that the overlimiting current with anion exchange membranes was associated with water dissociation.

It is reported that temperature increases were not negligible any more in the region beyond the limiting current density (Mavrov et al. 1993), and once the temperature increases, the degradation of sulfonate groups in the CEM may occur. As for the water splitting, it may lead to a pH change on both sides of the membrane. As a result scaling onto or into the membrane may occur, and the membrane may deteriorate under extreme pH conditions such as degradation of quaternary ammonium groups. Further, water dissociation is an energy-consuming process and reduces the current efficiency (Krol et al. 1999).

4.3 Membrane Degradation Under the Attack of Radicals

The membrane applied in the fuel cell is subjected to both a harsh chemically oxidizing environment on the cathode side and chemically reducing environment on the anode side. Additionally, peroxy and hydroperoxy radicals formed in the fuel cell attack the membrane. The chemical degradation of IEM membranes is mainly attributed to these attacks (Collier et al. 2006). Peroxy and hydroperoxy radicals are formed by a series reactions between hydrogen and oxygen via Pt catalyst (Xic et al. 2005). The peroxide may also form from incomplete reduction of oxygen on the platinum surface and then the intermediate reacted with trace metal ions to form peroxide radicals (Jingrong et al. 2003; Pozio et al. 2003).

Lindén et al. (1993) describe several reactions that polymers can undergo in the presence of peroxide radicals. It involves the abstraction of hydrogen from the polymer, leading to the degradation of the polymers. In membranes containing aromatic rings, such as polystyrene sulfonic acid (PSSA) membranes, the degradation is usually due to the addition of OH· to the aromatic ring. Since the para position is usually blocked, most of the attack is on the orthocarbon of the alkyl substituent. There is acid catalyzed water elimination leading to loss of chain scission and $-SO_3$ groups. Once a free radical is formed from the polymer, it will degrade further by reacting with oxygen (Panchenko 2004). The peroxide intermediates can attack hydrogen on the alpha-carbon, resulting in oxidative decomposition (Jingrong et al. 2003). O_2 has a strong tendency to be added directly to a cyclohexadienyl-like compounds to form a radical intermediate. This reaction can result in higher hydroxylation products and/or bond breaking reactions (Hübner and Roduner 1999).

5 The Possible Prevention Measures of Membrane Aging and Degradation

In response to the aging and degradation of ion-exchange membranes summarized in this view, the following possible measures could be insightful.

On the one hand, desirable membranes with better property like anti-fouling or anti-degradation (chemically inert) are urgently needed in practice. (1) Anti-fouling materials—For example, membrane surface modification method probably attach a protective layer to the surface of the membrane, which can reduce the adsorption-induced fouling of the membrane (Xu 2003). Grebenyuk et al. (Grebenyuk et al. 1998) demonstrated that the modification of AEM by high molecular mass surfactants enhanced the fouling resistance. The non-polar radicals of high molecular mass surfactants attached horizontally to the surface of the membranes form a protective layer on the membranes, making them less prone to a decrease of selectivity with respect to counterion transfer, or an acceleration of the voltage drop during electrodialysis. (2) Anti-degradation materials—Regarding the degradation possibly occurs in electrode compartment, a strong anti-oxidation membrane is recommended as the first membrane close to the electrode rinse solution. Rubinstein et al. (1984) found that replacing the AEM contains a quaternary ammonium group with an AEM contains a crown ether group can reduce water splitting, which to some degree ensure long lifetime of membrane by reducing the exposure of the membrane to the overlimiting current region.

On the other hand, the IEM-base processes should be operated with enough care in various aspects like membrane cleaning, pretreatment of the inflow or improving the process operation and setup/module design. (1) Membrane cleaning—For localized concentration polarization, measures such as online washing, periodic pickling, addition of anti-fouling agent, and reverse polarity operation are often used regarding to the process operation parameters (Jun et al. 2001). (2) Pretreatment—Improving the quality of the inflow water through pretreatment is widely used and proven successful in protecting the membrane in many other processes and definitely cannot be ignored in the IEM-based processes. (3) Operation improvement—Means to strengthen the mass transfer (or reduce the thickness of the boundary layer or mass transfer resistance) are meaningful to avoid the harsh conditions, like replacing the mesh with good agitation and ion conductive materials (Jun et al. 2001). During the operation of the fuel cell, a free radical scavenger is once suggested as it can reduce the concentration of free radicals, thereby reducing the rate of membrane degradation caused by free radical attack (Collier et al. 2006). (4) System improvement—A delicate setup/module modification could further protect the membrane used; for example, protective chamber consisting of an anti-oxidation inert membranes is once proposed.

6 Conclusion and Perspectives

Ion-exchange membranes (IEMs) are experiencing large popularity in various fields ranging from water treatment to energy production. Due to the aging and degradation of IEMs, many adverse consequences, such as decreased permselectivity, increased membrane resistance, more energy consumption yet lower efficiency as well as membrane damage, greatly reduce the membranes lifetime and deteriorate the membrane lifetime. But, there are as yet few studies on the aging and degradation of IEM materials during long-term use. In this view, we summarized the various aging and/or degradation observations reported in different IEM-based processes, such as (reverse) electrodialysis, bipolar electrodialysis, electrodeionization, fuel cell.

Since many membrane characteristics are highly interdependent for the IEM, the aging and degradation often occur with several characteristics' clear change, normally toward decreasing the permselectivity and efficiency of the IEM. Three major mechanisms are accordingly proposed, namely membrane fouling, membrane burning under overlimiting current region, and membrane degradation under extreme conditions like solution pH and temperature at the membrane interface. We also proposed prevention measures to deal with different kinds of aging and degradation of IEM.

The measures proposed in this review to prevent aging and degradation of IEM are of great significance to further improve its applications. Future work could be categorized as the two following domains: (1) To comprehensively optimize the systems from aspects of membrane materials (polymeric or inorganic or even mixed type), stack/cell design (channel dimension, flow geometry), water quality characteristics and process operation conditions. (2) To better understand the membrane degradation and aging phenomena occurring at the interface or the inside part of the membrane, including its reasons, dynamics, and influential factors, as well as the mechanism involved.

References

Choi, J.-H., & Moon, S.-H. (2003). Structural change of ion-exchange membrane surfaces under high electric fields and its effects on membrane properties. *Journal of Colloid and Interface Science, 265,* 93–100.

Collier, A., Wang, H., Ziyuan, X., Zhang, J., & Wilkinson, D. (2006). Degradation of polymer electrolyte membranes. *International Journal of Hydrogen Energy, 31,* 1838–1854.

Cope, A. C., & Mehta, A. S. (1963). Mechanism of the Hofmann elimination reaction: An ylide intermediate in the pyrolysis of a highly branched quaternary hydroxide. *Journal of the American Chemical Society, 85.*

Ghalloussi, R., Garcia-Vasquez, W., Chaabane, L., Dammak, L., Larchet, C., Deabate, S. V., et al. (2013). Ageing of ion-exchange membranes in electrodialysis: A structural and physicochemical investigation. *Journal of Membrane Science, 436,* 68–78.

Ghigo, G., Cagnina, S., Maranzana, A., & Tonachini, G. (2010). The mechanism of the Stevens and Sommelet-Hauser rearrangements. A theoretical study. *The Journal of organic chemistry, 75,* 3608–3617.

Grebenyuk, V. D., Chebotareva, R. D., Peters, S., & Linkov, V. (1998). Surface modification of anion-exchange electrodialysis membranes to enhance anti-fouling characteristics. *Desalination, 115,* 313–329.

Hirayama, K. (1993). Analysis of properties of advanced ion-exchange membranes: Neosepta CIMS and ACS-2. In *7th Symposium on Salt* (pp. 53–58).

Hübner, G., & Roduner, E. (1999). EPR investigation of HO' radical initiated degradation reactions of sulfonated aromatics as model compounds for fuel cell proton conducting membranes. *Journal of Materials Chemistry, 9,* 409–418.

Hui, W., Xuehao, W., Jia, G., Yongxing, C., Jingjing, G., Lijuan, L., & Xiulian, R. (2016). Present status and development trend of modification of ion-exchange membranes. *Inorganic Chemicals Industry, 48,* 1–4.

Hwang, U.-S., & Choi, J.-H. (2006). Changes in the electrochemical characteristics of a bipolar membrane immersed in high concentration of alkaline solutions. *Separation and Purification Technology, 48,* 16–23.

Jingrong, Y., Baolian, Y., Danmin, X., Fuqiang, L., Zhigang, S., Yongzhu, F., & Huamin, Z. (2003). Degradation mechanism of polystyrene sulfonic acid membrane and application of its composite membranes in fuel cells. *Physical Chemistry Chemical Physics, 5,* 611–615.

Jun, S., Quan, Y., Congjie, G. (2001). *Membrane technical manual* (Vol. 28, pp. 127).

Krol, J. J., Wessling, M., & Strathmann, H. (1999). Concentration polarization with monopolar ion exchange membranes: Current–voltage curves and water dissociation. *Journal of Membrane Science, 162,* 145–154.

Laconti, A. B., Hamdan, M., & McDonald, R. C. (2003). Mechanisms of membrane degradation.

Li, J., Wang, Y., Yang, C., Long, G., & Shen, H. (1998). Membrane catalytic deprotonation effects. *Journal of Membrane Science, 147,* 247–256.

Lindén, L. Å, Rabek, J. F., Kaczmarek, H., Kaminska, A., & Scoponi, M. (1993). Photooxidative degradation of polymers by HOxxx and HO2xxx radicals generated during the photolysis of H_2O_2, $FeCl_3$, and Fenton reagents. *Coordination Chemistry Reviews, 125,* 195–217.

Lu, L., Zhi-juan, Z., Ya, L., Shao-yuan, S., Xin-min, W., Li-ying, W., & Lin, L. (2015). Research progress in fouling of ion exchange membrane for electrodialysis desalination of industrial wastewater. *The Chinese Journal of Process Engineering, 15,* 881–891.

Mani, K. N. (1991). Electrodialysis water splitting technology. *Journal of Membrane Science, 58,* 117–138.

Mavrov, V., Pusch, W., Kominek, O., & Wheelwright, S. (1993). Concentration polarization and water splitting at electrodialysis membranes. *Desalination, 91,* 225–252.

Merle, G., Wessling, M., & Nijmeijer, K. (2011). Anion exchange membranes for alkaline fuel cells: A review. *Journal of Membrane Science, 377,* 1–35.

Mulder, B. M. (1998). Basic principles of membrane technology. *Zeitschrift Für Physikalische Chemie, 203,* 263–263.

Ngai Yin, Y., Vermaas, D. A., Kitty, N., & Menachem, E. (2014). Thermodynamic, energy efficiency, and power density analysis of reverse electrodialysis power generation with natural salinity gradients. *Environmental Science & Technology, 48,* 4925–4936.

Norcross, B. E. (1993). Advanced organic chemistry: Reactions, mechanisms, and structure (4th ed.) (March, Jerry). *Journal of Chemical Education, 70.*

Panchenko, D. (2004). MÖLler, Sixt, Roduner, In situ EPR investigation of polymer electrolyte membrane degradation in fuel cell applications. *Journal of Power Sources, 127,* 325–330.

Pozio, A., Silva, R. F., Francesco, M. D., & Giorgi, L. (2003). Nafion degradation in PEFCs from end plate iron contamination. *Electrochimica Acta, 48,* 1543–1549.

Raucq, D., Pourcelly, G., & Gavach, C. (1993). Production of sulphuric acid and caustic soda from sodium sulphate by electromembrane processes. Comparison between electro-electrodialysis and electrodialysis on bipolar membrane. *Desalination, 91,* 163–175.

Rubinstein, I. (1984). Effects of deviation from local electroneutrality upon electro-diffusional ionic transport across a cation-selective membrane. *Reactive Polymers Ion Exchangers Sorbents, 2,* 117–131.

Rubinstein, I., & Maletzki, F. (1991). Electroconvection at an electrically inhomogeneous permselective membrane surface. *Journal of the Chemical Society, Faraday Transactions, 87,* 2079–2087.

Rubinstein, I., & Segel, L. A. (1979). Breakdown of a stationary solution to the Nernst–Planck–Poisson equations. *Journal of the Chemical Society Faraday Transactions Molecular & Chemical Physics, 75,* 936–940.

Rubinstein, I., & Shtilman, L. (1979). Voltage against current curves of cation exchange membranes. *Journal of the Chemical Society Faraday Transactions, 75,* 231–246.

Rubinstein, I., Staude, E., & Kedem, O. (1988). Role of the membrane surface in concentration polarization at ion-exchange membrane. *Desalination, 69,* 101–114.

Rubinstein, I., Warshawsky, A., Schechtman, L., & Kedem, O. (1984). Elimination of acid-base generation ('water-splitting') in electrodialysis. *Desalination, 51,* 55–60.

Samms, S. R., Wasmus, S., & Savinell, R. F. (1996). Thermal stability of Nafion® in simulated fuel cell environments. *Journal of the Electrochemical Society, 143,* 1498–1504.

Sarapulova, V., Nevakshenova, E., Nebavskaya, X., Kozmai, A., Aleshkina, D., Pourcelly, G., & Pismenskaya, N. (2018). Characterization of bulk and surface properties of anion-exchange membranes in initial stages of fouling by red wine. *Journal of Membrane Science, 559,* 170–182.

Sata, T. (2004). Ion exchange membranes: Preparation, characterization, modification and application. *Royal Society of Chemistry.*

Sata, T., Tsujimoto, M., Yamaguchi, T., & Matsusaki, K. (1996). Change of anion exchange membranes in an aqueous sodium hydroxide solution at high temperature. *Journal of Membrane Science, 112,* 161–170.

Simons, R. (1984). Electric field effects on proton transfer between ionizable groups and water in ion exchange membranes. *Electrochimica Acta, 29,* 151–158.

Strathmann, H. (2010). Electrodialysis, a mature technology with a multitude of new applications. *Desalination, 264,* 268–288.

Tanaka, Y. (2015a). Fundamental properties of ion exchange membranes. In *Ion exchange membranes* (pp. 29–65).

Tanaka, Y. (2015b). Fuel cell. In *Ion exchange membranes* (pp. 459–470).

Tanaka, Y. (2015c). Bipolar membrane electrodialysis. In *Ion exchange membranes* (pp. 369–392).

Tanaka, Y. (2015d). Concentration polarization. In *Ion exchange membranes* (pp. 101–121).

Tanaka, N., Nagase, M., & Higa, M. (2011). Preparation of aliphatic-hydrocarbon-based anion-exchange membranes and their anti-organic-fouling properties. *Journal of Membrane Science, 384,* 27–36.

Taniguchi, A., Akita, T., Yasuda, K., & Miyazaki, Y. (2004). Analysis of electrocatalyst degradation in PEMFC caused by cell reversal during fuel starvation. *Journal of Power Sources, 130,* 42–49.

Ting-qing, L., Shan, G., Zong-heng, W., & Shi-huai, Z. (2014). Ion exchange membrane properties variation and its effect in electrodeionization process. *Journal of Tianjin Polytechnic University,* 10–14.

Tong, B. (2016). Preparation, characterization and application of polymeric ion exchange membranes. University of Science and Technology of China.

Wilkie, C. A., Thomsen, J. R., & Mittleman, M. L. (2010). Interaction of poly(methyl methacrylate) and Nafions. *Journal of Applied Polymer Science, 42,* 901–909.

Xia, C., Chenxiao, J., Yaoming, W., & Tongwen, X. (2018). Advances in reverse electrodialysis and its applications on renewable energy & environment protection. *CIESC Jorunal, 69,* 188–202.

Xie, J., Wood Iii, D. L., Wayne, D. M., Zawodzinski, T. A., Atanassov, P., & Borup, R. L. (2005). Durability of PEFCs at high humidity conditions. *Journal of the Electrochemical Society, 152,* A104–A113.

Xu, T. (2003). Chemistry and technology tutorial of membrane. University of Science and Technology of China.

Xu, T. (2005). Ion exchange membranes: State of their development and perspective. *Journal of Membrane Science, 263,* 1–29.

Xu, T., & Huang, C. (2008). *Preparation and application technology of ion exchange membrane.* Chemical Industry Press.

Zhang, W. (2017). Study on membrane electrochemical properties and fouling behaviors in reverse electrodialysis. Harbin Institute of Technology.

Recent Trends in Membrane Processes for Water Purification of Brackish Water

Muhammad Sarfraz

Abstract

Naturally occurring brackish water, normally containing 500–10,000 mg/L of total dissolved solids, is not safe for direct consumption due to its salinity. The salinity level needs to be reduced to a level below 500 mg/L to make it drinkable as per recommendations of the World Health Organization (WHO). Reverse osmosis (RO) process for water desalination purposes is currently considered to be the most effective, economical, efficient, and optimized method dominating the water purification market. An extensive research has been carried out in the field of membrane-based brackish water reverse osmosis (BWRO) process to improve its desalting performance. Various aspects of a BWRO process system such as nature and type of membrane material, module design parameters, process configuration, energy recovery devices, operating parameters, economical aspects are reviewed in this chapter. Theoretical background of a BWRO process, transport mechanism through BWRO membranes, and desalination performance of BWRO membranes are considered here. An updated review of different commercially available BWRO membranes, membrane modules, and process configurations is also provided. In addition, major components of a typical BWRO plant such as pretreatment unit, pumping system, membrane module section, and post-treatment unit are also described in this review. General process considerations, economic aspects, energy recovery options, and process optimization of a BWRO system are discussed here. High-performance BWRO membranes prepared from polymeric and thin-film composite materials are inserted in commercial spiral wound modules to make the desalting process economically efficient. Concentrated brine rejected from a BWRO plant can be economically treated by installing solar stills at sunlit places. A double-stage membrane process can enhance water recovery of BWRO plants from the usual range of 85–90% to about 95–98%. Brackish water can be purified by BWRO process at reduced cost by using high rejection membranes, installing larger pressure vessels, and adopting hybrid membrane design.

Keywords

Brackish water • Desalination • Reverse osmosis • Reverse osmosis membranes • Integrally skinned anisotropic membranes • Membrane modules • BWRO plant • Water flux • Salt rejection • Water pretreatment

Nomenclature

Abbreviations

BWRO	Brackish water reverse osmosis
CA	Cellulose acetate
ED	Electrodialysis
MD	Membrane distillation
MEE	Multi-effect evaporation
MSFD	Multistage flash distillation
MVCD	Mechanical vapor compression distillation
NF	Nanofiltration
PA	Polyamide
PVC	Polyvinyl chloride
RO	Reverse osmosis
SWRO	Seawater reverse osmosis
TVCD	Thermal vapor compression distillation
WHO	World Health Organization

Symbols

A	Membrane permeability coefficient for water
B	Membrane permeability coefficient for salt
C	Solute concentration (mol/L or mg/L)

M. Sarfraz (✉)
Department of Polymer and Process Engineering, University of Engineering and Technology, Lahore, 54890, Pakistan
e-mail: msarfraz@uet.edu.pk

© Springer Nature Switzerland AG 2021
Z. Zhang et al. (eds.), *Membrane Technology Enhancement for Environmental Protection and Sustainable Industrial Growth*, Advances in Science, Technology & Innovation,
https://doi.org/10.1007/978-3-030-41295-1_4

J	Flux (molar)
p	Pressure (externally applied)
q	Flow rate (volumetric)
R	Water recovery
SR	Solute rejection
SP	Salt passage

Greek Letters

π	Osmotic pressure (natural)
%	Percent

Subscripts

F	Feed
P	Permeate
s	Salt
w	Water

Superscript

°	Degree

1 Introduction

Every living creature on the globe essentially requires water for its survival and nurturing. Ample water is available on Earth as seawater (94%), brackish water (2%), and freshwater (4%). Main sources of freshwater are underground water (72%) and large masses of glacier ice (27%). Based on its salinity level, water is mainly classified into high-salinity seawater, medium-salinity brackish water, and low-salinity freshwater (El-Manharawy and Hafez 2001). High-salinity levels of brackish water limit its direct use as drinking water according to World Health Organization (WHO) guidelines; brackish water needs to be desalinated before it can be consumed as drinking water. Desalination is a technique used to obtain hygienic water appropriate for drinking, domestic, and industrial consumption by eliminating salts, minerals, and other contaminating agents from brackish water or seawater (Shenvi et al. 2015). Clean water finds vital applications in domestic, industrial, and farming sectors (Aquastat 2013). Inappropriate handling of clean water, exploitation of water resources, contamination of water reservoirs, disproportionate population boom, and changing meteorological conditions are the main reasons of water shortage. The situation of water scarceness is even more aggravated in distant localities.

Water desalination is believed to be the leading process to obtain freshwater from various brackish water sources (Prihasto et al. 2009). Legislative policies of some countries to maintain good quality of drinkable water require improvements in the efficiency of water desalination processes (Xia et al. 2007). Conventional thermal desalination processes to produce portable water are based on the principle of phase change. These expensive energy-intensive methods include multi-effect evaporation (MEE), multistage flash distillation (MSFD), thermal vapor compression distillation (TVCD), and mechanical vapor compression distillation (MVCD) processes (Ali et al. 2018; Youssef et al. 2014). Recent developments in membrane science and technology have resulted in the adoption of more economical and high-tech membrane-based desalination processes like reverse osmosis (RO), ultrafiltration (UF), nanofiltration (NF), electrodialysis (ED), and membrane distillation (MD) on account of their low energy requirements, high process efficiency, compact design, low space requirement, simple operation, and easy process control (Altaee et al. 2014; Cay-Durgun and Lind 2018; Werber et al. 2016a, b; Qasim et al. 2018).

Among the contemporary membrane-based water desalination processes, reverse osmosis is the most economical, mature, reliable, and state-of-the-minute technique to purify brackish water and seawater (Atab et al. 2016; Goh 2018). Depending upon salt concentration of raw feed water, reverse osmosis processes are broadly characterized as seawater reverse osmosis (SWRO) plants processing seawater having salinity of about 30,000 mg/L, and brackish water reverse osmosis (BWRO) plants operating on brackish water having salinity in the range of 500–10,000 mg/L. BWRO processes are further categorized as high-salinity BRWO process purifying feed water possessing salinity level in the range of 2500–10,000 mg/L and low-salinity BRWO process desalting feed water having salinity in the range of 500–2500 mg/L. Development of high-performance membrane materials, reduced energy requirements, and optimized process conditions have helped the reverse osmosis-based desalination technology to steadily capture the market place. At present, more than half of the desalted water accessible worldwide is being processed via reverse osmosis technology to produce portable water at competitively low price (Goh 2018). Desalination performance of BWRO process depends on permeation properties of semipermeable membrane, quality of feed water being processed, and working conditions of the process.

This chapter overviews the state-of-the-minute trends and numerous aspects of reverse osmosis membrane processes for water purification of brackish water. The fundamental governing principle, theoretical background, and various desalination performance measuring quantities associated with reverse osmosis membrane desalination process are comprehensively reviewed here. Recent advancements made in the field of membrane material development to prepare integrally skinned cellulose acetate membranes, non-cellulosic polymeric membranes, and thin-film interfacial composite membranes are also discussed in detail. Furthermore, a general description of a complete commercial BWRO plant along with different types of commercial membrane modules is thoroughly discussed. Lastly, general design principles of BWRO process, optimization strategies, and pertinent energy and economic considerations are also summarized in this chapter.

2 Theoretical Background of BWRO Process

2.1 Reverse Osmosis Process

Water molecules spontaneously permeate from a dilute solution to a concentrated one through a semipermeable membrane without any externally applied pressure in the naturally occurring osmosis phenomena (Fig. 1a). In this situation, the partially permeable membrane does not let the solute to pass through it. This normal process persists until the chemical potentials of brine solutions on either side of the membrane osmotically equilibrate (Fig. 1b). Permeation of water molecules can be halted or even reversed by exerting an external pressure on the concentrated side of the membrane (Chen et al. 2011). If the level of externally applied pressure (p) on the concentrated side of membrane is higher than natural osmotic pressure (π) of concentrated solution, water molecules would move from a concentrated solution to a dilute solution (Fig. 1c). This unnatural process known as reverse osmosis is used to produce clean water from brackish water.

2.2 Transport Through BWRO Membranes

Transport of water and salt permeating through BWRO membranes takes place via solution-diffusion mechanism (Wijmans and Baker 1995). According to this model, water flux (J_w) and solute flux (J_s) are, respectively, proportional to net transmembrane pressure difference ($\Delta p > \Delta \pi$) and solute concentration difference ($C_F - C_P$) for BWRO systems operating at steady state (Wang et al. 2014):

$$J_w = A(\Delta p - \Delta \pi)$$
$$J_s = B(C_F - C_P)$$

where A and B are membrane permeability coefficients for water and salt, respectively. Δp is the transmembrane applied pressure difference while $\Delta \pi$ is the transmembrane osmotic pressure difference between feed and permeate sides. C_F and C_P are feed-side and permeate-side solute concentrations, respectively.

2.3 Desalination Performance of BWRO Membranes

Recovery (R) of a BWRO desalination process is the fractional volume of feed water permeated through the membrane. Percentage recovery ($\%R$) of a BWRO system, in view of Fig. 2, can be estimated as follows:

$$R(\%) = \frac{q_P}{q_F} \times 100$$

where q_F and q_P are respective volumetric flow rates of feed and permeate streams. Recovery of a BWRO system generally depends on quality and salinity level of feed water, process design arrangement, pretreatment process, and concentrate handling (Greenlee et al. 2009). Typical recovery values of most BWRO systems vary from 50 to 85% (Kucera 2010).

Salt or solute rejection (SR) refers to the fractional rejection of a specific solute by BWRO membrane as given by following expression:

$$SR = \left(1 - \frac{C_P}{C_F}\right) \times 100\%$$

where C_F and C_P, respectively, denote feed and permeate solute concentrations normally expressed as mol/L or mg/L. Salt rejection is a strong function of molecular size, molecular weight, and nature of solute molecules, as well as solute–membrane interaction determined by the type of BWRO membrane (Kucera 2010).

Another relevant term, opposite to SR, called solute or salt passage (SP) expresses fraction of a certain permeating solute passed through BWRO membrane.

$$SP = \frac{C_P}{C_F} \times 100\%$$

Both the opposing terms SR and SP can be correlated as:

$$SP(\%) = 100\% - SR(\%)$$

3 BWRO Membranes

The heart of BWRO desalination process is the selective semipermeable membranes which preferentially permit water molecules to permeate through them while rejecting solutes (salts and other dissolved solids) to produce freshwater product. Desalination performance of BWRO plants measured in terms of permeate water flux and salt rejection strongly depends on chemical, morphological, and structural properties of the employed membrane (Kucera 2010). Continuous development of unique membrane materials is

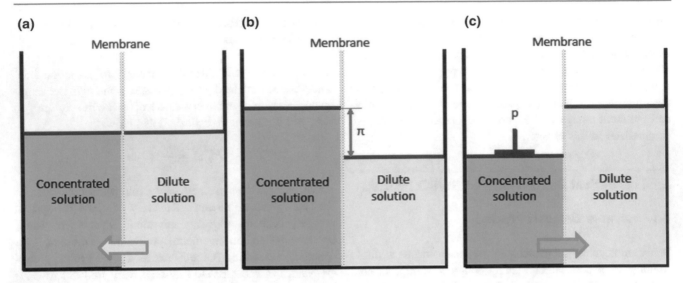

Fig. 1 Schematic illustration of **a** osmosis phenomena, **b** osmotic equilibrium, and **c** reverse osmosis process

Fig. 2 Schematic representation of a membrane-based brackish water reverse osmosis process

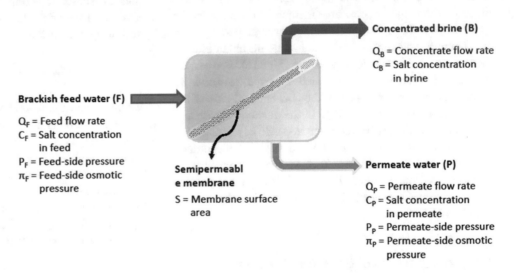

$\Delta P = P_F - P_P = $ *Transmembrane operating pressure difference between feed and permeate sides*
$\Delta \pi = \pi_F - \pi_P = $ *Transmembrane osmotic pressure difference between feed and permeate sides*

crucially important to improve desalination performance, process efficiency, technical progress, and overall economics of BWRO desalination plants (Kang and Cao 2012). Essential features of atypical membrane include high permeate water flux, high salt rejection, improved mechanical strength, high thermal stability, extended membrane life, low cost, low fouling tendency, and high resistance to biological and chemical degradation (Misdan et al. 2012; Li and Wang 2010).

A variety of BWRO membranes have been prepared from various polymeric and inorganic materials via different preparation techniques. BWRO membranes can systematically be classified into cellulosic membranes, non-cellulosic polymer membranes, interfacial composite membranes, and cross-linked membranes (Baker 2012). Commercially

dominating BWRO desalination membranes include conventional cellulose acetate (CA) and non-cellulosic aliphatic and aromatic polyamide (PA) membranes (Lee et al. 2011; Asadollahi et al. 2017; Curcio and Drioli 2009; Ghosh et al. 2011). Majority of polymer membranes consist of anisotropic (asymmetric) layered structures with varying permeation properties, porosity level, pore diameter, chemical composition, and sometimes different polymer materials throughout the matrix. Typical anisotropic membrane structure comprises a thin, dense, less permeable, and selective skin layer detained on a thicker, microporous, highly permeable, and less selective substrate layer (Pinnau and Freeman 1999). Top selective layer controls membrane permeation properties while porous sublayer renders mechanical strength needed to safely handle the membrane

(Ismail et al. 2019; Duarte and Bordado 2016). Asymmetric membranes can be categorized into integrally skinned asymmetric cellulosic membranes, non-cellulosic polymer membranes, and thin-film interfacial composite membranes depending on whether the membrane structure consists of the same or different materials (Wang and Wang 2019). In addition, several other BWRO membranes have also been prepared from some other materials during the last few years. Research advancements regarding the development and modification of commercially important BWRO desalination membranes to enhance their desalination performance are briefly described here.

3.1 Integrally Skinned Cellulosic Membranes

The earliest defect-free high-performance anisotropic BWRO desalination membrane was prepared from cellulose acetate via Loeb–Sourirajan method (Loeb and Sourirajan 1963; Kimura and Sourirajan 1967; Loeb 1981). Discovery of this first asymmetric membrane composed of layered structure proved to be a revolutionary achievement resulting into the commercialization of BWRO desalination process (Ulbricht 2006; Glater 1998). In spite of being one of the oldest BWRO membranes, cellulosic membranes still occupy a small size of the market due to its salient features including easy manufacturing, good mechanical toughness and comparatively high resistance against chlorine attack as compared to interfacial composite membranes. Sterilization of infected feed water by chlorine or other oxidizing agent can be easily accomplished via these membranes since they can resist perpetual exposure of 1 ppm chlorine.

Being highly sensitive to acetyl content contained by them, water flux and salt rejection characteristics of cellulose acetate membranes can be significantly improved by increasing the acetylation content of these membranes (Reid and Breton 1959; Rosenbaum et al. 1967; Lonsdale 1966). Cellulose diacetate and cellulose triacetate membranes exhibiting isotropic morphology were subsequently developed to improve permeate flux and salt rejection properties (Sidney and Srinivasa 1964; Kucera 2010; Loeb and Sourirajan 1963). An increase in degree of acetylation of cellulose acetate membranes generally improves water-to-salt flux ratio at the expense of reduced water flux (Lonsdale et al. 1965). In contrast to their diacetate counterpart, cellulose triacetate membranes are thermally, chemically, and biologically more stable. As compared to cellulose triacetate membrane, a blend membrane of cellulose diacetate and cellulose triacetate was prepared to further improve water flux, salt rejection and more importantly densification resistance when operated under high pressures (Sudak 1990; Lee et al. 2011). Advantageous features of cellulose acetate membranes encompass decent fouling resistance, reasonable chlorine resistance, and low preparation cost (Kucera 2010; El-Saied et al. 2003; Singh 2015). Practical limitations of cellulose acetate membranes include low silica rejection, degradation susceptibility towards bacteriological attack, and low thermal stability. Normally, cellulose acetate membranes are used below an operating temperature of 35 °C (Kucera 2010). Furthermore, these membranes can only be safely used in a narrow operating range of pH of feed water; acetate groups tend to gradually hydrolyze outside the pH range of 4–6 and leads to shortening of membrane lifetime (Vos et al. 1966). Key information of some of the commercially available cellulose acetate membranes is reported in Table 1.

3.2 Non-cellulosic Polymer Membranes

Integrally skinned isotropic membranes can also be commercially prepared from non-cellulosic polymers like aliphatic and aromatic polyamides. Asymmetric membranes prepared from aliphatic polyamides offered mild water fluxes and low salt rejections. Various asymmetric membranes composed of aromatic polyamide in the form of hollow fiber rendered low water fluxes and reasonable salt rejections (Endoh et al. 1977; McKinney and Rhodes 1971; Richter and Hoehn 1971). Aromatic polyamide membranes having trade name Permasep® B-9 were prepared by Du Pont to commercially desalinate brackish water on a large scale (Mehta and Loeb 1978; Hoehn and Richter 1973). As compared to cellulose membranes, separation performance of polyamide membranes was found to be better and remained a first priority of BWRO plants until early 1990s. Noticeable features of polyamide membranes include good permeate flux, high salt rejection, moderate resistance against hydrolysis and biological attack, good thermal and mechanical stability, and wide pH range operability (Soltanieh and Gill 1981). Du Pont, however, abandoned the production of polyamide membranes on account of their susceptibility to disinfectants on prolonged exposure to chlorine or ozone (Lee et al. 2011).

With the passage of time, other integrally skinned asymmetric BWRO desalination membranes were also prepared from polybenzimidazole and polybenzimidazoline materials (Sawyer and Jones 1984; Goldsmith et al. 1977). These membranes not only had inherently low salt rejection but were also susceptible to chlorine attack. Aiming to overcome this difficulty, membranologists prepared polypiperazinamides membranes but could not be commercialized owing to their low salt rejection (Credali et al. 1974; Parrini 1983; Credali and Parrini 1971). Other integrally skinned asymmetric membranes giving low permeability and low salt rejection were prepared from polyimide, polyoxadiazole, polyvinylchloride/cellulose acetate blend, and cellulose

Table 1 Desalination properties of commercially available cellulosic membranes

Base polymer	Trade name	Permeate flux (m^3/m^2 day)	Salt rejection (%)	Test conditions	References
Cellulose triacetate	HOLLOSEP® HA 8130	0.52	94	1500 ppm NaCl, 29.4 bar, 25 °C, 75% recovery	Toyobo Co. Ltd. (2006)
	HOLLOSEP® HJ 9155	0.18	99.6	35,000 ppm NaCl, 53.9 bar, 25 °C, 30% recovery	
	M-C4040A	0.12	96.1	500 ppm tap water, 16.0 bar, 25 °C, 15% recovery, pH 7–8	Applied Membranes (2018)
Blend of cellulose diacetate and cellulose triacetate	CD series	0.37	98.5	2000 ppm NaCl, 29.3 bar, 25 °C, pH 6.5, 15% recovery	Suez Water Technologies (2018a)
	CE series	0.40	97.5		Suez Water Technologies (2018b)

acetate/polyvinyl pyrrolidone blend (Fox 1980; Sekiguchi et al. 1978; El-Gendi et al. 2017; Saljoughi and Mohammadi 2009). Desalination performance of important asymmetric BWRO membranes is summarized in Table 2.

3.3 Thin-Film Interfacial Composite Membranes

Development of thin-film interfacial composite membranes by Cadotte et al. via interfacial polymerization in 1981 was a breakthrough in the field of water desalination technology via reverse osmosis process (Cadotte 1981, 1985; Larson et al. 1981). Intensive research carried out on thin-film asymmetric membranes led to successful commercialization and patenting of these membranes on industrial scale (Sasaki et al. 1988, 1989; Hachisuka and Ikeda 2002). These membranes consist of three layers: (1) woven polymer fabric for membrane handling, (2) a porous polymer substratum cast on the fabric to provide mechanical toughness, and (3) a permselective dense layer of crosslinked polyamide deposited on the substrate via polymerization occurring at the interface (Fig. 3). Commercial preparation technique of interfacial composite membranes is almost same except that the process is continuous and fast (Petersen 1993). Water flux, salt elimination, fouling propensity, and operating conditions of thin-film composite membranes can be optimized via separately choosing and modifying a proper microporous substrate, barrier dense layer, reactive monomers, and suitable additives. Detailed description of some of the commercially available polyamide-based thin-film composite membranes is enlisted in Table 3.

In contrast to cellulose acetate membranes, composite membranes render substantially higher water fluxes along with higher salt and silica rejections. Flux and selectivity of these membranes are determined by the porosity of support layer and crosslink density of selective layer, respectively. Owing to their inherently thin selective layer deposited on a microporous support layer, composite membranes can function well under low operating pressure while withstanding relatively broader ranges of acidity scale (pH: 2–12) and feed water temperature (up to 45 °C). These membranes, however, are susceptible to fouling and disinfectants attack (Kucera 2010).

4 BWRO Membrane Modules

BWRO plants consist of membrane modules containing efficient and economical packaging of huge surface areas (thousands of square meters) of membranes to accomplish desalination on large industrial scale. Effective module designs guarantee process compactness, low capital cost, easy installation, comfortable cleaning, and replacement of membranes (Belfort 1984). Fabrication of inexpensive membrane modules was a technological breakthrough to launch membrane desalting process on a commercial scale in 1970s. Since membrane modules have been confidentially fabricated inside private companies, lots of membranologists are unacquainted about the issues of module design, fabrication, and operation.

Depending on their configuration, two basic types of membrane module designs are flat and cylindrical (Mulder 1996). Flat shape membranes are implanted in plate and frame and spiral wound modules, whereas membranes of cylindrical profile are inserted in tubular, hollow fiber, and capillary modules. In addition, some patents have reported

Table 2 Desalination properties of integrally skinned asymmetric membranes for brackish water purification

Base polymer	Permeate flux (m³/m² day)	Salt rejection (%)	Test conditions	References
Polyamide	0.05	99	5000 ppm NaCl, 27.6 bar	Mehta and Loeb (1978)
Polybenzimidazole	0.13	95	1050 ppm NaCl, 5.89 bar	Senoo et al. (1971)
Polybenzimidazoline	0.61–0.65	90.0–95.0	5000 ppm NaCl, 18.3 bar	Goldsmith et al. (1977)
Polyimide	0.49	97	5000 ppm NaCl, 28.6 bar	Fox (1980)
Polyoxadiazole	1.50×10^{-7}	92	5000 ppm NaCl, 44.1 bar	Sekiguchi et al. (1978)
Polypiperazinamide	0.67	97.2	3600 ppm NaCl, 80.0 bar	Lee et al. (2011), Credali et al. (1975)
Blend of cellulose acetate and polyvinylpyrrolidone	0.60–1.56	–	Pure water, 0.35 bar	Saljoughi and Mohammadi (2009)

Fig. 3 Schematic diagram of thin-film interfacial composite membrane

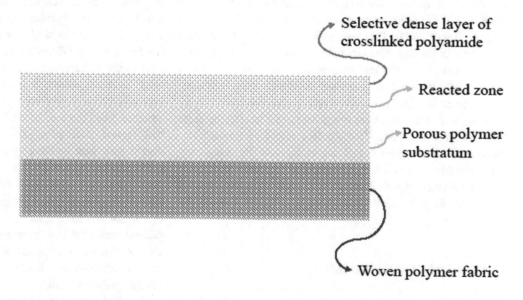

Selective dense layer of crosslinked polyamide

Reacted zone

Porous polymer substratum

Woven polymer fabric

several modified forms of these modules (Eckman 1995; Solomon 1989; Casey 1983, 1984). Early BWRO plants employed low fouling plate and frame and tubular module designs. Both of these modules, however, were gradually abandoned and substituted with more economical, compact and high throughput spiral wound and hollow fiber modules (Kucera 2014; Dupont et al. 1982). A short comparison and selection criteria for main types of BWRO module designs are overviewed in Table 4. Critically important parameters include manufacturing cost, packing density, membrane material specificity, membrane fouling, concentration polarization, ease of cleaning, mechanical sustainability

against high pressure, and magnitude of permeate-side pressure drop (Baker 2012; Kucera 2010).

4.1 Plate and Frame Modules

The initial desalination membrane systems were constructed on the basis of plate and frame module designs (Günther et al. 1996). In such designs, numerous grooved flat sheet membrane envelopes along with feed and product spacers arranged in series are stacked together between two rigid endplates made of glass fiber, reinforced paper, or plastic

Table 3 Desalination properties of commercially available polyamide-based thin-film composite membranes

Brand name	Permeate flux (m^3/ m^2 day)	Salt rejection (%)	Test conditions	References
HF1 (Axeon)	0.29	99	550 ppm NaCl, 10.34 bar, 25 °C, 15% recovery	Axeon (2017)
FILMTE™ BW30-365 (Dow)	0.56	99	2000 ppm NaCl, 15.5 bar, 25 °C, pH 8, 15% recovery	Dow (2015)
Osmo HR (Suez Water Technologies)	0.57	99	2000 ppm NaCl, 15.5 bar, 25 °C, pH 7.5, 15% recovery	Suez Water Technologies (2015)
TM700 (Toray Industries, Inc.)	0.65	99.7	2000 ppm NaCl, 15.5 bar, 25 °C, pH 7, 15% recovery	Toray Industries Inc. (2014)
CPA series (Hydranautics)	0.59	99.7	1500 ppm NaCl, 15.5 bar, 25 °C, pH 6.5–7, 15% recovery	Hydranautics (2018)
FLUID SYSTEMS® TFC® HR (Koch Membrane Systems)	0.64	97.6	2000 ppm NaCl, 15.5 bar, 25 °C, pH 7.5, 15% recovery	Koch Membrane Systems (2018)

material (MacNeil 1988; Belfort 1984). The whole structure is housed inside a pressure shell. The schematic diagram of such a BWRO membrane module system is illustrated in Fig. 4 (Baker 2004). Brackish water being introduced from shell side of the module is distributed across each plate, pure water permeates through membrane envelopes and is received in the central permeate collecting pipe. Concentrated brine is taken from other end of the module.

Plate and frame modules are pricey units having tiresome fabrication design and low surface area-to-volume ratio. Formation of stagnant zones within module assembly may lead to fouling, which can, however, be cleaned easily. Ease of maintenance makes them a good choice for desalinating highly fouling brackish water feed streams on a relatively small scale.

4.2 Spiral Wound Modules

The Gulf General Atomic with financial support of the Office of Saline Water initiated the development of spiral wound modules and commercialized them to carry out membrane separation processes on large industrial scale (Westmoreland 1968; Bray 1968; Kremen 1977). Later on, design and configuration of spiral wound modules have been revised to reduce cost and enhance process efficiency of these systems (Schneider 1989; Mannapperuma 1994; Doll 1984; Ng et al. 2008). Currently, these type of units are the most popular forms of industrial membrane module systems being used in BWRO desalination plants to obtain clean water.

A spiral wound module comprises a number of polyamide-based thin-film composite membrane leaves/envelopes and feed/concentrate mesh spacers spirally wrapped around a central pierced permeate receiving pipe as shown in Fig. 5 (Westmoreland 1968; Kucera 2010; MacNeil 1988; Kucera 2014). A typical membrane envelope is made by sandwiching a permeate weaved spacer between two membrane sheets followed by gluing the sandwiched structure from three sides while leaving the fourth side of the leaf opened. A number of such leaves are alternately positioned with feed webbed spacers to provide flow paths for feed/concentrate streams, promote turbulence, and control concentration polarization effects. In order to minimize undue pressure drop in a single-envelope module, a number of membrane envelopes are attached to the central axial pierced permeate collecting tube in a multi-envelope spiral wound module as depicted in Fig. 6. Normally, 4–6 spiral wound modules are housed in series inside a particular tubular pressure vessel.

Brackish feed water being entered from one side of the module moves in axial direction along the surface of membrane envelope. Major portion of feed water permeated through the membrane surface travels towards the module center in a spiral direction and is collected as clean water in the central perforated collection pipe. The concentrated brine traveling axially along the membrane surface exits the unit at the other end of the module.

Advantageous factors of hollow fiber modules such as higher membrane surface to volume ratio, better packing density, and higher production rate are offset by inherently

Table 4 A short comparison and selection criteria for main types of BWRO module designs (Baker 2012; Kucera 2010)

Parameter	Plate and frame	Spiral wound	Tubular	Hollow fiber
Manufacturing cost ($/m²)	High (50–200)	Moderate (5–100)	Moderate (10–50)	Low (5–20)
Packing density (m²/m³)	150–500	500–1200	20–400	500–5000
Limited to specific types of membrane material	No	No	Yes	Yes
Fouling tendency	Moderate	High	Low	Very high
Ease of cleaning	Good	Poor	Excellent	Poor
Suitability for high-pressure operation	Yes	Yes	No	Yes
Permeate-side pressure drop	Low	Moderate	Moderate	High

Fig. 4 Schematic diagram of a BWRO plate and frame module

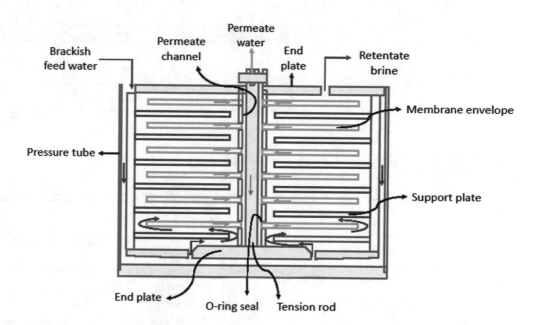

low fouling susceptibility, low cost, provision for elevated mass transfer rates, and minimal feed pretreatment requirements of spiral wound modules. Owing to these valuable characteristics, spiral wound modules are commonly used for large-scale treatment of brackish water rendering a clean water recovery up to 90% (Lee et al. 2011; Joshi et al. 2004). Insufficient feed pretreatment may, however, promote module fouling entailing frequent cleaning. Single-envelope spiral wound modules may lead to high feed-side pressure drop; the excessive pressure drop may be overcome by using multiple-envelope modules (Fritzmann et al. 2007).

4.3 Tubular Modules

Tubular modules, being simple in design and fabrication, encompass a vast membrane area in a smaller-sized module housing. They comprise several smaller-diameter selective membrane-containing porous tubes fitted inside a single larger supporting pressure vessel as shown in Fig. 7. The nested polymer membranes are usually cast on plastic, fiberglass, ceramic, carbon, or paper tubes while the support tube is made of plastic, fiberglass, or steel (Baker 2004; Dupont et al. 1982; Casey 1984). High-pressure saline feed water goes into the tube from its one end, radially permeated product water through each tube is collected in permeate collection header, while the axially flowing rejected concentrated brine is expelled from other tube end. In a typical larger BWRO plant, multiple tube bundles can be assembled in series or parallel configuration in a larger casing in order to enhance system production capacity as shown in Fig. 8 (Dupont et al. 1982).

Manufacturing cost of tubular membranes and their modules is generally high. They also depict low package density owing to their low surface area-to-volume ratio (Kucera 2010). However, improved liquid hydrokinetics

Fig. 5 Exploded view of spiral wound module. Some portion of the brackish feed water flowing along the axis of membrane module permeates through the membrane surface and travels spirally inwards to the perforated central collection pipe. The retentate concentrated brine axially moving through the spacers is collected at the other module end opposite to the feed end

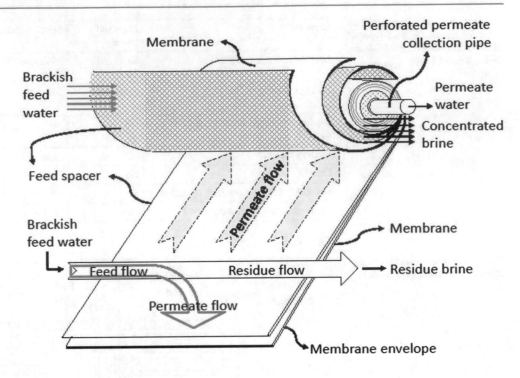

Fig. 6 Multi-envelope spiral wound module having 6 envelopes. There may be dozens of envelopes in larger diameter modules

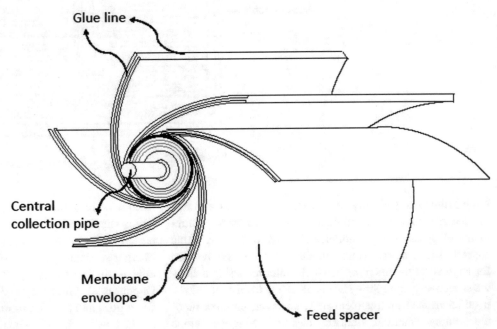

through these wide-bored tubes makes them fouling resistant. Tubular membranes can be easily cleaned by chemical and mechanical methods after dismantling the module assembly (MacNeil 1988). Tubular membrane modules find limited applications in BWRO desalination processes. They are being commonly applied in ultrafiltration and microfiltration applications (McCutchan and Goel 1974; Wiley et al. 1985; Tsuge et al. 1977).

4.4 Hollow Fiber Modules

After its development by Dow Chemical in 1966 for the first time, cellulosic hollow fiber modules have been commercially produced, promoted and also patented by other companies too (Mahon 1966; Kumano and Fujiwara 2008). Hollow fiber membrane modules for brackish water desalination are often manufactured in shell-side feed design as

Fig. 7 Schematic diagram of a typical tubular module nesting many membrane tubes inside a plastic casing

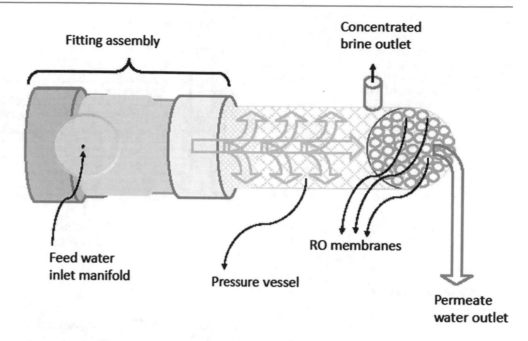

Fig. 8 Exploded view of tubular module containing multiple tube bundles assembled in series or parallel arrangement in a larger cylindrical tube sheet

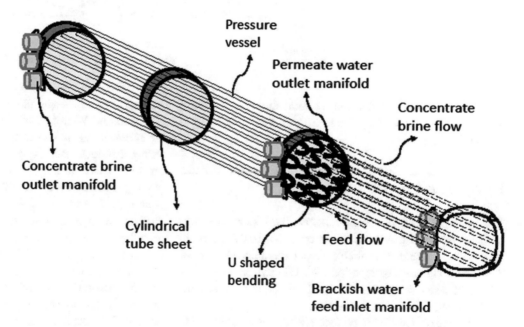

displayed in Fig. 9 (Kumano et al. 2008). Such a module is made by containing a bundle of thousands of small-diameter hollow fibers in a pressure vessel. The bundle of fibers is closed from one end while kept opened from other end for the permeate flow. A perforated central pipe is inserted all along the module length to uniformly distribute brackish feed water (MacNeil 1988). A portion of pressurized feed water entering the module via core tube is forced to permeate radially into the walls of bundled hollow fibers from shell side. Desalted water leaves the tubes to form their open ends, whereas the concentrated brine exits the module from its shell side.

These types of modules are cost-effective, render high membrane area-to-volume ratio, uniformly high permeate flow, and high recovery (Marcovecchio et al. 2010; Nakayama and Sano 2013; Senthilmurugan and Gupta 2006). The production rate of clean water of hollow fiber modules is almost one order of magnitude higher than that obtained from spiral wound modules. Notable drawbacks of these modules, however, include low fouling resistance and difficult cleaning on account of compact fiber arrangement and formation of stagnant zones (Kucera 2010; MacNeil 1988; Dupont et al. 1982). In addition, thick-walled fibers should be made to tolerate high shell-side hydrostatic

Fig. 9 A hollow fiber membrane module ensuring improved flow distribution and reduced fouling using cross-flow pattern. Feed entering through perforated core tube moves towards vessel shell

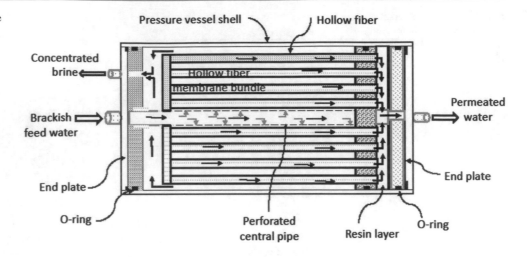

Fig. 10 Key components of a typical BWRO system

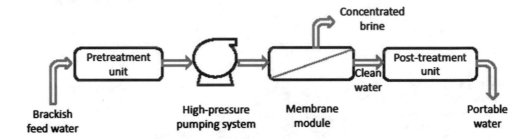

pressures. Typical internal and external diameters of these hollow fibers, respectively, are 50 μm and 100–200 μm (Baker 2012).

5 A Typical BWRO Plant

The brininess level of typical brackish feed waters needs to be lowered from a normal range of 2000–10,000 mg/L to 500 mg/L to make it drinkable as per recommendations of the World Health Organization (WHO). Typical salt rejections and water recoveries of such feeds lie in the ranges of 75–95% and 85–90%, respectively.

Key components of a typical BWRO system comprise pretreatment unit, high-pressure pumping system, desalting membrane module, and post-treatment section arranged as shown in Fig. 10 (Khawaji et al. 2008; Sauvet-Goichon 2007). A streamlined flow diagram of a commercial BWRO plant is schematically illustrated in Fig. 11.

5.1 Pretreatment

Pretreatment, the first process of a BWRO system, is applied to combat fouling and scaling of reverse osmosis membranes by removing suspended solids, dissolved organic matter, and

biological moieties via addition of chemicals and filtration (Durham and Walton 1999; Isaias 2001). Suspended solids are removed by adding a flocculating agent to feed water, letting the flocs settle down in gravity settlers and filtering brackish water through sand and cartridge filters. After passing it through the sand filter, acidity level of feed water is attuned. In addition, a disinfectant and an antiscalant are also added to feed water for the prevention of microbial growth and precipitation suppression of multivalent ions on membrane surface, respectively. As a matter of precaution, sodium sulfite can be added to eradicate surplus chlorine if chlorine-susceptible thin-film interfacial composite membranes are employed. In general, BWRO plants utilizing hollow fiber membrane modules demand stringent pretreatment actions than those employing spiral wound modules.

5.2 High-Pressure Pumps

Contaminants-free pretreated brackish feed water is pressurized far beyond its osmotic pressure via heavy-duty stainless steel pumps to move it through subsequent desalination membrane modules. Pumps exerting high pressure are essential to facilitate reverse osmosis phenomena to occur across the semipermeable membrane to obtain freshwater on permeate side and concentrated brine on retentate side.

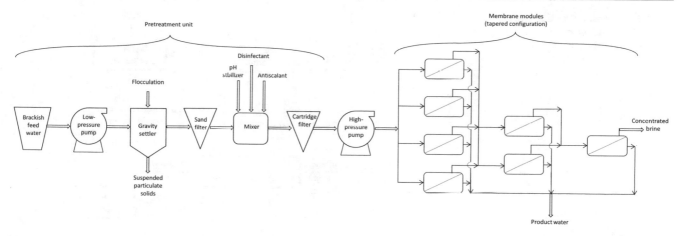

Fig. 11 Schematic flow diagram of a typical brackish water reverse osmosis plant

5.3 Membrane Module

The main component of a BWRO plant is the staggered array of desalting membrane modules containing selective semipermeable membranes which preferentially allow permeation of water molecules through them while retaining salt back. Insignificant amount of salt may, however, also pass through the membrane at high operating pressure owing to imperfections and flaws inherently present in membrane structure. Minor quantities of dissolved gases, if present in feed water, may also permeate along with water; these gases need to be eliminated from product water at post-treatment stage. Currently, spiral wound and hollow fiber modules containing thin-film polyamide and cellulosic membranes are the dominating configurations of commercial BWRO desalination plants. Pressure vessels are frequently arranged in tapered module configuration to ensure almost uniform and high average velocity of permeating water through each module. Typical operating pressure of brackish water desalination plants have steadily lowered down to the range of 8–12 bar.

Two separate streams having different salt concentrations leave the membrane module: (i) a desalted permeate water stream flowing to post-treatment unit for further processing and (ii) a high-salinity concentrated brine stream for further treatment before being discharged into the environment.

5.4 Post-Treatment

Desalted water leaving the membrane module is still insecure for drinking; it needs further treatment to improve its taste and to fulfill the requirements of portable drinking water. Main post-treatment processes include pH neutralization, H_2S removal via aeration, recarbonation, CO_2 degasification, sterilization by Cl_2 gas, and remineralization through filtration or chemical addition (Sauvet-Goichon 2007; Withers 2005; Muramoto and Nishino 1994; Delion et al. 2004; Birnhack et al. 2008). Desalted water is healthy and safe for drinking after passing through post-treatment processes.

Reject stream leaving the membrane module contains high-salinity brine and other harmful chemicals (ions of calcium, sulfate, silica, etc.) which need treatment before their ejection into surroundings since they can affect the ecosystem (Malaeb and Ayoub 2011; Voutchkov and Semiat 2008; Lattemann and Höpner 2008). One of the proposed solutions to simply, effectively and economically treat rejected concentrated brine makes use of solar stills at any sunny location (Hasnain and Alajlan 1998).

Water recovery achieved in a single-stage BWRO membrane system usually varies from 85–90%. Discarding the residual 10–15% retentate concentrated brine not only poses an environmental problem but also results into the wastage of a huge quantity of water in water-scarce desert regions. These factors have stimulated researchers to propose a two-stage membrane process for enhancing water recovery of BWRO plants in the range of 95–98% as schematically represented in Fig. 12 (Rahardianto et al. 2007; Kurihara et al. 1999; Ahmed et al. 2003). Usual water recovery of 85–90% is achieved in a normal way from the first stage membrane module. Anticipated scale-forming compounds present in the concentrated brine exiting the first stage are precipitated by adding a suitable precipitant into the retentate brine stream. Concentrated brine is further treated before introducing it to the second stage of BWRO membrane module to obtain a total water recovery of about 98% from the whole process. Small concentrated brine stream leaving the second stage of BWRO plant is finally disposed-off to evaporation pond to obtain edible salt.

Fig. 12 Schematic flow diagram of a double-stage membrane desalination process to increase water recovery of BWRO plants

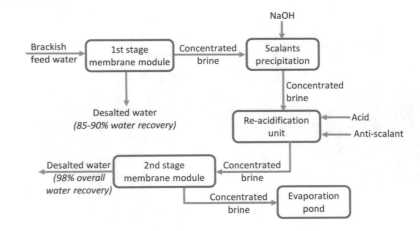

6 General BWRO Process Considerations

6.1 Process Configurations

Process configuration of membrane modules in a BWRO system can be arranged either in a single stage or multiple stages (Burn and Gray 2015). A stage is termed as the number of membrane module elements fitted in one pressure vessel used for concentrate treatment as determined by plant design output. For instance, single- and double-stage process configurations are represented in Figs. 11 and 12, respectively. Selecting the best optimized BWRO process configuration entails adequate design experience and awareness of sustainable marketing aspects (Lu et al. 2007; Almulla et al. 2002). Standardized operational parameters comprise design specifications, energy consumption, operational cost, maintenance cost, cartridge filter replacement frequency, plant operational availability, membrane cleaning frequency, and membrane replacement rate (Voutchkov 2014).

6.2 Effect of Operational Parameters

Pressure, temperature, and salt concentration of entering brackish feed water stream along with active surface area, selectivity and fouling propensity of the employed membrane are the imperative parameters affecting desalination performance of a BWRO plant; their effect on water flux and salt rejection is schematically demonstrated in Fig. 13 (Greenlee et al. 2009; Cadotte et al. 1980). Some practical trade-offs worth of considering while designing a BWRO process system are discussed as follows:

- Water permeability and water-to-salt selectivity of a membrane are inversely related with each other: BWRO membranes generally provide high water flux and low salt rejection subjected to a given set of feed water and

operating conditions and vice versa (Greenlee et al. 2009).

- A tradeoff exists between membrane size (capital cost) and feed pressure (operating cost): An increase in water recovery would result in a corresponding reduction in membrane area required to handle a specific feed stream (Wilf and Bartels 2005).
- Water flux increases linearly while salt rejection rapidly escalates with an increase in feed pressure of brackish water (Cadotte et al. 1980). Both the water flux and salt rejection become zero at an operating pressure equal to osmotic pressure of the feed water.
- An increase in feed temperature exponentially increases water flux while slightly decreases salt rejection (Cadotte et al. 1980). A temperature elevation of 30 °C almost doubles the water flux with a minute drop in salt rejection. In addition, elevated temperature promotes biofouling due to an increase in osmotic pressure of feed water (Greenlee et al. 2009; Voutchkov 2014).
- Membrane desalting performance is seriously affected by introducing a concentrated feed solution to the BWRO membrane system. Both the water flux and salt rejection are declined by raising the salt concentration of entering feed solution at a fixed feed pressure (Lu et al. 2007).

6.3 Energy Recovery of BWRO Process

Considerable amount of energy delivered to the feed water by the high-pressure pumping system of the BWRO plant is carried by the exiting concentrated brine stream. Some portion of the leaving energy with the brine stream can be recovered by installing either centrifugal energy recovery devices or isobaric systems to the BWRO plant (Voutchkov 2012). Energy recovery devices of former type, working on the principle of small watermills, are driven by jetting concentrated brine onto them. These devices comprise Pelton turbines, reversible pumps, hydraulic turbo boosters, etc. Energy recovery devices of latter type, capable of recovering 93–96% energy of the brine, comprise Dual Work Exchanger Energy Recovery System, Pressure exchanger, Sal Tec device, etc.

Materials of construction to manufacture energy recovery devices, piping and fittings, valves, and other associated accessories should be mechanically strong as well as sustainable in chlorinated and corrosive environments (Olsson 2005; Olsson and Snis 2007). Commonly used materials to manufacture BWRO plants include polyvinyl chloride (PVC) and various grades of stainless steel like 316 L, 904 L, 254 SMO, and SAF 2507 (Voutchkov 2014).

6.4 Economic Aspects of BWRO Process

Owing to its lowest energy requirement, water purification via reverse osmosis membrane technology is the most economical method among all the desalination processes (Wilf and Bartels 2005). Overall water desalination cost of reverse osmosis membrane process has heavily fallen over the last few years on account of significant developments made in membrane technology (Ghaffour et al. 2013). Notable improvements in this area comprise increased water flux, enhanced salt rejection, improved plant efficiency, better energy recovery, reduced energy consumption, preparation of high boron rejection membranes, and manufacturing of fouling-resistant membranes (Lee et al. 2011; Amy et al. 2017; Zhao et al. 2013; Kang and Cao 2012).

6.5 Optimization of BWRO Process

Cost of producing drinking water through BWRO membrane process can be further reduced by using larger pressure vessels, employing high productivity and/or high rejection membranes, utilizing membranes made from nanomaterials, adopting hybrid membrane inter-stage design, and making use of subsurface intakes instead of open intakes.

- Capital costs can be significantly reduced by shifting from contemporary smaller 8 in. diameter to larger 16 in. diameter pressure vessels to meet industrial standards as per recommendations of membrane suppliers. Switching to larger pressure vessels results in cost savings of fabrication and life cycle by 30% and 10%, respectively (Bartels et al. 2005).
- High throughput and/or highly selective membranes can also result in cost savings of BWRO membrane process by optimizing water flux and salt rejection properties of the membrane (Voutchkov 2018; Werber et al. 2016a, b).

- A tradeoff must be made between water recovery and membrane area since they directly affect the working and capital costs of membrane desalination process (Gordon and Hui 2016; Elimelech and Phillip 2011; Zhu et al. 2009).
- Desalting performance of BWRO process can be considerably improved by employing high selectivity next generation membranes made of nanomaterials such as carbon nanotubes and graphene nanosheets (Werber et al. 2016a, b).
- Both the capital and operational costs of BWRO plant can be reduced by using hybrid membrane inter-stage design in which membrane elements of varying throughput and rejection are mounted in a single pressure vessel to achieve uniform flux distribution (Voutchkov 2018; Peñate and García-Rodríguez 2011; Voutchkov 2014).
- Utilization of subsurface intakes in place of open intakes decreases working cost, increases feed water quality, and prolongs membrane life expectancy (Rachman et al. 2014). Good quality feed water does not require laborious pre-treatment, thus reducing capital and operational costs (Missimer et al. 2013).

7 Concluding Remarks

Salinity level and composition of regional brackish water helps to determine appropriate process conditions, pretreatment specifications, best type of membrane material and module, and economic aspects of a brackish water reverse osmosis (BWRO) plant. Although originated few decades earlier, membrane-based reverse osmosis desalination technology has now matured and surpassed conventional thermal desalination processes. A well-designed BWRO purification plant should safely and economically produce good quality drinking water meeting the standards of the World Health Organization (WHO). In spite of their high chlorine-resistant properties, conventional cellulosic membranes now find a little market space due to the availability of contemporary high-performance thin-film composite and polymeric membranes. Commercial membranes available in flat sheet or cylindrical shape are inserted in various modules to make the BWRO desalination process more efficient and economical.

Owing to their low manufacturing cost, low fouling tendency, and low operating pressure, spiral wound modules are the most commonly used types of BWRO membrane modules setting aside other module types like plate and frame, tubular, and hollow fiber. Main components of a classic BWRO process system comprise pretreatment, high-pressure pumping, membrane module, and post-treatment units. Process parameters which need to be optimized include high permeate water flux, high solute rejection, high water recovery, low fouling membrane material, low operating pressure, high energy recovery, process configuration, and low energy utilization. The operating cost of a typical BWRO system can be heavily shortened by making use of efficient energy recovery systems.

References

Ahmed, M., Arakel, A., Hoey, D., Thumarukudy, M. R., Goosen, M. F. A., Al-Haddabi, M., & Al-Belushi, A. (2003). Feasibility of salt production from inland RO desalination plant reject brine: A case study. *Desalination, 158*, 109–117. https://doi.org/10.1016/S0011-9164(03)00441-7.

Ali, A., Tufa, R. A., Macedonio, F., Curcio, E., & Drioli, E. (2018). Membrane technology in renewable-energy-driven desalination. *Renewable & Sustainable Energy Reviews, 81*, 1–21. https://doi.org/10.1016/j.rser.2017.07.047.

Almulla, A., Eid, M., Cote, P., & Coburn, J. (2002). Development in high recovery brackish water desalination plants as part of the solution to water quantity problems. *Desalination, 153*, 237–243. https://doi.org/10.1016/S0011-9164(02)01142-6.

Altaee, A., Zaragoza, G., & van Tonningen, H. R. (2014). Comparison between forward osmosis-reverse osmosis and reverse osmosis processes for seawater desalination. *Desalination, 336*, 50–57. https://doi.org/10.1016/j.desal.2014.01.002.

Amy, G., Ghaffour, N., Li, Z., Francis, L., Linares, R. V., Missimer, T., & Lattemann, S. (2017). Membrane-based seawater desalination: Present and future prospects. *Desalination, 401*, 16–21. https://doi.org/10.1016/j.desal.2016.10.002.

Applied Membranes Inc. (2018). CTA Commercial/Industrial RO Membranes. https://www.appliedmembranes.com/cta-ro-membranes.html. Accessed date: October 12, 2018.

Aquastat. (2013). Water Uses. Food and Agriculture Organization of the United Nations.

Asadollahi, M., Bastani, D., & Musavi, S. A. (2017). Enhancement of surface properties and performance of reverse osmosis membranes after surface modification: A review. *Desalination, 420*, 330–383. https://doi.org/10.1016/j.desal.2017.05.027.

Atab, M. S., Smallbone, A. J., & Roskilly, A. P. (2016). An operational and economic study of a reverse osmosis desalination system for potable water and land irrigation. *Desalination, 397*, 174–184. https://doi.org/10.1016/j.desal.2016.06.020.

Axeon. (2017). HF1—Series Membrane Elements. https://www.axeonwater.com/skin/common_files/admin/overview/MKTF_133_E_HF1_MEMBRANE_ELEMENT_SPEC_SHEET.pdf. Accessed date: November 1, 2018.

Baker, R. W. (2004). *Membrane technology and applications* (2nd ed.). New York: Wiley. ISBN: 978-0-470-02039-5.

Baker, R. W. (2012). *Membrane technology and applications* (3rd ed.). New York: Wiley. https://doi.org/10.1002/9781118359686.

Bartels, C., Bergman, R., Hallan, M., Henthorne, L., Knappe, P., Lozier, J., et al. (2005). *Industry consortium analysis of large reverse osmosis/nanofiltration element diameters*. Desalination and Water Purification Research and Development Report No. 114, U.S. Department of Interior Bureau of Reclamation, September 2004.

Belfort, G. (1984). Desalting experience by hyperfiltration (reverse osmosis) in the United States. In G. Belfort (Ed.), *Synthetic membrane process: Fundamentals and water applications* (Chap. 7,

pp. 221–280). Cambridge: Academic Press. https://doi.org/10.1016/B978-0-12-085480-6.50013-X.

Birnhack, L., Penn, R., & Lahav, O. (2008). Quality criteria for desalinated water and introduction of a novel, cost effective and advantageous post treatment process. *Desalination, 221*, 70–83. https://doi.org/10.1016/j.desal.2007.01.068.

Bray, D. T. (1968). Reverse osmosis purification apparatus. US Patent US3417870.

Burn, S., & Gray, S. (2015). *Efficient desalination by reverse osmosis: A best practice guide to RO*. London, UK: IWA Publishing. https://www.abebooks.co.uk/9781780405056/Efficient-Desalination-Reverse-Osmosis-Best-1780405057/plp. ISBN 13: 9781780405056.

Cadotte, J. E. (1981). Interfacially synthesized reverse osmosis membrane. US Patent US4277344A.

Cadotte, J. E. (1985). Evolution of composite reverse osmosis membranes. In D. R. Lloyd (Ed.), *Materials science of synthetic membranes. ACS Symposium Series Number* (Vol. 269, pp. 273–294). Washington, DC: American Chemical Society. https://doi.org/10.1021/bk-1985-0269.ch012.

Cadotte, J. E., Petersen, R. J., Larson, R. E., & Erickson, E. E. (1980). A new thin film sea water reverse osmosis membrane. *Desalination, 32*, 25–31. https://doi.org/10.1016/S0011-9164(00)86003-8.

Casey, W. P. (1983). Reverse osmosis water purification element and cartridge.US Patent US4715952A.

Casey, W. P. (1984). Tubular element for reverse osmosis water purification. US Patent US4874514A.

Cay-Durgun, P., & Lind, M. L. (2018). Nanoporous materials in polymeric membranes for desalination. *Current Opinion in Chemical Engineering, 20*, 19–27. https://doi.org/10.1016/j.coche.2018.01.001.

Chen, J. P., Chian, E. S. K., Sheng, P.-X., Nanayakkara, K. G. N., Wang, L. K., & Ting, Y.-P. (2011). Desalination of seawater by reverse osmosis. In *Membrane desalination technologies* (Vol. 13, pp. 559–601). https://doi.org/10.1007/978-1-59745-278-6_13.

Credali, L., Baruzzi, G., & Guidotti, V. (1975). Reverse osmosis anisotropic membranes based on polypiperazine amides. US Patent US4129559A.

Credali, L., Chiolle, A., & Parrini, P. (1974). New polymer materials for reverse osmosis membranes. *Desalination, 14*, 137–150. https://doi.org/10.1016/S0011-9164(00)82047-0.

Credali, L., & Parrini, P. (1971). Properties of piperazine homopolyamide films. *Polymer (Guildf), 12*, 717–729. https://doi.org/10.1016/0032-3861(71)90087-5.

Curcio, E., & Drioli, E. (2009). Membranes for desalination. In *Seawater desalination: Conventional and renewable energy processes* (Chap. 3, pp. 41–75). Berlin, Heidelberg: Springer. https://doi.org/10.1007/978-3-642-01150-4_3.

Delion, N., Mauguin, G., & Corsin, P. (2004). Importance and impact of post treatments on design and operation of SWRO plants. *Desalination, 165*, 323–334. https://doi.org/10.1016/j.desal.2004.06.037.

Doll, D. W. (1984). Spirally wrapped reverse osmosis membrane cell. US Patent US4476022A.

Dow. (2015). DOW FILMTEC™ BW30-365 Element. https://www.dupont.com/content/dam/Dupont2.0/Products/water/literature/609-00153.pdf. Accessed date: November 1, 2018.

Duarte, A. P., & Bordado, J. C. (2016). 12—Smart composite reverse-osmosis membranes for energy generation and water desalination processes. In *Smart composite coatings and membranes. Woodhead Publishing Series in Composites Science and Engineering* (pp. 329–350). Sawston: Woodhead Publishing. https://doi.org/10.1016/B978-1-78242-283-9.00012-9.

Dupont, R. R., Eisenberg, T. N., & Middlebrooks, E. J. (1982). *Reverse osmosis in the treatment of drinking water*. Reports, Utah Water Research Laboratory, Utah State University.

Durham, B., & Walton, A. (1999). Membrane pretreatment of reverse osmosis: Long-term experience on difficult waters. *Desalination, 122*, 157–170. https://doi.org/10.1016/S0011-9164(99)00037-5.

Eckman, T. J. (1995). Hollow fiber cartridge. US Patent US5470469A.

El-Gendi, A., Abdallah, H., Amin, A., & Amin, S. K. (2017). Investigation of polyvinyl chloride and cellulose acetate blend membranes for desalination. *Journal of Molecular Structure, 1146*, 14–22. https://doi.org/10.1016/j.molstruc.2017.05.122.

Elimelech, M., & Phillip, W. A. (2011). The future of seawater desalination: Energy, technology, and the environment. *Science, 80* (333), 712–717. https://doi.org/10.1126/science.1200488.

El-Manharawy, S., & Hafez, A. (2001). Water type and guidelines for RO system design. *Desalination, 139*, 97–113. https://doi.org/10.1016/S0011-9164(01)00298-3.

El-Saied, H., Basta, A. H., Barsoum, B. N., & Elberry, M. M. (2003). Cellulose membranes for reverse osmosis part I. RO cellulose acetate membranes including a composite with polypropylene. *Desalination, 159*, 171–181. https://doi.org/10.1016/S0011-9164(03)90069-5.

Endoh, R., Tanaka, T., Kurihara, M., & Ikeda, K. (1977). New polymeric materials for reverse osmosis membranes. *Desalination, 21*, 35–44. https://doi.org/10.1016/S0011-9164(00)84107-7.

Fox, R. L. (1980). Process for preparing an asymmetric permselective membrane. US Patent US4307135A.

Fritzmann, C., Löwenberg, J., Wintgens, T., & Melin, T. (2007). State-of-the-art of reverse osmosis desalination. *Desalination, 216*, 1–76. https://doi.org/10.1016/j.desal.2006.12.009.

Ghaffour, N., Missimer, T. M., & Amy, G. L. (2013). Technical review and evaluation of the economics of water desalination: Current and future challenges for better water supply sustainability. *Desalination, 309*, 197–207. https://doi.org/10.1016/j.desal.2012.10.015.

Ghosh, A. K., Bindal, R., Prabhakar, S., & Tewari, P. K. (2011). Composite polyamide reverse osmosis (RO) membranes—Recent developments and future directions. *BARC Newsletter, 321*, 43–51.

Glater, J. (1998). The early history of reverse osmosis membrane development. *Desalination, 117*, 297–309. https://doi.org/10.1016/S0011-9164(98)00122-2.

Goldsmith, R. L., Wechsler, B. A., Hara, S., Mori, K., & Taketani, Y. (1977). Development of PBIL low pressure brackish-water reverse osmosis membranes. *Desalination, 22*, 311–333. https://doi.org/10.1016/S0011-9164(00)88387-3.

Goh, P. S., Lau, W. J., Othman, M. H. D., & Ismail, A. F. (2018). Membrane fouling in desalination and its mitigation strategies. *Desalination, 425*, 130–155. https://doi.org/10.1016/j.desal.2017.10.018 .

Gordon, J. M., & Hui, T. C. (2016). Thermodynamic perspective for the specific energy consumption of seawater desalination. *Desalination, 386*, 13–18. https://doi.org/10.1016/j.desal.2016.02.030.

Greenlee, L. F., Lawler, D. F., Freeman, B. D., Marrot, B., & Moulin, P. (2009). Reverse osmosis desalination: Water sources, technology, and today's challenges. *Water Research, 43*, 2317–2348. https://doi.org/10.1016/j.watres.2009.03.010.

Günther, R., Perschall, B., Reese, D., & Hapke, J. (1996). Engineering for high pressure reverse osmosis. *Journal of Membrane Science, 121*, 95–107. https://doi.org/10.1016/0376-7388(96)00161-5.

Hachisuka, H., & Ikeda, K. (2002). Reverse osmosis composite membrane and reverse osmosis treatment method for water using the same. US Patent US6177011B1.

Hasnain, S. M., & Alajlan, S. A. (1998). Coupling of PV-powered RO brackish water desalination plant with solar stills. *Desalination, 116*, 57–64. https://doi.org/10.1016/S0011-9164(98)00057-5.

Hoehn, H. H., & Richter, J. W. (1973). Aromatic polyimide, polyester and polyamide separation membranes. US Patent USRE30351E.

Hydranautics. (2018). CPA2. Nitto Group Company. https://membranes.com/wp-content/uploads/2017/03/CPA2.pdf. Accessed date: November 1, 2018.

Isaias, N. P. (2001). Experience in reverse osmosis pretreatment. *Desalination, 139*, 57–64. https://doi.org/10.1016/S0011-9164(01)00294-6.

Ismail, F., Khulbe, K. C., & Matsuura, T. (2019). *Reverse osmosis* (1st ed.). Amsterdam: Elsevier. Paperback ISBN: 9780128114681; eBook ISBN: 9780128115398.

Joshi, S. V., Ghosh, P. K., Shah, V. J., Devmurari, C. V., Trivedi, J. J., & Rao, P. (2004). CSMCRI experience with reverse osmosis membranes and desalination: Case studies. *Desalination, 165*, 201–208. https://doi.org/10.1016/j.desal.2004.06.023.

Kang, G. D., & Cao, Y. M. (2012). Development of antifouling reverse osmosis membranes for water treatment: A review. *Water Research, 46*, 584–600. https://doi.org/10.1016/j.watres.2011.11.041.

Khawaji, A. D., Kutubkhanah, I. K., & Wie, J.-M. (2008). Advances in seawater desalination technologies. *Desalination, 221*, 47–69. https://doi.org/10.1016/j.desal.2007.01.067.

Kimura, S., & Sourirajan, S. (1967). Analysis of data in reverse osmosis with porous cellulose acetate membranes used. *AIChE Journal, 13*, 497–503. https://doi.org/10.1002/aic.690130319.

Koch Membrane Systems. (2018). FLUID SYSTEMS® TFC® HR 8″ Elements. https://www.kochmembrane.com/KochMembrane Solutions/media/Product-Datasheets/Spiral. Accessed date: November 1, 2018.

Kremen, S. S. (1977). Technology and engineering of ROGA spiral-wound reverse osmosis membrane modules. In S. Sourirajan (Ed.) *Reverse osmosis and synthetic membranes* (pp. 371–386). Ottawa: National Research Council Canada. https://doi.org/10.1002/pol.1977.130151011.

Kucera, J. (2010). Reverse osmosis principles. In: *Reverse osmosis: Design, processes, and applications for engineers*. Hoboken, NJ, USA: Wiley. https://doi.org/10.1002/9780470882634.ch2.

Kucera, J. (2014). *Desalination: Water from water*. Salem: Wiley-Scrivener. https://doi.org/10.1002/9781118904855.

Kumano, A., & Fujiwara, N. (2008). Cellulose triacetate membranes for reverse osmosis. In N. N. Li, A. G. Fane, W. S. W. Ho, & T. Matsuura (Eds.), *Advanced membrane technology and applications* (Chap. 2, pp. 21–45). New Jersey: Wiley. https://doi.org/10.1002/9780470276280.ch2.

Kumano, A., Sekino, M., Matsui, Y., Fujiwara, N., & Matsuyama, H. (2008). Study of mass transfer characteristics for a hollow fiber reverse osmosis module. *Journal of Membrane Science, 324*, 136–141. https://doi.org/10.1016/j.memsci.2008.07.011.

Kurihara, M., Yamamura, H., & Nakanishi, T. (1999). High recovery/high pressure membranes for brine conversion SWRO process development and its performance data. *Desalination, 125*, 9–15. https://doi.org/10.1016/S0011-9164(99)00119-8.

Larson, R. E., Cadotte, J. E., & Petersen, R. J. (1981). The FT-30 seawater reverse osmosis membrane-element test results. *Desalination, 38*, 473–483. https://doi.org/10.1016/S0011-9164(00)86092-0.

Lattemann, S., & Höpner, T. (2008). Environmental impact and impact assessment of seawater desalination. *Desalination, 220*, 1–15. https://doi.org/10.1016/j.desal.2007.03.009.

Lee, K. P., Arnot, T. C., & Mattia, D. (2011). A review of reverse osmosis membrane materials for desalination—Development to date and future potential. *Journal of Membrane Science, 370*, 1–22. https://doi.org/10.1016/j.memsci.2010.12.036.

Li, D., & Wang, H. (2010). Recent developments in reverse osmosis desalination membranes. *Journal of Materials Chemistry, 20*, 4551–4566. https://doi.org/10.1039/b924553g.

Loeb, S. (1981). The Loeb-Sourirajan membrane: How it came about. In *Synthetic Membranes* (Chap. 1, pp. 1–9). American Chemical Society. https://doi.org/10.1021/bk-1981-0153.ch001.

Loeb, S., & Sourirajan, S. (1963). Sea water demineralization by means of an osmotic membrane. In *Saline Water Conversion-II* (Chap. 9,

Vol. 38, pp. 117–132 SE-9). American Chemical Society. https://doi.org/10.1021/ba-1963-0038.ch009.

Lonsdale, H. K. (1966). Properties of cellulose acetate membranes. In M. Merten (Ed.), *Desalination by reverse osmosis* (pp. 93–160). Cambridge, MA: MIT Press. ASIN: B0000CNLGY.

Lonsdale, H. K., Merten, U., & Riley, R. L. (1965). Transport properties of cellulose acetate osmotic membranes. *Journal of Applied Polymer Science, 9,* 1341–1362. https://doi.org/10.1002/app.1965.070090413.

Lu, Y. Y., Hu, Y. D., Zhang, X. L., Wu, L. Y., & Liu, Q. Z. (2007). Optimum design of reverse osmosis system under different feed concentration and product specification. *Journal of Membrane Science, 287,* 219–229. https://doi.org/10.1016/j.memsci.2006.10.037.

MacNeil, C. J. (1988). Membrane separation technologies for treatment of hazardous wastes. *Critical Reviews in Environmental Control, 18,* 91–131. https://doi.org/10.1080/10643388809388344.

Mahon, H. I. (1966). Permeability separatory apparatus, permeability separator membrane element, method of making the same and process utilizing the same. US Patent US3228876A.

Malaeb, L., & Ayoub, G. M. (2011). Reverse osmosis technology for water treatment: State of the art review. *Desalination, 267,* 1–8. https://doi.org/10.1016/j.desal.2010.09.001.

Mannapperuma, J. D. (1994). Corrugated spiral membrane module. US Patent US5458774A.

Marcovecchio, M. G., Scenna, N. J., & Aguirre, P. A. (2010). Improvements of a hollow fiber reverse osmosis desalination model: Analysis of numerical results. *Chemical Engineering Research and Design, 88,* 789–802. https://doi.org/10.1016/j.cherd.2009.12.003.

McCutchan, J. W., & Goel, V. (1974). Systems analysis of a multi-stage tubular module reverse osmosis plant for sea water desalination. *Desalination, 14,* 57–76. https://doi.org/10.1016/S0011-9164(00)80047-8.

McKinney, R., & Rhodes, J. H. (1971). Aromatic polyamide membranes for reverse osmosis separations. *Macromolecules, 4,* 633–637. https://doi.org/10.1021/ma60023a025.

Mehta, G. D., & Loeb, S. (1978). Performance of permasep B-9 and B-10 membranes in various osmotic regions and at high osmotic pressures. *Journal of Membrane Science, 4,* 335–349. https://doi.org/10.1016/S0376-7388(00)83312-8.

Misdan, N., Lau, W. J., & Ismail, A. F. (2012). Seawater reverse osmosis (SWRO) desalination by thin-film composite membrane-current development, challenges and future prospects. *Desalination, 287,* 228–237. https://doi.org/10.1016/j.desal.2011.11.001.

Missimer, T. M., Ghaffour, N., Dehwah, A. H. A., Rachman, R., Maliva, R. G., & Amy, G. (2013). Subsurface intakes for seawater reverse osmosis facilities: Capacity limitation, water quality improvement, and economics. *Desalination, 322,* 37–51. https://doi.org/10.1016/j.desal.2013.04.021.

Mulder, M. (1996). Module and process design. In *Basic principles of membrane technology* (2nd ed., pp. 312–351). The Netherlands: Kluwer Academic Publishers. https://doi.org/10.1007/978-94-017-0835-7_8.

Muramoto, S., & Nishino, J. (1994). An advanced purification system combining ozonation and BAC, developed by The Bureau of Waterworks, Tokyo Metropolitan Government. *Desalination, 98,* 207–215. https://doi.org/10.1016/0011-9164(94)00145-6.

Nakayama, A., & Sano, Y. (2013). An application of the Sano-Nakayama membrane transport model in hollow fiber reverse osmosis desalination systems. *Desalination, 311,* 95–102. https://doi.org/10.1016/j.desal.2012.11.012.

Ng, H. Y., Tay, K. G., Chua, S. C., & Seah, H. (2008). Novel 16-inch spiral-wound RO systems for water reclamation—A quantum leap in water reclamation technology. *Desalination, 225,* 274–287. https://doi.org/10.1016/j.desal.2007.02.097.

Olsson, J. (2005). Stainless steels for desalination plants. *Desalination, 183,* 217–225. https://doi.org/10.1016/j.desal.2005.02.050.

Olsson, J., & Snis, M. (2007). Duplex—A new generation of stainless steels for desalination plants. *Desalination, 205,* 104–113. https://doi.org/10.1016/j.desal.2006.02.051.

Parrini, P. (1983). Polypiperazinamides: New polymers useful for membrane processes. *Desalination, 48,* 67–78. https://doi.org/10.1016/0011-9164(83)80006-X.

Peñate, B., & García-Rodríguez, L. (2011). Reverse osmosis hybrid membrane inter-stage design: A comparative performance assessment. *Desalination, 281,* 354–363. https://doi.org/10.1016/j.desal.2011.08.010.

Petersen, R. J. (1993). Composite reverse osmosis and nanofiltration membranes. *Journal of Membrane Science, 83,* 81–150. https://doi.org/10.1016/0376-7388(93)80014-O.

Pinnau, I., & Freeman, B. D. (1999). Formation and modification of polymeric membranes: Overview. In *ACS Symposium Series* (Chap. 1, Vol. 744, pp. 1–22). https://doi.org/10.1021/bk-2000-0744.ch001.

Prihasto, N., Liu, Q. F., & Kim, S. H. (2009). Pre-treatment strategies for seawater desalination by reverse osmosis system. *Desalination, 249,* 308–316. https://doi.org/10.1016/j.desal.2008.09.010.

Qasim, M., Darwish, N. N., Mhiyo, S., Darwish, N. A., & Hilal, N. (2018). The use of ultrasound to mitigate membrane fouling in desalination and water treatment. *Desalination, 443,* 143–164. https://doi.org/10.1016/j.desal.2018.04.007.

Rachman, R. M., Li, S., & Missimer, T. M. (2014). SWRO feed water quality improvement using subsurface intakes in Oman, Spain, Turks and Caicos Islands, and Saudi Arabia. *Desalination, 351,* 88–100. https://doi.org/10.1016/j.desal.2014.07.032.

Rahardianto, A., Gao, J., Gabelich, C. J., Williams, M. D., & Cohen, Y. (2007). High recovery membrane desalting of low-salinity brackish water: Integration of accelerated precipitation softening with membrane RO. *Journal of Membrane Science, 289,* 123–137. https://doi.org/10.1016/j.memsci.2006.11.043.

Reid, C. E., & Breton, E. J. (1959). Water and ion flow across cellulosic membranes. *Journal of Applied Polymer Science, 1,* 133–143. https://doi.org/10.1002/app.1959.070010202.

Richter, J. W., & Hoehn, H. H. (1971). Selective aromatic nitrogen-containing polymeric membranes. US Patent US3567632A.

Rosenbaum, S., Mahon, H. I., & Cotton, O. (1967). Permeation of water and sodium chloride through cellulose acetate. *Journal of Applied Polymer Science, 11,* 2041–2065. https://doi.org/10.1002/app.1967.070111021.

Saljoughi, E., & Mohammadi, T. (2009). Cellulose acetate (CA)/polyvinylpyrrolidone (PVP) blend asymmetric membranes: Preparation, morphology and performance. *Desalination, 249,* 850–854. https://doi.org/10.1016/j.desal.2008.12.066.

Sasaki, T., Fujimaki, H., Uemura, T., & Kurihara, M. (1988). Interfacially synthesized reverse osmosis membrane. US Patent US4758343A.

Sasaki, T., Fujimaki, H., Uemura, T., & Kurihara, M. (1989). Process for preparation of semipermeable composite membrane. US Patent US4857363A.

Sauvet-Goichon, B. (2007). Ashkelon desalination plant—A successful challenge. *Desalination, 203,* 75–81. https://doi.org/10.1016/j.desal.2006.03.525.

Sawyer, L. C., & Jones, R. S. (1984). Observations on the structure of first generation polybenzimidazole reverse osmosis membranes. *Journal of Membrane Science, 20,* 147–166. https://doi.org/10.1016/S0376-7388(00)81329-0.

Schneider, B. M. (1989). Spirally wrapped reverse osmosis membrane cell. US Patent US4814079A.

Sekiguchi, H., Sato, F., Sadamitsu, K., & Yoshida, K. (1978). Solute-separating membrane. US Patent US4067804A.

Senoo, M., Hara, S., & Ozawa, S. (1971). Permselective polymeric membrane prepared from polybenzimidazoles. US Patent US3951920A.

Senthilmurugan, S., & Gupta, S. K. (2006). Separation of inorganic and organic compounds by using a radial flow hollow-fiber reverse osmosis module. *Desalination, 196,* 221–236. https://doi.org/10.1016/j.desal.2006.02.001.

Shenvi, S. S., Isloor, A. M., & Ismail, A. F. (2015). A review on RO membrane technology: Developments and challenges. *Desalination, 368,* 10–26. https://doi.org/10.1016/j.desal.2014.12.042 .

Sidney, L., & Srinivasa, S. (1964). High flow porous membranes for separating water from saline solutions. US Patent US3133132A.

Singh, R. (2015). Introduction to membrane technology. In R. Singh (Ed.), *Membrane technology and engineering for water purification* (2nd ed., pp. 1–80). Oxford: Butterworth-Heinemann. https://doi.org/10.1016/B978-0-444-63362-0.00001-X.

Solomon, D. F. (1989). Reverse osmosis element. US Patent US4844805A.

Soltanieh, M., & Gill, W. N. (1981). Review of reverse osmosis membranes and transport models. *Chemical Engineering Communications, 12,* 279–363. https://doi.org/10.1080/00986448108910843.

Sudak, R. G. (1990). Reverse osmosis. In M. C. Poter (Ed.), *Handbook of industrial membrane technology* (pp. 260–305). New Jersey: Noyes Publication. eBook ISBN: 9780815517559.

Suez Water Technologies. (2015). OSMO HR (PA) series. https://www.suezwatertechnologies.com/kcpguest/documents/Fact. Accessed date: November 1, 2018.

Suez Water Technologies. (2018a). CD series high rejection brackish water RO elements (cellulose acetate). https://www.suezwatertechnologies.com/kcpguest/documents/Fact. Accessed date: October 15, 2018.

Suez Water Technologies. (2018b). CE series brackish water RO elements (cellulose acetate). https://www.suezwatertechnologies.com/kcpguest/documents/Fact. Accessed date: October 15, 2018.

Toray Industries Inc. (2014). Standard BWRO TM 700. https://www.toraywater.com/products/ro/pdf/TM700.pdf. Accessed date: November 1, 2018.

Toyobo Co. Ltd. (2006). HOLLOSEP®. https://www.toyobo-global.com/seihin/ro/tokucho.htm. Accessed date: October 3.

Tsuge, H., Yanagi, C., & Mori, K. (1977). Desalination of sea water by reverse osmosis using tubular module. *Desalination, 23,* 235–243. https://doi.org/10.1016/S0011-9164(00)82526-6.

Ulbricht, M. (2006). Advanced functional polymer membranes. *Polymer (Guildf), 47,* 2217–2262. https://doi.org/10.1016/j.polymer.2006.01.084.

Vos, K. D., Burris, F. O., Jr., & Riley, R. L. (1966). Kinetic study of the hydrolysis of cellulose acetate in the pH range of 2–10. *Journal of Applied Polymer Science, 10,* 825–832. https://doi.org/10.1002/app.1966.070100515.

Voutchkov, N. (2012). *Desalination engineering: Planning and design.* New York: McGraw Hill. ISBN: 9780071777155 0071777156.

Voutchkov, N. (2014). *Desalination engineering: Operation and maintenance.* New York: McGraw-Hill. ISBN: 9780071804219 0071804218.

Voutchkov, N. (2018). Energy use for membrane seawater desalination—Current status and trends. *Desalination, 431,* 2–14. https://doi.org/10.1016/j.desal.2017.10.033.

Voutchkov, N., & Semiat, R. (2008). Seawater desalination. In N. N. Li, A. G. Fane, W. S. W. Ho, & T. Matsuura (Eds.), *Advanced membrane technology and applications* (pp. 47–85). New Jersey: Wiley. https://doi.org/10.1002/9780470276280; ISBN: 9780470276280.

Wang, J., Dlamini, D. S., Mishra, A. K., Pendergast, M. T. M., Wong, M. C. Y., Mamba, B. B., et al. (2014). A critical review of transport through osmotic membranes. *Journal of Membrane Science, 454,* 516–537. https://doi.org/10.1016/j.memsci.2013.12.034.

Wang, Y.-N., & Wang, R. (2019). Reverse osmosis membrane separation technology. In A. F. Ismail, M. A. Rahman, M. H. D. Othman, & T. Matsuura (Eds.), *Membrane separation principles and applications. Handbooks in Separation Science* (Chap. 1, pp. 1–45). Amsterdam: Elsevier. https://doi.org/10.1016/B978-0-12-812815-2.00001-6.

Werber, J. R., Deshmukh, A., & Elimelech, M. (2016a). The critical need for increased selectivity, not increased water permeability, for desalination membranes. *Environmental Science & Technology Letters, 3,* 112–120. https://doi.org/10.1021/acs.estlett.6b00050.

Werber, J. R., Osuji, C. O., & Elimelech, M. (2016b). Materials for next-generation desalination and water purification membranes. *Nature Reviews Materials, 1,* 16018. https://doi.org/10.1038/natrevmats.2016.18.

Westmoreland, J. C. (1968). Spirally wrapped reverse osmosis membrane cell. US Patent US3367504A.

Wijmans, J. G., & Baker, R. W. (1995). The solution-diffusion model: A review. *Journal of Membrane Science, 107,* 1–21. https://doi.org/10.1016/0376-7388(95)00102-I.

Wiley, D. E., Fell, C. J. D., & Fane, A. G. (1985). Optimisation of membrane module design for brackish water desalination. *Desalination, 52,* 249–265. https://doi.org/10.1016/0011-9164(85)80036-9.

Wilf, M., & Bartels, C. (2005). Optimization of seawater RO systems design. *Desalination, 173,* 1–12. https://doi.org/10.1016/j.desal.2004.06.206.

Withers, A. (2005). Options for recarbonation, remineralisation and disinfection for desalination plants. *Desalination, 179,* 11–24. https://doi.org/10.1016/j.desal.2004.11.051.

Xia, S., Li, X., Zhang, Q., Xu, B., & Li, G. (2007). Ultrafiltration of surface water with coagulation pretreatment by streaming current control. *Desalination, 204,* 351–358. https://doi.org/10.1016/j.desal.2006.03.544.

Youssef, P. G., Al-Dadah, R. K., & Mahmoud, S. M. (2014). Comparative analysis of desalination technologies. *Energy Procedia, 61,* 2604–2607. https://doi.org/10.1016/j.egypro.2014.12.258.

Zhao, L., Chang, P. C. Y., & Ho, W. S. W. (2013). High-flux reverse osmosis membranes incorporated with hydrophilic additives for brackish water desalination. *Desalination, 308,* 225–232. https://doi.org/10.1016/j.desal.2012.07.020.

Zhu, A., Christofides, P. D., & Cohen, Y. (2009). On RO membrane and energy costs and associated incentives for future enhancements of membrane permeability. *Journal of Membrane Science, 344,* 1–5. https://doi.org/10.1016/j.memsci.2009.08.006.

High Performance Membrane for Natural Gas Sweetening Plants

Imran Ullah Khan, Mohd Hafiz Dzarfan Othman, and Asim Jilani

Abstract

The use of membrane technology in natural gas sweetening process has grown rapidly in recent years compared to other traditional purification processes. Carbon dioxide (CO_2) is the main culprit of greenhouse gases produced through the combustion of fossil fuel. In this chapter, various principles and mechanisms of gas separations on the basis of solubility and diffusion are systematically reviewed. Furthermore, the development and performance of various polymeric and inorganic materials in light of the recent challenges, advantages, and disadvantages for natural gas purification are described. Development of mixed matrix membrane for natural gas purification is one of the potential solutions for the use of expensive, brittle, and highly selective inorganic materials with low cost and stable polymers. At the end of this chapter, an attempt is also made to show the research development and future direction of natural gas purification process through highly selective membrane materials. The potential of the hybrid process for natural gas purification is more advantageous and economical compared to a single membrane process.

Keywords

Membrane • Gas separation • Performance • Mechanisms • Solubility • Diffusion

1 Introduction

Gas separation by the membrane is based on the interaction of specific gases with various membrane materials by physical or chemical interaction. This technology is considered to be visible and advantageous for the purification of raw natural gas at the industrial level. Natural gas sweetening is the process to clean the raw natural gas from impurities to meet the specifications of the natural gas grid. Raw natural gas is mainly composed of methane (CH_4), some other light gases such as ethane (C_2H_6), propane (C_3H_8), butane (C_4H_{10}), and acid gases such as carbon dioxide (CO_2) and hydrogen sulfide (H_2S) (Rezakazemi et al. 2014). The typical composition of raw natural gas is shown in Table 1.

CO_2 is the main corrosive component of natural gas which reduces the calorific value and increases the maintenance and operational cost (Khan et al. 2018a, b, c). It must be reduced to less than 2% to decrease the corrosion of the pipelines and equipment and also increase the heating value of the gas (Hwang et al. 2015). Traditional technologies which are available on the industrial level to purify the raw natural gas include adsorption, membrane separation, absorption (physical and chemical), and cryogenic technique. Chemical absorption is well known and has been commercially used for CO_2 removal in various processes and considered as a state-of-the-art technology. Although these technologies offer many advantages such as high separation performance and purity, they suffer from economics and environmental impacts that have led researchers to find a more economical and environmental-friendly

I. U. Khan (✉) · M. H. D. Othman (✉) · A. Jilani
Faculty of Engineering, School of Chemical and Energy Engineering, Advanced Membrane Technology Research Centre (AMTEC), Universiti Teknologi Malaysia, 81310 Skudai, Johor Bahru, Malaysia
e-mail: imran.khan@fcm3.paf-iast.edu.pk

M. H. D. Othman
e-mail: hafiz@petroleum.utm.my

I. U. Khan
Department of Chemical and Energy Engineering, Pak-Austria Fachhochshule, Institute of Applied Sciences &Technology, Khanpur Road, Mang, 22650 Haripur, Pakistan

A. Jilani
Center of Nanotechnology, King Abdul-Aziz University, Jeddah, 21589, Saudi Arabia

Z. Zhang et al. (eds.), *Membrane Technology Enhancement for Environmental Protection and Sustainable Industrial Growth*, Advances in Science, Technology & Innovation,
https://doi.org/10.1007/978-3-030-41295-1_5

Table 1 Typical composition of raw natural gas (Rezakazemi et al. 2014)

Component	Chemical formula	Composition (%)
Methane	CH_4	70–90
Ethane	C_2H_6	3–8
Propane	C_3H_8	1–2
Butane	C_4H_{10}	< 1
Carbon dioxide	CO_2	0–8
Oxygen	O_2	0–0.02
Nitrogen	N_2	0–5
Hydrogen sulfide	H_2S	0–5
Rare gases	He, Ar, Xe, Ne	Trace

technology for cleaning of natural gas efficiently (Lampinen 2014; Sun et al. 2015; Zhao et al. 2010).

Membrane system shows great potential for natural gas sweetening since it possesses numerous environmental advantages such as (1) low operational and investment cost with high CH_4 production up to >96% (Sun et al. 2015), (2) fewer space requirements (Basu et al. 2011), (3) easy maintenance without hazardous chemicals (Khan et al. 2017), (4) low maintenance cost (Song et al. 2012a, b), (5) simplicity in the operational and environmentally friendly process without chemical additives (Ahn et al. 2010), and (6) simple and easy process with low energy needs (Mohshim et al. 2013).

Membrane technology became the part of gas separation when Thomas Graham measured the permeation rates of all the gases and gave the first description of the solution diffusion model, and his work on porous membranes led to Graham's law of diffusion in 1850 (Weller and Steiner 1950). To date, the membrane technology is dominated by polymeric materials owing to the low cost and easy processing (Dong et al. 2013). Although polymeric materials showed encouraging results for gas separation, they suffer from some drawbacks (Sridhar et al. 2007a). Low membrane selectivity is the major inconvenient loss in gas separation that demands a multi-stage separation system which likely to impart higher capital cost. In addition, polymeric membranes commonly could not maintain their performance and thus deteriorate in extreme environmental operating conditions of high temperature and pressure. The chain swelling in the presence of highly corrosive components in the feed, plasticization, compaction, and aging of membranes are the main reasons of problematic phenomena (Liu et al. 2002). Moreover, all conventional polymeric membranes are inevitably bounded by Robeson's upper boundary. Since the development of membrane materials exceeding, the upper bound limit has become the major challenge for new researchers. Currently, inorganic membrane materials have gained much attention owing to their excellent separation performance; improve chemical and thermal resistance to the

harsh environment and long operational life. The selectivity of inorganic membranes is exceptionally high so as to surpass the Robeson's upper boundary. Unfortunately, its high cost and complex fabrication process, intrinsic fragility, and the low surface to volume ratio have hindered their further industrial application in the separation process. Furthermore, despite the availability of highly selective polymer materials, membrane fabrication process was not sufficiently advanced to make useful membranes (Aroon et al. 2010).

Therefore, many attempts have been made to fabricate a mixed matrix membrane (MMM) for gas separation and proved to be an alternative to overcome this limitation of polymeric membranes. To date, various potential MMMs have been explored and well documented in literature highlighting the advantages and limitations suffered by the resulted membranes (Adams et al. 2010; Aroon et al. 2010; Chaidou et al. 2012; Dorosti et al. 2014; Li et al. 2013; Ordoñez and Balkus 2010; Shahid and Nijmeijer 2014). Generally, the incorporation of inorganic particles into the polymer matrix would surely improves the membrane properties due to the superiority of the dispersed phase itself. In actual practice, MMMs development is often encountered with the deterioration of membrane performance. The incompatibility between the polymer and inorganic material is regarded as the main factor that leads to the defective interface. The development of MMM is a surely interesting approach with robustness, moderate cost, and high thermal, chemical, and mechanical stability. Selection of fillers remains as the heart of MMM development to ensure good compatibility with the polymer matrix, consequently, boost the membrane performance exceeding the Robeson upper bound (Khan et al. 2018a, b, c). The further main challenges for natural gas processing using MMM are material cost, CH_4 loss, and plasticization at high pressure operation. In this contribution, the main challenges, advantages, and limitations of the membrane separation technology for natural gas purification are thoroughly discussed. Lastly, the future research developments and directions for raw natural gas processing through membrane technology are also presented.

2 Separation Principles and Mechanisms

Membranes are thin films used to transport gases due to differences in their permeability rates. The permeability of gases in a membrane material is highly reliant on its structure, size, shape, polarization, and the interaction of membrane materials with permeant species (Lalia et al. 2013). The membrane is a permeable barrier which controls the permeability of various gases due to the applied driving forces of pressure, temperature, concentration, and electric charges difference of each gas. The solubility strength of any gas species in the membrane materials determines the compatibility of membrane for permeant. The permeability or permeability coefficient of gas passing through membrane is the product of thermodynamic parameter, i.e., the solubility coefficient and kinetic parameter, i.e., diffusion coefficient. The solubility coefficient is the amount of gas absorbed on the membrane surface at given pressure and temperature, while the diffusion coefficient shows how fast the gas is passed through the membrane (Shekhawat et al. 2003).

The pore flow and solution diffusion are the two phenomena used to describe the membrane separation process. In the solution diffusion, the concentration difference is used to dissolve permeates in the membrane materials and then diffuse them. While, in the pore flow, the pressure-driven convective flow in the pores is used to separate permeates (Rongwong et al. 2012). Mostly, the solution diffusion is suggested for gas transports through polymeric membranes (Kentish 2008). Figure 1 illustrates the process of natural gas purification using membrane separation technique.

The following mechanisms are used when natural gas transport through any membrane materials. A brief description of these commonly occurring mechanisms is given below:

2.1 Molecular Diffusion

The molecular diffusion occurs mainly through molecule-molecule collisions when the mean free path of the gas molecules is smaller than the pore size. The driving force for the separation of gases is the concentration difference. Furthermore, if a pressure gradient is applied in such pore regimes, the laminar flow occurs, as identified by Poiseuille's equation. Such flow is often known as *Poiseuille flow* or viscous flow (Javaid 2005).

2.2 Knudsen Diffusion

Knudsen diffusion transport is significant when the mean free path of the gas molecules is greater than the pore size. In these conditions, the collisions of the molecules with the pore wall are more prominent than the collisions between molecules. The separation factor (selectivity) is proportional to the ratio of the inverse square root of the molecular weights. This mechanism is leading to macroporous and mesoporous membranes (Vinoba et al. 2017).

2.3 Surface Diffusion

In surface diffusion, the permeating species exhibit a strong affinity on the membrane surface and adsorb along the pore walls. The difference in the adsorbed amount of the permeating species is the driving force in this mechanism. It also occurs with other transport mechanisms such as Knudsen diffusion (Fain 1994).

2.4 Capillary Condensation

In this transport mode, one of the gases should be condensable gas so the pores get completely filled by the condensed gas at a specific pressure. Meniscus formed at both ends of the pores and flow can only take place by the capillary pressure difference between the two ends. This mechanism is used to attain high separation factor, as the formation of the meniscus will prevent the flow of the non-condensable gases (Tomita et al. 2004).

2.5 Micropore Diffusion

Micropore diffusion is the same as surface diffusion in the limit where the pore size becomes comparable to the molecular size. In this mechanism, separation occurs due to molecular shape and size, pore size, and interactions between the pore wall and gas molecules (Javaid 2005).

2.6 Solution Diffusion

In dense polymeric materials, solution diffusion is considered as the major mechanism of transport. Firstly, the gas molecules are absorbed on the membrane surface and then

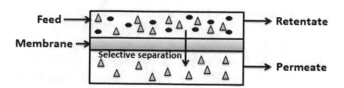

Fig. 1 Various membrane materials for natural gas purification (Khan et al. 2017)

followed by the diffusion of the gas molecules through the polymer matrix. Finally, the gas molecules evaporate through the downstream end (Stern 1994).

3 Membrane Materials for Natural Gas Purification

Nowadays, there have been growing interests in the development and applications of membrane-based natural gas separation process to tackle serious environmental issues and challenges. Researchers have been made a remarkable improvement in the gas separation membranes based on both polymeric and inorganic materials for industrial application (Centeno and Fuertes 1999; Ho et al. 2008; Mahajan et al. 2002; Noble 2011). Interestingly, new polymers, copolymers, and advanced materials such as metal–organic frameworks (MOF) and composite materials have been successfully used for natural gas sweetening process. Furthermore, nano-sized range inorganic materials have been merged in polymeric structures to introduce a new class of mixed matrix membranes (MMMs) with higher separation performance to surpass the Robeson upper boundary limits (Khan et al. 2018a, b, c).

Furthermore, the membrane technology has now reached its initial stage of maturity. The following important factors are considered for the selection of membrane materials for gas separation such as (1) intrinsic membrane selectivity (Thomas et al. 2017), (2) having high plasticization resistance (Iwasa et al. 2018), (3) having good mechanical and thermal strength (Khan et al. 2018a, b, c), (4) ability to convert materials into membrane morphology with excellent gas separation performance under adversarial feed mixture conditions (Zornoza et al. 2011a, b), and (5) having excellent interaction and sorption with one of the species of the mixture for efficient gas separation (Ismail et al. 2009). Additionally, molecular structure, polarity, and the presence of chemical groups in the membrane materials are other parameters of interest as it affects the natural gas separation performance.

Generally, the gas permeability and selectivity of membrane material are a decisive factor for high-gas separation performance and attracted the attention of future researchers from industries as well as academia sector. Moreover, the scientific aspects such as thermodynamics, mass, and heat transfer phenomena and surface chemistry at various feed conditions are equally important to encourage natural gas separation process. Polymeric, inorganic, and mixed matrix membranes are used for purification of natural gas as shown in Fig. 2. The following sections further explain the current challenges of the aforementioned membrane materials for natural gas purification.

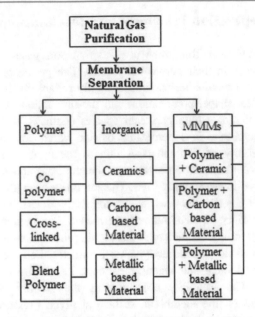

Fig. 2 Various membrane materials for natural gas purification

3.1 Polymeric Membranes

Polymeric membranes are commercially available for various gas mixture separations such as separation of nitrogen and oxygen from air and purification of natural gas. In the process of natural gas sweetening, compression at feed gas stream and low pressure of CO_2 after separation is required to offer the driving force for permeation. Generally, glassy and rubbery polymeric materials are used for gas separation (Zornoza et al. 2011a, b). Glassy materials are glass-like rigid materials and perform below their glass transition temperatures (T_g). On the other hand, rubbery materials are flexible, soft, and perform above their T_g (Stern 1994). Typically, the rubbery polymers give high permeability with low selectivity while glassy polymers display vice versa. Glassy polymeric membranes are dominating in industrial separation processes due to their high-gas separation performance and excellent thermal and mechanical properties (Samarasinghe et al. 2018). These membranes are further classified on the basis of structure such as nonporous dense membrane and microporous membrane. Microporous membranes have a rigid morphology with connected pores on the surface. While, the nonporous dense membranes have a dense film through which permeants pass by diffusion under the driving force of pressure, electrical potential, and concentration gradient (Kanehashi et al. 2015). The separation of different components of a gas mixture depends on the relative transport rate within the membrane, which is determined by their diffusivity and solubility in the membrane material.

Commercially available polymeric materials for gas separation are polyimide (PI), polycarbonate (PC), polysulfone (PSF), cellulose acetate (CA), and polydimethylsiloxane (PDMS). They have exceptional mechanical strength with low cost and high permeability. CA is the first polymeric material commercialized for gas separation (Scholz et al. 2013). It is relatively low priced due to abundant resources of cellulose with excellent separation properties. However, the CA is prone to plasticization ($P_{Plasticization}$ = 8 bar) (Bos et al. 1999) owing to its −OH rich functional groups that dissolve CO_2 during the process. PI is another crystalline polymer material with high permeability and selectivity used for gas separation. Matrimid® is a commercially available PI which is very rigid and stable with stiff polymer backbone which can be used in harsh conditions. But, it is very expensive and susceptible to plasticization ($P_{Plasticization}$ = 17 bar) (Bos et al. 1999). PSF is another important polymeric material with excellent mechanical and thermal strength, high rigidity, and acceptable gas pair selectivity. Although the PSF is still lacking in separation properties compared to PI, it is inexpensive with acceptable plasticization resistance ($P_{Plasticization}$ = 34 bar) (Bos et al. 1999).

Although polymeric membranes showed convincing results in gas separation, they suffer from drawbacks (Vinoba et al. 2017). Low membrane selectivity was the major inconvenient loss in gas separation that demands a multi-stage separation system which likely to impart higher capital cost. In addition, polymeric membranes commonly could not maintain their performance and thus deteriorate in extreme environmental operating conditions of high temperature and pressure. The chain swelling in the presence of highly corrosive components in the feed, plasticization, compaction, and aging of membranes were the main reasons for problematic phenomena (Vinoba et al. 2017). The performance of the polymeric materials for gas separation is also challenged by Robeson's upper bound trade-off limit. Highly permeable membrane materials are commonly accompanied by low gas pair selectivity and vice versa (Robeson 2008) as shown in Fig. 3. The performance of

various polymeric membranes for CO_2/CH_4 separation is summarized in Table 2.

3.2 Inorganic Membranes

Inorganic materials are used for gas separation due to their exceptional properties such as higher thermal and mechanical stability, excellent erosion resistance, insensitivity to bacterial action, and extensive operative life. Inorganic membranes show high permeability and selectivity, exceeding the Robeson's upper bound trade-off limit. It also exhibits good resistance to harsh chemical conditions and can withstand high pressures and temperatures (Khan et al. 2017). Zeolite, activated carbon, silica, carbon nanotubes, and metal–organic frameworks are different inorganic membrane materials. Table 3 presents the separation performance of different inorganic membranes for CO_2/CH_4.

Generally, the dense, nonporous inorganic membranes are impermeable to the majority of gases except for a very limited number of gases that can permeate and transport through them (Javaid 2005). Therefore, they are rarely used in natural gas separation processes. On the other hand, microporous inorganic membranes are used in the industries for gas separation (Fain 1994; Javaid 2005; Shekhawat et al. 2003). Although inorganic membranes have exhibited high-gas separation performance, their performance tends to be a strong function of operating conditions such as temperature, pressure, and mole fraction of the condensable species in the feed. It requires high cost for membrane fabrication due to their fragile structure (Baker and Lokhandwala 2008). The intrinsic fragility and low surface to volume ratio have also hindered their further application in the separation process (Rezakazemi et al. 2014). The limitations of these materials have motivated researchers to develop new membrane materials. Comparison in terms of the advantages and limitations of inorganic membranes are given in Table 4.

3.3 Mixed Matrix Membranes (MMMs)

Limitations suffered by both polymeric and inorganic membrane have motivated the researchers to develop a new class of membranes, i.e., mixed matrix membranes (MMMs). The polymeric material is used as a continuous phase while inorganic particles are homogeneously dispersed in the membrane as shown in Fig. 4. Improved gas separation properties are the driving force in the development of mixed matrix membrane for gas separation (Zornoza et al. 2011a, b). The addition of inorganic filler into the polymer matrix improves the membrane performances. For example, the zeolite provides a molecular sieving

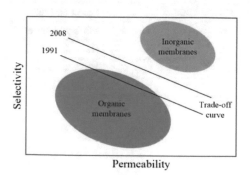

Fig. 3 Robeson trade-off limit between gas permeability and gas pair selectivity (Robeson 2008)

Table 2 Separation performance of different polymeric membranes for CO_2/CH_4

Membrane material	Conditions		Permeability (barrer)[a]		Selectivity (α) P_{CO_2}/P_{CH_4}	References
	Pressure (bar)	Temperature (°C)	P_{CO_2}	P_{CH_4}		
Polydimethyl siloxane	4	23	3800	1187.5	3.2	José et al. (2004)
Polycarbonate	5	75	10	0.77	13	Costello and Koros (1992)
Polyamide	2	35	11	0.303	36.3	Ghosali and Freeman (1995)
Polycarbonate	20	30	2	0.074	27.2[b]	Sridhar et al. (2007a)
Matrimid	10	25	6.72	0.22	30.5[b]	Jiang et al. (2006)
Matrimid	34.5	35	10	0.28	35.7	Vu et al. (2003b)
Matrimid	2.7	35	11	0.26	42.14[b]	Ordoñez et al. (2010)
6FDA-Durene	2	25	1468	64.96	22.6	Nafisi and Hagg (2014)
6FDA-Durene	10	35	541	41.27	13.11	Askari and Chung (2013)
6FDA-Durene	10	35	456	28.32	16.1	Liu et al. (2002)
6FDA-Durene/DABA	10	35	519	27.30	19.01	Askari and Chung (2013)
6FDA-DAM	2	25	390	16.25	24[b]	Bae et al. (2010)
6FDA-1,5-NDA	10	35	22.6	0.47	48.1	Chan et al. (2003)
Natural rubber	10	25	134	28.5	4.7	Sridhar et al. (2007b)
Poly(4-methyl-1-pentene)	10	25	83	13.2	6.3	Sridhar et al. (2007b)
Polysulfone	10	30	5.60	0.25	22.4	Sridhar et al. (2007a)
Polyetherimide	10	25	0.63	0.0175	36	Sridhar et al. (2007b)

[a] 1 barrer = 846 mL (gas) mm per cm^2 per day and per bar or 10^{-10} cm^3 (STP) cm cm^{-2} s^{-1} cm Hg^{-1}

[b] Mixed gas permeation test

Table 3 Performance of different inorganic membranes for CO_2/CH_4 separation

Membrane material	Conditions		Permeability (GPU)[a]		Selectivity (α) P_{CO_2}/P_{CH_4}	References
	Pressure (bar)	Temperature (°C)	P_{CO_2}	P_{CH_4}		
CMS	1	25	7	0.044	160[b]	Centeno and Fuertes (1999)
SAPO-34 zeolite	2.2	22	1045.9	8.72	120[b]	Li et al. (2006a, b)
Silicalite	1	30	589.8	137.2	4.3[b]	Zhu et al. (2006)
H-ZSM-5	0.3	27	5667.8	1030	5.5	Poshusta et al. (1999)
Silicalite	1	30	5755.2	2398	2.4[b]	van den Broeke et al. (1999)
Silica	1.4	25	2570	210.7	12.2	Raman and Brinker (1995)
DDR zeolite	2	100	99.6	0.996	100[b]	Tomita et al. (2004)
Zeolite-T	1	35	137.4	0.344	400	Cui et al. (2004)
Zeolite-T	5	30	30	0.424	70.8	Mirfendereski et al. (2008)
Carbon	2	30	30	0.288	104[c]	He et al. (2011)
CMS 550-2	3.4	35	1250	19.84	63[c]	Vu et al. (2003a)
CMS 800-2	3.4	35	43.5	0.218	200[c]	Vu et al. (2003b)

[a] 1 GPU = 10^{-6} cm^3 (STP) cm^{-2} s^{-1} cm Hg^{-1}

[b] Mixed gas permeation test

[c] Unit in barrer

Table 4 Advantages and limitations of inorganic membranes (Javaid 2005)

Advantages for inorganic membranes	Limitation for inorganic membranes
High stability at high temperatures and pressure	High capital costs
High resistance to harsh environments	Brittleness (membrane cracking due to the brittleness and high sensitivity to the temperature gradient
High resistance to microbiological degradation	Low membrane surface area per module volume
Ease of cleanability after fouling	Difficult to achieve high selectivities at large scale
Ease of catalytic activation	Challenging for sealing of membrane into the module at high temperatures

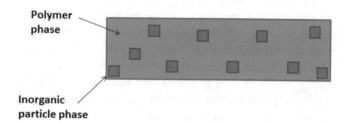

Polymer phase

Inorganic particle phase

Fig. 4 Schematic of a mixed matrix membrane (MMM)

mechanism to stop CH_4 while its interaction with CO_2 gives better surface diffusion across the membrane (Thompson et al. 2014).

Increase in membrane tensile strength indicates good dispersion of particles in the polymer matrix and good interaction between polymer and filler (Basu et al. 2011). Increase in membrane thermal stability results from an increase in rigidity of polymer chains and thus restricts motion of polymer chain due to particle disruption (Zornoza et al. 2011a, b; Basu et al. 2011). The key factors in developing ideal MMM are high intrinsic properties of MMM materials and good interaction between polymer and filler. For continuous phase, glassy polymers are the preferable choice since it is highly selective with moderate gas permeability, while rubbery polymers have low intrinsic selectivity (Aroon et al. 2010). As for the dispersed phase, the selection of filler is dependent on the specific properties to be improved, either permeability or selectivity. Table 5 compares the properties of polymeric, inorganic, and MMM membranes. MMMs have the properties of both polymeric and inorganic membranes with better performance and stability.

3.3.1 Progress in MMM Development

To date, various potential MMMs have been explored and well-documented in the literature highlighting the advantages and limitations suffered by the resulted membranes. Zeolites as inorganic filler have been in the highlights due to high intrinsic separation properties, high thermal, and mechanical strength, as well as high chemical stability.

There are 218 different types of zeolite, only a few have shown promising fillers in MMM for CO_2/CH_4 separation (Chen et al. 2015). Among them, zeolite 4A, silicoaluminophosphate-34 (SAPO-34), faujasite (FAU), and zeolite socony mobil-5 (ZSM-5) have been widely reported to improve the gas separation properties (Zhang et al. 2008). For example, boost CO_2 in permeability up to 140%, while providing 60% in CO_2/CH_4 selectivity when utilizing FAU/EMT zeolite into a polyimide matrix (Chen et al. 2012). Similarly, Junaidi et al. (2014) reported that CO_2 permeance increased from 105 to 706 GPU, while CO_2/CH_4 selectivity from 15 to 30.7 for asymmetric PSF/SAPO-34 membrane. It is expected since slower diffusion of large gases (e.g., CH_4, N_2) in zeolite channels hindered its permeation, whereas smaller gas (e.g., CO_2) is unaffected thus increasing both the CO_2 permeance and CO_2/CH_4 selectivity. Zeolite preference toward quadrupole moment of CO_2 has hindered CH_4 permeation and increase the gas pair selectivity. Combinations of preferential adsorption and molecular sieving have emerged MMM/zeolite as a potential pair for CO_2/CH_4 separation.

Incorporation of carbon molecular sieve (CMS) into polymer matrices has also been reported in past years. It is expected that CMS to have good affinity with glassy polymers compared to other class of fillers due to carbon-rich materials, thus allowing good adhesion at the interface without leading to interfacial defects (Vu et al. 2003a). In a series of their works, CMS was prepared using dense Matrimid® 5218 as a precursor before undergoing pyrolysis at a final temperature of 800 °C for 2 h under vacuum. The resulted MMM with CMS showed promising separation properties with the CO_2 permeability of 43.5 barrer and CO_2/CH_4 selectivity of 200. The increase in separation properties was attributed to the highly selective and permeable of the incorporated CMS (Vu et al. 2003b).

Despite having poor intrinsic CO_2/CH_4 separation properties compared to zeolites and CMS, utilizing silica as fillers MMM have also been investigated for CO_2 removal membrane. The embodiment of silica into polymer matrices influence membrane performance distinctly compared to

Table 5 Comparison of polymeric, inorganic, and MMM properties (Ismail et al. 2009)

Properties	Polymeric	Inorganic	MMM
Cost	Economical	High	Moderate
Chemical and thermal stability	Moderate	High	High
Mechanical Strength	Good	Poor	Excellent
Compatibility to solvent	Limited	Wide range	Limited
Separation performance	Moderate	High	Moderate
Handling	Robust	Brittle	Robust

zeolites and CMS. MMM/silica tends to improve the CO_2/CH_4 selectivity by restricting polymer chain mobility and hindered larger molecules to permeate (Ahn et al. 2010). CNT possesses unique characteristics such as superior mechanical and thermal stability with high electrical properties and is thus widely regarded as versatile materials. As gas adsorption separation media, CNT tends to adsorb polar molecules such as H_2S (dipole moment of 0.97 Debye), while discriminating non-polar molecules, despite CNT being non-polar materials (Nour et al. 2013; Vinoba et al. 2017). CNT alone as gas separation medium is not considered to be attractive due to its non-polar nature; non-selective sorption; and large pores do not provide molecular sieving. Nevertheless, the incorporation of CNT into polymer matrix has been reported to be beneficial to gas separation properties. The limitations of CNT alone as gas separation media are compensated by its relatively low density materials. Compared to other class of fillers, a large number of CNT materials can be incorporated into a polymer matrix at similar mass loading with other fillers. As a result, even low loading of CNT embedded in the membrane can provide a significant contribution toward overall membrane properties and hence separation performance (Moghadassi and Rajabi 2014).

Application of metal–organic frameworks (MOFs) for gas separation is not well developed comparing other applications. Most of the studies on MOF focused as a catalyst, gas storage, and gas adsorbent, while as a gas separation media is discouraging despite their highly permeable properties due to poor gas pair selectivity. Nevertheless, MOF has higher BET surface area, no dead volume, high uptake capacity, and low desorption energy compared to zeolites and CMS. Most importantly, the presence of organic linker within its structure has provided good interaction between polymer matrices which could minimize the interfacial defects. The development of MMM is a surely interesting approach with robustness, moderate cost, and high thermal, chemical, and mechanical stability. Selection of fillers remains as the heart of MMM development to ensure good compatibility with the polymer matrix, consequently, boost the membrane performance exceeding the Robeson upper bound.

3.3.2 Challenges in MMM

Generally, the incorporation of inorganic particles into the polymer matrix would surely improve the membrane properties due to the superiority of the dispersed phase itself. In actual practice, MMMs development is often encountered with the deterioration of membrane performance (Chung et al. 2007). The incompatibility between polymer-inorganic is regarded as the main factor that leads to the defective interface. The defective MMM can be identified through its gas separation performance relative to its neat polymer membrane. Unselective voids would be formed due to incompatibility between polymer and filler. Besides the incompatibility between polymer and filler, the formation of the voids is due to interfacial cracking at high loading, elongation stress during spinning hollow fiber, and repulsive force between polymer and filler (Xing and Ho 2009). Consequently, the presence of voids interface would lead to poor gas pair selectivity. Formation of voids can be reduced by using the treatment on filler before dispersed into the polymer solution. To address the issue, Zornoza et al. (2011a, b) proposed filler treatment on mesoporous silica spheres by using calcinations and chemical extraction.

Filler surface modification is also helpful to enhance its affinity toward polymer. This approach acts as adhesive agent at filler-polymer interface hence minimizes the formation of unselective voids. The common surface modification methods were octadecylamine (Hashemifard et al. 2011), silane coupling agent (Nik et al. 2011), diethanolamine (Clarizia et al. 2008), and amine modification (Nordin, et al. 2014). It was also reported that the incorporation of unmodified zeolite has diminished CO_2/CH_4 selectivity up to 80% due to the formation of unselective voids. Whereas, the incorporated modified zeolites have boosted CO_2/CH_4 selectivity, 50% better than the neat membrane. With the absent of unselective voids after surface modification, the larger penetrant, CH_4, traveled in longer permeation path while CO_2 can access through the filler and give an improvement in CO_2/CH_4 selectivity. Particle distribution is one of the important factors to be considered in fabricating MMM. Poor particle distribution would provoke particles to agglomerate with each other thus deteriorating the membrane performance. Poor interaction between polymer matrices and the filler is the main factor for particles to agglomerate. The agglomeration caused the filler to be inaccessible by penetrant and cause severe CO_2/CH_4 reduction, thus hindering the potential of MMM. The addition of excess particle is also increased the chance of particles to agglomerate. Car et al. (2006) demonstrated that the CO_2/CH_4 selectivity began to decrease when $Cu_3(BTC)_2$ loading exceeds 30 wt% in the PDMS membrane. At higher filler loading, the filler began to agglomerate and disrupt the polymer chain severely. Several methods have been proposed to avoid particle to agglomerate. Ge et al. (2011) applied the pretreatment of CNT with acid and phase transition method which has improved the particle distribution in the polymer matrix. Using H_2SO_4/HNO_3 and fast phase inversion, the absent of particle agglomeration at 1–5 wt% of CNT was observed. However, a further increase in CNT loading up to 10 wt%, particle agglomerations took place as a cluster of nanotubes formed due to intermolecular force in the polymer matrix.

Pore blockage can be categorized into the partial blockage and total blockage. Partial blockage allows molecules smaller than blocking pores to pass through while total blockage acts as an impermeable filler. Pore blockage can be caused by the sorbent, solvent, contaminant, a component in the polymer chain before, during or after membrane fabrication. The pore blockage phenomenon often resulted from excessive filler loading that creates inaccessible pores after agglomeration occurs. Moreover, low compatibility between filler and polymer also lead to pores blockage. By functionalization or alteration of polymer or filler with introducing mutually interactive functional group will enhance its compatibility and minimize pore blockage (Bhuwania et al. 2014). Another important phenomenon was the plasticization of the membrane during the gas separation process. CO_2 has been shown to plasticize a wide range of glassy polymers at elevated feed pressure (Scholes et al. 2010). Plasticization of glassy polymers occurred due to a reduction in the interaction between polymer and filler which decline the gas separation performance of the resulted membranes. The CO_2 increases the mobility of polymer chain segments, thereby increasing the diffusion coefficients of all penetrants in the membrane. Generally, cross-linking or modification methods were widely applied for improving the plasticization resistance (Adewole et al. 2013). Cross-linking process leads to a reduction in chain mobility by improving adhesion between polymer and inorganic filler. Subsequently, the gas separation performance and long-term stability could be achieved by overcoming CO_2-induced plasticization.

3.3.3 Promising Nanoparticles for Mixed Matrix Membranes

Metal–organic frameworks (MOFs) have recently emerged as important materials for the catalytic application, gas storage, adsorption, and gas separation. MOFs are crystalline compounds consisting of metal ions and secondary building units (SBUs) or organic ligands (Fig. 5). Interesting characteristics of MOF's such as high micropore volume, large pore sizes, high phase crystallinity, and high metal content offering valuable active sites are the key features of this new and emerging class of porous materials (Schlichte et al. 2004). Large surface area and pore volume of MOF give advantages over other porous materials like activated carbon and zeolite. The high surface area gives more contact with the targeted species thus increasing the effectiveness of the particles. Previous researchers had synthesized MOF with high BET surface areas such as MOF-5 (3000 m^2/g) (Perez and Balkus 2009), Co-MOF-74 (1314 m^2/g) (Cho et al. 2012), Mg-MOF-74 (1332 m^2/g) (Bao et al. 2011), higher than commercial zeolite Y (900 m^2/g), and zeolite beta (710 m^2/g). The role of MOF's in enhancing MMM performance can be summarized as follows; (1) enhancement of CO_2 diffusivity coefficient through polymer chain disruption exceeding CH_4 (Song et al. 2012a, b); (2) interaction between quadrupole moment of CO_2 with the weak electrostatic field of MOF (Basu et al. 2011); and (3) molecular sieving induced by MOF pores (Ordoñez and Balkus 2010).

One of the attractive properties of MOFs is their adsorption capability. The high surface area of the materials with open metal sites act as a large platform for adsorption of specific gases (Bao et al. 2011). The HKUST-1 was able to adsorb 12.5 mmol CO_2/g at 15 bar and still did not reach its saturation point due to a large surface area and pore volume (Liang et al. 2009). Other notable MOF's with high CO_2 adsorption capacities were Cu-MOF with 10.9 mmol/g (Lincke et al. 2011) and Mg-MOF-74 with 15 mmol/g (Choi et al. 2012). In additions, due to the structural flexibility of MOF's, their adsorption capacity has increased even at high pressure while retaining its structure (Li et al. 2011b). It was reported that the adsorption capacity of ZIF-69 increases up to 40% when changing to larger pores at high pressure (Li et al. 2011a) and able to retain 82.6 L CO_2 for every liter of ZIF-69 (Banerjee et al. 2008). MOFs for gas separation were also reported in recent years. Since MOF's are highly porous materials with high surface area, gases are able to permeate across the membrane with high permeation rates.

Although MOF possesses a unique characteristic, it should be noted that MOF is the relatively new class of materials. Thus, several insights such as influences of different gases and vapors and the cost of the materials are necessary before it can be used for various industrial applications (Tagliabue and Farrusseng 2009). Besides, the selective sorption of the MOF is relatively low compared to other classes of materials (Yeo et al. 2014). Although the versatility of its organic ligands allowed vast modification approaches to compensate its limitations, an intensive study regarding this matter is necessary. Table 6 summarizes the

Fig. 5 Schematic representations of the general classification of porous solids and a typical construction procedure of MOF (Li et al. 2011b)

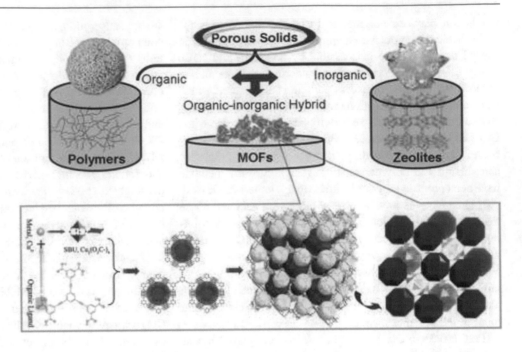

Table 6 Performance of various MOF-based MMMs for CO_2/CH_4 separation

MMM materials		Filler content (%)	Permeability (barrer)[a]		Selectivity (α) P_{CO_2}/P_{CH_4}	References
Polymer	MOF		P_{CO_2}	P_{CH_4}		
PSF	$Cu_3(BTC)_2$	10	3034	902.9	3.36	Anja et al. (2006)
PSF	$Mn(HCOO)_2$	10	6.83	0.75	9.16	Anja et al. (2006)
Matrimid	Cu-BPY-HFS	20	9.88	0.358	27.62	Yanfeng et al. (2008)
Matrimid	MOF-5	30	20.2	0.45	44.89[b]	Edson et al. (2009)
PSF	MIL-101	19	32	1.36	23.50	Jeazet et al. (2013)
PSF	HKUST-1	16	8.80	0.54	16.20[b]	Beatriz et al. (2011)
6FDA-ODA	NH_2-MOF-199	25	26.6	0.45	59.6	Nik et al. (2012)
Matrimid	MIL-53	15	12.4	0.24	51.8	Dorosti et al. (2014)
6FDA-ODA	UiO-66	25	50.4	1.09	46.1	Nik et al. (2012)
Matrimid	Fe(BTC)	30	13.5	0.45	30	Shahid and Nijmeijer (2014)
Matrimid	MOP-18	44	15.6	0.947	16.47	Perez et al. (2014)
Matrimid	$Cu_3(BTC)_2$	20	24.8	0.656	37.8	Duan et al. (2014)

[a]1 barrer = 846 mL (gas) mm per cm^2 per day and per bar or 10^{-10} cm^3 (STP) cm cm^{-2} s^{-1} cm Hg^{-1}
[b]Mixed gas permeation test

CO_2/CH_4 separation performance for MOF-based mixed matrix membranes. It is seen that PSF and Matrimid have compatibility with various MOFs.

4 Research and Development Prospects for Membrane-Based Natural Gas Purification

Future research opportunities, directions, and development of membrane technology have many driving forces such as economic and environmental reasons and growing interest of

industries and markets. This section shows the prospective research and development activities in the field of natural gas purification via membrane separation process.

MMMs prepared are still lacking in high CO_2/CH_4 selectivity and plasticized at high feed pressure. In the near future, the research for the purification of natural gas using membrane process will extend in more directions such as: synthesize new membrane materials (polymer and inorganic materials) with high separation performance, materials could perform in a harsh environment with high resistance, plasticization reduction with improved performance, enhanced CO_2 selectivity with longer service life and hollow fiber mixed matrix membrane with improved contact area and subsequently high permeance and selectivity. The use of membrane technology for natural gas processing should be further researched and focused to develop a simple process with low energy requirements.

5 Conclusions and Future Directions

This chapter attempted to address the main challenges of membrane technology encountered during the natural gas sweetening processes. Nowadays, the membrane separation technology has a vital role in the industries due to environmental and economic driving forces. Among impurities, the removal of acid gas CO_2 from the raw natural gas is more crucial due to its corrosive nature. The aim is to improve the heating value, reduce corrosion of the equipment, and decrease the hazards for human health. Membrane gas separation method is the most economical, simple, and environmentally friendly process. Deficiencies in both the polymeric and inorganic membranes have suggested the need for the development of novel mixed matrix membrane with superior gas separation performance. In addition, it may offer better thermal, chemical, and mechanical properties for aggressive and harsh conditions. It is further suggested that the future research must be focused on the development of new membrane materials that overcome the current challenges. Earlier work has shown that a huge potential exists for the improvement of membrane technology for natural gas processing. Further in the long term, the hybrid processes for natural gas purification process are more effective and advantageous, where it could be combined with other traditional techniques. This collective performance can reduce the operational cost of the process. The potential of the hybrid processes over the single process needs to be fully explored by considering all thermodynamic, kinetic, and modeling parameters. In view of the current situation, mixed membranes (MMMs) are considered the most practical alternative approach for commercial application. It is highly desirable to produce a cost-effective membrane that should operate at high temperature and pressure conditions. Furthermore, the membrane plasticization and long-term operation capacity will be other challenges and typical future research directions. Moreover, the modeling of the membrane processes for high performance can be recommended for further research direction.

References

Adams, R., Carson, C., Ward, J., Tannenbaum, R., & Koros, W. (2010). Metal organic framework mixed matrix membranes for gas separations. *Microporous and Mesoporous Materials, 131*(1–3), 13–20.

Adewole, J. K., Ahmad, A. L., Ismail, S., & Leo, C. P. (2013). Current challenges in membrane separation of CO_2 from natural gas: A review. *International Journal of Greenhouse Gas Control, 17*, 46–65.

Ahn, J, Chung, W. J., Pinnau, I., & Song, J. (2010). Gas transport behavior of mixed-matrix membranes composed of silica nanoparticles in a polymer of intrinsic microporosity (PIM-1). *Journal of Membrane Science, 346*, 280–287.

Anja, Stropnik, C., & Peinemann, K. V. (2006). Hybrid membrane materials with different metal-organic frameworks (MOFs) for gas separation. *Desalination, 200*(1-3), 424–426.

Aroon, M. A., Ismail, A. F., Matsuura, T., & Montazer-Rahmati, M. M. (2010). Performance studies of mixed matrix membranes for gas separation: A review. *Separation and Purification Technology, 75*(3), 229–242.

Askari, M., & Chung, T.-S. (2013). Natural gas purification and olefin/paraffin separation using thermal cross-linkable co-polyimide/ZIF-8 mixed matrix membranes. *Journal of Membrane Science, 444*, 173–183.

Bae, T. H. (2010). A high-performance gas-separation membrane containing submicrometer-sized metal-organic framework crystals. *Angewandte Chemie - International Edition, 49*(51), 9863–9866.

Baker, R. W., & Lokhandwala, K. (2008). Natural gas processing with membranes: An overview. *Industrial & Engineering Chemistry Research, 47*(7), 2109–2121.

Banerjee, R., Phan, A., Wang, B., Knobler, C., Furukawa, H., O'Keeffe, M., & Yaghi, O. M. (2008). High-throughput synthesis of zeolitic imidazolate frameworks and application to CO_2 capture. *Science, 319*(5865), 939–943.

Bao, Z., Yu, L., Ren, Q., Lu, X., & Deng, S. (2011). Adsorption of CO_2 and CH_4 on a magnesium-based metal organic framework. *Journal of Colloid and Interface Science, 353*(2), 549–556.

Basu, S., Cano-Odena, A., & Vankelecom, I. F. J. (2011). MOF-containing mixed-matrix membranes for CO_2/CH_4 and CO_2/N_2 binary gas mixture separations. *Separation and Purification Technology, 81*, 31–40.

Bhuwania, N., Labreche, Y., & Achoundong, C. S. K. (2014). Engineering substructure morphology of asymmetric carbon molecular sieve hollow fiber membranes. *Carbon, 76*, 417–434.

Bos, A., Pünt, I. G. M., Wessling, M., & Strathmann, H. (1999). CO_2-induced plasticization phenomena in glassy polymers. *Journal of Membrane Science, 155*(1), 67–78.

Centeno, T. A., & Fuertes, A. B. (1999). Supported carbon molecular sieve membranes based on a phenolic resin. *Journal of Membrane Science, 160*(2), 201–211.

Chaidou, C. I., Pantoleontos, G., Koutsonikolas, D. E., Kaldis, S. P., & Sakellaropoulos, G. P. (2012). Gas separation properties of polyimide-zeolite mixed matrix membranes. *Separation Science and Technology, 47*(7), 950–962.

Chan, S. S., Chung, T.-S., Liu, Y., & Wang, R. (2003). Gas and hydrocarbon (C2 and C3) transport properties of co-polyimides synthesized from 6FDA and 1, 5-NDA (naphthalene)/durene diamines. *Journal of Membrane Science 218*(1), 235–245.

Chen, X. Y., Nik, O. G., Rodrigue, D., & Kaliaguine, S. (2012). Mixed matrix membranes of aminosilanes grafted FAU/EMT zeolite and cross-linked polyimide for CO_2/CH_4 separation. *Polymer, 53*, 3269–3280.

Chen, X.-Y., Vinh, H., Ramirez, A. A., Kaliaguine, S., & Rodrigue, D. (2015). Membrane gas separation technologies for biogas upgrading. *RSC Advances, 5*(31), 24399–23448.

Cho, H.-Y., Yang, D.-A., Kim, J., Jeong, S.-Y., & Ahn, W.-S. (2012). CO_2 adsorption and catalytic application of Co-MOF-74 synthesized by microwave heating. *Catalysis Today, 185*(1), 35–40.

Choi, S., Watanabe, T., & Bae, T. H. (2012). Modification of the Mg/DOBDC MOF with amines to enhance CO_2 adsorption from ultradilute gases. *The Journal of Physical Chemistry Letters, 3*(9), 1136–1141.

Chung, T. S., Jiang, L. Y., Li, Y., & Kulprathipanja, S. (2007). Mixed matrix membranes (MMMs) comprising organic polymers with dispersed inorganic fillers for gas separation. *Progress in Polymer Science, 32*(4), 483–507.

Clarizia, G., Algieri, C., Regina, A., & Drioli, E. (2008). Zeolite-based composite PEEK-WC membranes: Gas transport and surface properties. *Microporous and Mesoporous Materials, 115*, 67–74.

Costello, L. M., & Koros, W. J. (1992). Temperature dependence of gas sorption and transport properties in polymers: Measurement and applications. *Industrial & Engineering Chemistry Research, 31*(12), 2708–2714.

Cui, Y., Kita, H., & Okamoto, K. I. (2004). Zeolite T membrane: Preparation, characterization, pervaporation of water/organic liquid mixtures and acid stability. *Journal of Membrane Science, 236*(1–2), 17–27.

Dong, G., Li, H., & Chen, V. (2013). Challenges and opportunities for mixed-matrix membranes for gas separation. *Journal of Materials Chemistry A, 1*(15), 4610–4630.

Dorosti, F., Omidkhah, M., & Abedini, R. (2014). Fabrication and characterization of matrimid/MIL-53 mixed matrix membrane for CO_2/CH_4 separation. *Chemical Engineering Research and Design, 92*(11), 2439–2448.

Duan, C., Jie, X., Liu, D., Cao, Y., & Yuan, Q. (2014). Post-treatment effect on gas separation property of mixed matrix membranes containing metal organic frameworks. *Journal of Membrane Science, 466*, 92–102.

Fain, D. E. (1994). Membrane gas separation principles. *MRS Bulletin, 19*(4), 40–43.

Ge, L., Zhu, Z., & Rudolph, V. (2011). Enhanced gas permeability by fabricating functionalized multi-walled carbon nanotubes and polyethersulfone nanocomposite membrane. *Separation and Purification Technology, 78*, 76–82.

Ghosali, K., & Freeman, B. D. (1995). Gas separation properties of aromatic polyamides with sulfone groups. *Polymer, 36*(4), 793–800.

Hashemifard, S. A., Ismail, A. F., & Matsuura, T. (2011). Effects of montmorillonite nano-clay fillers on PEI mixed matrix membrane for CO_2 removal. *Chemical Engineering Journal, 170*, 316–325.

He, X., Lie, J. A., Sheridan, E., & Hagg, M. B. (2011). Preparation and characterization of hollow fiber carbon membranes from cellulose acetate precursors. *Industrial and Engineering Chemistry Research, 50*(4), 2080–2087.

Ho, M. T., Allinson, G. W., & Wiley, D. E. (2008). Reducing the cost of CO_2 capture from flue gases using pressure swing adsorption. *Industrial & Engineering Chemistry Research, 47*(14), 4883–4890.

Hwang, S., Chi, W. S., Lee, S. J., Im, S. H., Kim, J. K., & Kim, J. (2015). Hollow ZIF-8 nanoparticles improve the permeability of mixed matrix membranes for CO_2/CH_4 gas separation. *Journal of Membrane Science, 480*, 11–19.

Ismail, A. F., Goh, P. S., Sanip, S. M., & Aziz, M. (2009). Transport and separation properties of carbon nanotube-mixed matrix membrane. *Separation and Purification Technology, 70*, 12–26.

Iwasa, R., Suizu, T., Yamaji, H., Yoshioka, T., & Nagai, K. (2018). Gas separation in polyimide membranes with molecular sieve-like chemical/physical dual crosslink elements onto the top of surface. *Journal of Membrane Science, 550*, 80–90.

Javaid, A. (2005). Membranes for solubility-based gas separation applications. *Chemical Engineering Journal, 112*(1–3), 219–226.

Jeazet, H. B. T., Koschine, T., Staudt, C., Raetzke, K., & Janiak, C. (2013). Correlation of gas permeability in a metal-organic framework MIL-101(Cr)—Polysulfone mixed-matrix membrane with free volume measurements by positron annihilation lifetime spectroscopy (PALS). *Membranes, 3*(4), 331–353.

Jiang, L. Y., Chung, T. S., & Kulprathipanja, S. (2006). Fabrication of mixed matrix hollow fibers with intimate polymer-zeolite interface for gas separation. *AIChE Journal, 52*(8), 2898–2908.

José, N. M., Prado, L. A. S. A., & Yoshida, I. V. P. (2004). Synthesis, characterization, and permeability evaluation of hybrid organic-inorganic films. *Journal of Polymer Science, Part B: Polymer Physics, 42*(23), 4281–4292.

Junaidi, M. U. M., Khoo, C. P., Leo, C. P., & Ahmad, A. L. (2014). The effects of solvents on the modification of SAPO-34 zeolite using 3-aminopropyl trimethoxy silane for the preparation of asymmetric polysulfone mixed. *Microporous and Mesoporous Materials, 192*(52–59).

Kanehashi, S., Chen, G. Q., Scholes, C. A., Ozcelik, B., Hua, C., Ciddor, L., et al. (2015). Enhancing gas permeability in mixed matrix membranes through tuning the nanoparticle properties. *Journal of Membrane Science, 482*, 49–55.

Kentish, S. E. (2008). Carbon dioxide separation through polymeric membrane systems for flue gas applications. *Recent Patents on Chemical Engineering, 1*(1), 52–66.

Khan, I. U., Othman, M. H. D., Hashim, H., Matsuura, T., Ismail, A. F., Arzhandi, M. R. D., & Azelee, I. W. (2017). Biogas as a renewable energy fuel—A review of biogas upgrading, utilisation and storage. *Energy Conversion and Management, 150*, 277–294.

Khan, I. U., Othman, M. H. D., Ismail, A. F., Matsuura, T., Hashim, H., Nordin, N. A. H., et al. (2018). Status and improvement of dual-layer hollow fiber membranes via co-extrusion process for gas separation: A review. *Journal of Natural Gas Science and Engineering, 52*, 215–234.

Khan, I. U., Othman, M. H. D., Jilani, A., Ismail, A. F., Hashim, H., Jaafar, J., et al. (2018a). Economical, environmental friendly synthesis, characterization for the production of zeolitic imidazolate framework-8 (ZIF-8) nanoparticles with enhanced CO_2 adsorption. *Arabian Journal of Chemistry, 11*(7), 1072–1083.

Khan, I. U., Othman, M. H. D., Jaafar, J., Hashim, H., Ismail, A. F., Rahman, M. A., & Ismail, N. (2018b). Rapid synthesis and characterization of leaf-like zeolitic imidazolate framework. *Malaysian Journal of Analytical Sciences, 22*(3), 553–560.

Khan, I. U., Othman, M. H. D., Ismail, A. F., Ismail, N., Jaafar, J., Hashim, H., et al. (2018c). Structural transition from two-dimensional ZIF-L to three-dimensional ZIF-8 nanoparticles in aqueous room temperature synthesis with improved CO_2 adsorption. *Materials Characterization, 136*, 407–416.

Lalia, B. S., Kochkodan, V., Hashaikeh, R., & Hilal, N. (2013). A review on membrane fabrication: Structure, properties and performance relationship. *Desalination, 326*, 77–95.

Lampinen, A. (2014). Sweden: Role of municipal policy in renewable energy use in transportation in Sweden. *Renewable Energy Law and Policy Review, 5*, 179–190.

Li, B. S., Falconer, J. L., & Noble, R. D. (2006). Improved SAPO-34 membranes for CO_2/CH_4 separations. *Advanced Materials, 18*(19), 2601–2603.

Li, J. R., Ma, Y., McCarthy, M. C., & Sculley, J. (2011a). Carbon dioxide capture-related gas adsorption and separation in metal-organic frameworks. *Coordination Chemistry Reviews, 255*(15), 1791–1823.

Li, J. R., Sculley, J., & Zhou, H. C. (2011b). Metal-organic frameworks for separations. *Chemical Reviews, 112*(2), 869–932.

Li, T., Pan, Y., Peinemann, K.-V., & Lai, Z. (2013). Carbon dioxide selective mixed matrix composite membrane containing ZIF-7 nano-fillers. *Journal of Membrane Science, 425–426*(2013), 235–242.

Li, Y., Chung, T. S., Huang, Z., & Kulprathipanja, S. (2006). Dual-layer polyethersulfone (PES)/BTDA-TDI/MDI co-polyimide (P84) hollow fiber membranes with a submicron PES–zeolite beta mixed matrix dense-selective layer for gas separation. *Journal of Membrane Science, 277*, 28–37.

Liang, Z., Marshall, M., & Chaffee, A. L. (2009). Comparison of Cu-BTC and zeolite 13X for adsorbent based CO_2 separation. *Energy Procedia, 1*, 1265–1271.

Lincke, J., Lässig, D., & Moellmer, J. (2011). A novel copper-based MOF material: Synthesis, characterization and adsorption studies. *Microporous and Mesoporous Materials, 142*(1), 62–69.

Liu, S., Wang, R., Chung, T., Chng, M. L., Liu, Y., & Vora, R. H. (2002). Effect of diamine composition on the gas transport properties in 6FDA-durene/3, 3-diaminodiphenyl sulfone copolyimides. *Journal of Membrane Science, 202*, 165–176.

Mahajan, R., Burns, R., Schaeffer, M., & Koros, W. J. (2002). Challenges in forming successful mixed matrix membranes with rigid polymeric materials. *Journal of Applied Polymer Science, 86* (4), 881–890.

Mirfendereski, S. M., Mazaheri, T., Sadrzadeh, M., & Mohammadi, T. (2008). CO_2 and CH_4 permeation through T-type zeolite membranes: Effect of synthesis parameters and feed pressure. *Separation and Purification Technology, 61*(3), 317–323.

Moghadassi, A. R., & Rajabi, Z. (2014). Fabrication and modification of cellulose acetate based mixed matrix membrane: Gas separation and physical properties. *Journal of Industrial and Engineering Chemistry, 20*(3), 1050–1060.

Mohshim, D. F., Mukhtar, H. B., Man, Z., & Nasir, R. (2013). Latest development on membrane fabrication for natural gas purification: A review. *Journal of Engineering, 2013*, 101746.

Nafisi, V., & Hagg, M.-B. (2014). Gas separation properties of ZIF-8/6FDA-durene diamine mixed matrix membrane. *Separation and Purification Technology, 128*, 31–38.

Nik, O. G., Chen, X. Y., & Kaliaguine, S. (2011). Amine-functionalized zeolite FAU/EMT-polyimide mixed matrix membranes for CO_2/CH_4 separation. *Journal of Membrane Science, 379*, 468–478.

Nik, O. G., Chen, X. Y., & Kaliaguine, S. (2012). Functionalized metal organic framework-polyimide mixed matrix membranes for CO_2/CH_4 separation. *Journal of Membrane Science, 414*, 48–61.

Noble, R. D. (2011). Perspectives on mixed matrix membranes. *Journal of Membrane Science, 378*(1–2), 393–397.

Nordin, N. A. H. M., Ismail, A. F., Mustafa, A. B., Goh, P., Matsuura, T., & Rana, D. (2014). Aqueous room temperature synthesis of zeolitic imidazole framework 8 (ZIF-8) with various concentrations of triethylamine. *RSC Advances, 4*(63), 33292–33300.

Nour, M., Berean, K., Balendhran, S., Ou, J. Z., Plessis, J. D., McSweeney, C., et al. (2013). CNT/PDMS composite membranes for H_2 and CH_4 gas separation. *International Journal of Hydrogen Energy, 38*(25), 10494–10501.

Ordoñez, M. J. C., & Balkus, K. J. (2010). Molecular sieving realized with ZIF-8/Matrimid® mixed-matrix membranes. *Journal of Membrane Science, 361*(1), 28–37.

Ordoñez, M. J. C., Balkus, K. J., Ferraris, J. P., & Musselman, I. H. (2010). Molecular sieving realized with ZIF-8/Matrimid® mixed-matrix membranes. *Journal of Membrane Science, 361*(1–2), 28–37.

Perez, E. V., Balkus, K. J., Ferraris, J. P., & Musselman, I. H. (2009). Mixed-matrix membranes containing MOF-5 for gas separations. *Journal of Membrane Science, 328*(1-2), 165–173.

Perez, E. V., Balkus, K. J., Ferraris, J. P., & Musselman, I. H. (2014). Metal-organic polyhedra 18 mixed-matrix membranes for gas separation. *Journal of Membrane Science, 463*, 82–93.

Poshusta, J. C., Noble, R. D., & Falconer, J. L. (1999). Temperature and pressure effects on CO_2 and CH_4 permeation through MFI zeolite membranes. *Journal of Membrane Science, 160*(1), 115–125.

Raman, N. K., & Brinker, C. J. (1995). Organic 'Template' approach to molecular sieving silica membranes. *Journal of Membrane Science, 105*(3), 273–279.

Rezakazemi, M., Amooghin, A. E., Montazer-Rahmati, M. M., Ismail, A. F., & Matsuura, T. (2014). State-of-the-art membrane based CO_2 separation using mixed matrix membranes (MMMs): An overview on current status and future directions. *Progress in Polymer Science, 39*(5), 817–861.

Robeson, L. M. (2008). The upper bound revisited. *Journal of Membrane Science, 320*, 390–400.

Rongwong, W., Boributh, S., Assabumrungrat, S., Laosiripojana, N., & Jiraratananon, R. (2012). Simultaneous absorption of CO_2 and H_2S from biogas by capillary membrane contactor. *Journal of Membrane Science, 392–393*, 38–47.

Samarasinghe, S. A. S. C., Chuah, C. Y., Yang, Y., & Bae, T. H. (2018). Tailoring CO_2/CH_4 separation properties of mixed-matrix membranes via combined use of two- and three-dimensional metal-organic frameworks. *Journal of Membrane Science, 557*, 30–37.

Schlichte, K., Kratzke, T., & Kaskel, S. (2004). Improved synthesis, thermal stability and catalytic properties of the metal-organic framework compound $Cu_3(BTC)_2$. *Microporous and Mesoporous Materials, 73*(1), 81–88.

Scholes, C. A., Chen, G. Q., Stevens, G. W., & Kentish, S. E. (2010). Plasticization of ultra-thin polysulfone membranes by carbon dioxide. *Journal of Membrane Science, 346*(1), 208–214.

Scholz, M., Melin, T., & Wessling, M. (2013). Transforming biogas into biomethane using membrane technology. *Renewable and Sustainable Energy Reviews, 17*, 199–212.

Shahid, S., & Nijmeijer, K. (2014). High pressure gas separation performance of mixed-matrix polymer membranes containing mesoporous Fe(BTC). *Journal of Membrane Science, 459*, 33–44.

Shekhawat, D., Luebke, D. R., & Pennline, H. W. (2003). *A review of carbon dioxide selective membranes.* DOE/NETL-2003/1200. Pittsburgh, PA, and Morgantown, WV (United States).

Song, Q., Nataraj, S. K., & Roussenova, M. V. (2012a). Zeolitic imidazolate framework (ZIF-8) based polymer nanocomposite membranes for gas separation. *Energy & Environmental Science, 5*, 8359–8369.

Song, Q., Nataraj, S. K., Roussenova, M. V., Tan, J.-C., Hughes, D. J., Li, W., et al. (2012b). Zeolitic imidazolate framework (ZIF-8) based polymer nanocomposite membranes for gas separation. *Energy & Environmental Science, 5*(8), 8359.

Sridhar, S., Aminabhavi, T. M., & Ramakrishna, M. (2007a). Separation of binary mixtures of carbon dioxide and methane through sulfonated polycarbonate membranes. *Journal of Applied Polymer Science, 105*(4), 1749–1756.

Sridhar, S., Smitha, B., & Aminabhavi, T. M. (2007b). Separation of carbon dioxide from natural gas mixtures through polymeric membranes—A review. *Separation and Purification Reviews, 36* (2), 113–174.

Stern, S. A. (1994). Polymers for gas separations: The next decade. *Journal of Membrane Science, 94*(1), 1–65.

Sun, Q., Li, H., Yan, J., Liu, L., Yu, Z., & Yu, X. (2015). Selection of appropriate biogas upgrading technology—A review of biogas cleaning, upgrading and utilisation. *Renewable and Sustainable Energy Reviews, 51*, 521–532.

Tagliabue, M., & Farrusseng, D. (2009). Natural gas treating by selective adsorption: Material science and chemical engineering interplay. *Chemical Engineering Journal, 155*(3), 553–566.

Thomas, N., Mavukkandy, M. O., Loutatidou, S., & Arafat, H. A. (2017). Membrane distillation research & implementation: Lessons from the past five decades. *Separation and Purification Technology, 189*, 108–127.

Thompson, J. A., Vaughn, J. T., Brunelli, N. A., Koros, W. J., Jones, C. W., & Nair, S. (2014). Mixed-linker zeolitic imidazolate framework mixed-matrix membranes for aggressive CO_2 separation from natural gas. *Microporous and Mesoporous Materials, 192*, 43–51.

Tomita, T., Nakayama, K., & Sakai, H. (2004). Gas separation characteristics of DDR type zeolite membrane. *Microporous and Mesoporous Materials, 68*, 71–75.

van den Broeke, L. J. P., Bakker, W. J. W., Kapteijn, F., & Moulijn, J. A. (1999). Transport and separation properties of a silicalite-1 membrane—I. Operating conditions. *Chemical Engineering Science, 54*(2), 245–258.

Vinoba, M., Bhagiyalakshmi, M., Alqaheem, Y., Alomair, A. A., Pérez, A., & Rana, M. S. (2017). Recent progress of fillers in mixed matrix membranes for CO_2 separation: A review. *Separation and Purification Technology, 188*, 431–450.

Vu, D. Q., Koros, W. J., & Miller, S. J. (2003a). Effect of condensable impurities in CO_2/CH_4 gas feeds on carbon molecular sieve hollow-fiber membranes. *Industrial & Engineering Chemistry Research, 42*(5), 1064–1075.

Vu, D. Q., Koros, W. J., & Miller, S. J. (2003b). Mixed matrix membranes using carbon molecular sieves: I. Preparation and experimental results. *Journal of Membrane Science, 211*(2), 311–334.

Weller, S., & Steiner, W. A. (1950). Separation of gases by fractional permeation through membranes. *Journal of Applied Physics, 21*(4), 279–283.

Xing, R., & Ho, W. S. W. (2009). Synthesis and characterization of crosslinked polyvinylalcohol/polyethyleneglycol blend membranes for CO_2/CH_4 separation. *Journal of the Taiwan Institute of Chemical Engineers, 40*, 654–662.

Yeo, Z. Y., Chai, S. P., Zhu, P. W., & Mohamed, A. R. (2014). Development of a hybrid membrane through coupling of high selectivity zeolite T on ZIF-8 intermediate layer and its performance in carbon dioxide and methane gas separation. *Microporous and Mesoporous Materials, 196*, 79–88.

Zhang, Y., Musselman, I. H., Ferraris, J. P., & Balkus, K. J. (2008). Gas permeability properties of Matrimid® membranes containing the metal-organic framework Cu–BPY–HFS. *Journal of Membrane Science, 313*(1–2), 170–181.

Zhao, Q., Leonhardt, E., & MacConnell, C. (2010). Purification technologies for biogas generated by anaerobic digestion. *CSANR Research*. https://www.build-a-biogas-plant.com/PDF/BiogasPurificationTech2010.PDF (August 24, 2016).

Zhu, W. Hrabanek, P., Gora, L., Kapteijn, F., & Moulijn, J. A. (2006). Role of adsorption in the permeation of CH_4 and CO_2 through a silicalite-1 membrane. *Industrial and Engineering Chemistry Research, 45*(2), 767–776.

Zornoza, B., Seoane, B., Zamaro, J. M., Téllez, C., & Coronas, J. (2011a). Combination of MOFs and zeolites for mixed-matrix membranes. *ChemPhysChem, 12*(15), 2781–2785.

Zornoza, B., Téllez, C., & Coronas, J. (2011b). Mixed matrix membranes comprising glassy polymers and dispersed mesoporous silica spheres for gas separation. *Journal of Membrane Science, 368*(1), 100–109.

Zornoza, B., Martinez-Joaristi, A., Serra-Crespo, P., Tellez, C., Coronas, J., Gascon, J., & Kapteijn, F. (2011c). Functionalized flexible MOFs as fillers in mixed matrix membranes for highly selective separation of CO_2 from CH_4 at elevated pressures. *Chemical Communications, 47*(33), 9522.

Hydrocarbon Separation and Removal Using Membranes

Mohammad Arif Budiman Pauzan, Mazlinda Abd Rahman, and Mohd Hafiz Dzarfan Othman

Abstract

This chapter describes the implementation of membrane in separation and removal of hydrocarbon. The treatment of hydrocarbon using membrane normally started from produced water (PW) which later continued to petrochemical refinery wastewater for natural gas, and other chemicals recovery has been enhanced throughout the years. Generally, the applications of using membrane technologies such as the use of microfiltration (MF), ultrafiltration (UF), followed by nanofiltration (NF), forward and reverse osmosis (FO and RO) have been implemented in many oil refineries plant. Recently, with the advance of technologies, focus nowadays has been put on the use of these membranes in oil refinery membrane systems was being made according to its application, i.e. thermal, electric and biological driven and mixed technologies. Finally, the future developments and challenges that needed to counter are briefly discussed. Throughout this chapter, the use of membrane technologies for hydrocarbon separation and removal in petrochemical and produced water wastewater will be explained and the impact of the studies in this application.

Keywords

Hydrocarbon separation · Membrane technology · Produced water · Filtration

1 Introduction

In modern day life, fossil fuel has been one of the most important energy sources to humankind compared other energy sources. For centuries, the technology especially in transportation and industries, humankind still uses this energy in their life and the demand still increasing even other technology from other source of energy such as electric sources. Derivatives from hydrocarbon such as oil, gas, and polymers from fossil fuel are among the main products from this source.

Hydrocarbon or fossil fuels are basically deep drilled from the earth using hydraulic fracturing of shale oil or also known as crude oil. Recent technology has brought this process to another level due to high demand, and it does not show any sign of stopping. Engineers and scientists around the world have been working to improvize and improve this particular process to achieve the demand. Generally, there is concern from by-product from this industry which the production of produced water (PW) also known as co-produce water or flowback water. This water can be referred to as the water that came out from the well due to the drilling of the oil. This water is vital to assist the drilling process where the freshwater will be used to pressurize later the crude oil/gas was pumped out from the well. Due to this matter, the amount of PW from oil drilling can be considered in vast quantity compared to gas drilling even though it can be varied from well to well depending on the location, geological formation and age of the well. Typically, the volume of PW will make up more than 90% from the total fluid volume of the reservoir due to the process of pumping or drilling. Figure 1 depicts the general schematic of produced water (PW) in the (a) oil and gas reservoir and (b) refinery process plant. The PW can be described as the water produced naturally in the well or reservoir. Hence, the acidity or the content of minerals ions/heavy metals inside the water may vary corresponding to the geology of that respective well.

M. A. B. Pauzan · M. A. Rahman · M. H. D. Othman (✉)
Faculty of Engineering, School of Chemical and Energy Engineering, Advanced Membrane Technology Research Centre (AMTEC), Universiti Teknologi Malaysia (UTM), 81310 Johor Bahru, Johor, Malaysia
e-mail: hafiz@petroleum.utm.my

© Springer Nature Switzerland AG 2021
Z. Zhang et al. (eds.), *Membrane Technology Enhancement for Environmental Protection and Sustainable Industrial Growth*, Advances in Science, Technology & Innovation,
https://doi.org/10.1007/978-3-030-41295-1_6

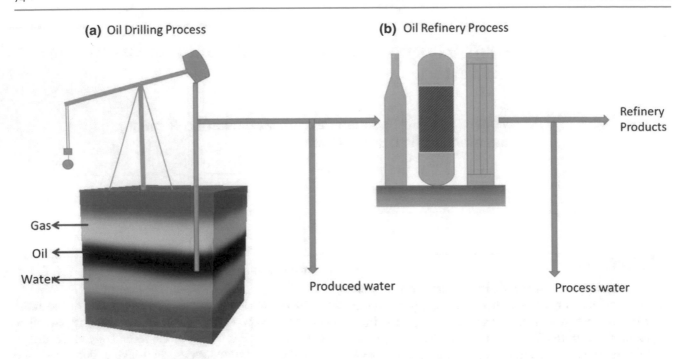

Fig. 1 General schematic of produced water (PW) in the **a** oil and gas reservoir and **b** from oil refinery process water (Munirasu et al. 2016). Lic. No 4616931488464

According to Chang et al. (2019), by using the combination of horizontal drilling and hydraulic fracturing techniques had made the extraction of shale/crude oil became more feasible, economically and technically. It was discovered that this technique requires up to 34,900 m^3 of water per horizontal well where it will limit the production of the crude oil from the well due to high baseline water stress which many countries with crude oil producer are facing. Hence, this technique allows more water to be taken out. Almost all crude oil producer countries facing this problem due to the baseline levels of water stress were high. The water needs to be taken out to allow the gas production once the hydraulic fracturing process was completed which resulting in flowback water (FW), and then, the PW was taken place, and the process continued. Here come the critical challenges faced by most refinery process plant to manage this FW and PW where protecting human health and the environment happened to be obeyed with respect to preserve the extraction. At first, treating the water in wastewater treatment plants (WWTP) has been one of the attempts for this matter. However, it was found that the quality of the receiving surface waters was reduced due to failure to the remove the dissolved impurities from the oil. Hence, it brought to another attempt of reuse and deep well injection technique. The famously used method to extract oil and gas is by deep well injection but limited not to all reservoirs. In addition to concerns on earthquakes, a smaller number of operating and constraint on the deep well injection, it was suggested that recycling or reusing and

discharging the waters after proper treatment has been the best solution for the industry.

On the other hand, there is another water produced from the refinery processing which is process water. Generated as the by-product after the refinery process of crude oil and gas, the mineral ions/heavy metals content from this water considerably lower than produced water due to the refinery process which taken place has removed almost half of the constituents of the minerals in the water. There are other chemicals broken down during the refinery process that should be taken note. Major constituents such as ammonia (NH_3), phenols, hydrogen sulphide (H_2S), toluene, xylene, benzene and ethylbenzene are among the chemical compounds that broken down which either loss or even absence after the transformation process in PW. Hence, it also can be considered that produced and process water possesses significant difference in term of chemicals content as generally, PW contains high salt directly from crude oil/gas while high organic matters containing in process water.

Even though the quality and quantity of PW may vary depending on its location and geological sources, previous studies have shown that there are no big differences in term of treatment using microfiltration (MF) and ultrafiltration (UF) membrane. In fact, even the results for PW and process water also have no significant differences even comparable. Therefore, in this chapter, the treatment for produce water and process water can be reviewed as single progression. In addition, there are only few literatures which stated where the original source of the crude oil was taken, and hence, the study of hydrocarbon separation using membrane separation will emerge the sources of the water, i.e. produced water and process water will be known generally as PW.

Normally, the process of refining crude oil/gas to simpler form involves three minors process depending on its size of molecules and level of separation:

1. Organic matter removal such as emulsified, dispersed and dissolved grease, oil and gases.
2. Dissolved inorganic matters removal.
3. Particulates removal which include clay, suspended particles and sand.

If along the process there is natural radioactive chemicals present, caution should be taken to remove those chemicals. The chemicals used for the sweetening or additives during the three steps process also should be removed without changing the original constituent.

In the process of PW treatment, the treatment can be divided to several steps of treatments started with primary treatment and followed by secondary treatment prior to membrane treatment as shown in Fig. 2. There are some situations where biological treatment was also involved at tertiary and/or secondary treatment. Currently, membrane technology application was notoriously studied after secondary treatment or tertiary treatment as this treatment was the closest to the end user before discharging to ground water.

Physical separation by gravity was commonly method used to separate oil from PW. However, policies regarding this matter has been progressively stringent due to the discharge of this PW was suggested to bring bad impact to the environment

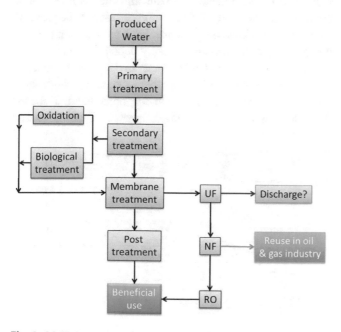

Fig. 2 Multiple treatments in PW treatment processes (Munirasu et al. 2016). Lic. No 4616931488464

(Witze 2015). Most countries have regulated that oil-in-water (OIW) content should not exceed 40 ppm prior to be discharged to the sea; however, nowadays the regulation has become stringent where the volume has becoming lesser (Igunnu and Chen 2012). This has led many countries to impose "zero-discharge" of where no harmful mixtures disposed. Approximately, 80% of PW was re-injected or recycle to the well to maintain the stability of the oil and water pressure as well as to optimize the usage of PW. It is considered beneficial economically to utilize the PW since normally oil producing areas are located near to water stressed area.

Thus, in order to meet with effective PW treatment, cleaning of polluted water together with recovering the purified water was imposed simultaneously. Several treatment methods were proposed or developed in order to recycle and for discharging such as membrane filtration, oxidation, basic separation technology, desalination technology as well as adsorption which hugely contributed by the components present in the PW. Apparently, Mulder (1996), Baker (2004) stated that membrane could be one of the effective ways for PW treatment in term of cost and practical compared to other available water purification technologies. However, the main issue of using membrane technology in PW treatment is to reduce and solve the presence of high total dissolved solids (TDS) which formed bunch of sediment on the membrane surface thus leads to membrane fouling. Therefore, research and study to counter this fact should be made by reducing the any organic and inorganic compounds to the lowest percent or concentration especially at pre-treatment stages in order to have a feasible practical membrane treatment as well as economically practicable process. Shaffer et al. (2013) suggested several membrane technologies that can be used in PW treatment such as membrane distillation (MD), reverse and forward osmosis (RO and FO), mechanical vapour compression, etc. However, for the use of microfiltration (MF), ultrafiltration (UF) and nanofiltration (NF) where the pore size played the main role in the treatment, extra care should be taken due to high pressure required to operate the system will damage the membrane and lead to formation of layers on the membrane's surface.

Rapid industrial growth, such as in oil and gas, petrochemical, pharmaceutical, metallurgical and food industries, has led to the large production of oily wastewater from hydrocarbon/shale/crude oil refinery process plant. It is very important for wastewater treatment in refinery plant not only to humankind but also to the environment, and the challenge still keeps growing. The increase of world population and economy has witnessed the increasing demand of safe and clean water especially at water-stressed areas. Thus, this non-renewable source of energy will be running out, and present consumption needs to come to some limitation if we intent to hold for future generation. Most interesting solution to the current problem will be to recycle and reuse the water with the implementation of membrane technology as this technology has become one of rapid growing industries for filtration with various applications could be found. Continuous studies and researches have been made to develop the best membrane not only academically but also in private industry. The introducing of membrane technology in hydrocarbon separation is not new discoveries as membrane technology has become one of the most effective ways to be used in this petrochemical industry. With the ability to separate oil from water, membrane also can be used to separate gas from the mixture and used in produce and process water treatment before being discharged or injected to the well. Hence, this chapter provides the information on the use of membrane technology in hydrocarbon separation by previous study.

2 Membrane Treatment Processes in Hydrocarbon Separation

The use of membrane technology for hydrocarbon waste from chemical and petrochemical/petroleum industries has been started as early 1950s where the first experiment was notated. Membrane technology is a simple and safe techniques where the separation requires small energy, and it is not exaggerated to mention it is an energy-saving technology as well as the materials used also non-harmful to humankind and environment (Munirasu et al. 2016).

Basically, a membrane is a barrier used to separate two phases in a selective manner and controlled by the carrying of various chemicals. Liquid or solid in form, the structure can be asymmetric or symmetric, heterogeneous or homogenous layers, as well as possess a negative or positive charge. The membrane transport mechanism can be influenced by diffusion or convection by respective molecules, depending on the concentration, pressure or temperature of the permeate and filtrate. The membrane can be classified by thickness, pore and particle size which vary from micro to nano scale. Driving force or pressure difference is required to run the membrane, and the performance of the membrane was evaluated based on any changes of concentration or interaction between two phases of the membrane. The parameters can be varied depending on the applications of the membrane where famously used parameters are concentration, pressure or temperature. Electrical or zeta potential difference also can be another driving force in membrane separations which only influenced by the moving of the membranes' charged molecules or particles (Ravanchi et al. 2009).

Generally, there are four types of methods were applied to investigate the mechanism of the membrane which are absorption, adsorption, cryogenics and normal filtration. To determine the methods going to use, the properties of the materials used will be defined depending to its specific applications. Compared to other methods, membranes offer versatility and simplicity where normally all membranes possessed similar separation processes regarding any materials and techniques. Among advantageous of membrane include compact, low labour intensity and maintenance where moving parts are not required, light in weight, operated at low capacity, minimum energy required, simple and safe module configuration for a simple permeation and last but not least low cost. Figure 3 depicts the example of membrane technology systems applications applied in industry.

In order to be used in wastewater treatment, few aspects should be considered such as the membrane operation mechanism, driving force, materials or even the structure of the membranes before being applied according to its applications. In next sub-chapters, the information of the membrane to be used in hydrocarbon or crude oil separation of PW will be categorized into various applications. Nowadays, the advancement of membrane technologies has brought to various improvement of membrane application systems which can be categorized according to the driving force which are low and high pressure, osmotic, thermal, electric and biological-driven membranes. Figure 4 shows the location where the membrane technologies were applied in refinery plant.

Typically, polymers and copolymers are used to fabricate the membrane and either in the shape of flat film or hollow fibre or even tubular depending on its applications. For gas separation, there are two types of membranes to be used either it is porous or nonporous. These two types of membrane have very significant different of transport mechanisms. Known as Knudsen flow, the mechanism stated that no separation of gas flow passing through a porous membrane due to small mean free path. In order to give a great pathway, the pore size of the membrane needs to be reduced. Correspond to Knudsen flow mechanism, the separation of two gas molecules was significantly depended on the square root of the molecular weights for a low separation factor.

Only by connecting to a number of modules, high separation can be accomplished. On the other hand, Knudsen flow does not involve if the transport of gases passed

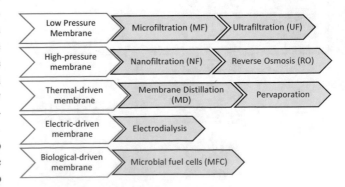

Fig. 3 Example of membrane technology system applications in refinery plant

Fig. 4 Schematic diagram of membrane-based technologies used in crude oil refinery process plant (Chang et al. 2019). Lic No. 4616931191915

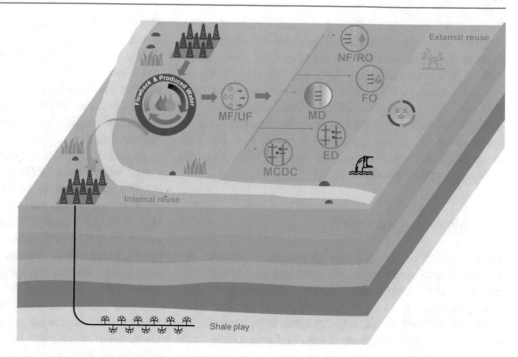

through a nonporous membrane. But in some cases, in a composite membrane with a porous substructure was supported by a dense top layer, Knudsen flow may contribute to the total flow if and only if the pore sizes in the following layer were considered.

In order to accomplish a perfect separation of gas in membrane, selectivity and high flux were used as the key parameters, and by using nonporous membranes, it will be depended on the differences in the permeabilities of various gases through a given membrane. The assessment of the membrane can be investigated based on high permeabilities or fluxes. High permeable materials are not required if high selectivities was used, but low permeable materials based on glassy polymers will be required for a moderate selectivity to create a balance between permeability and selectivity (Scoth 1999). According to Ravanchi et al. (2009), approximately more than 20 companies have commercially equipped membrane with gas permeation process due to its relatively simple process since it was first introduced in 1979. The applications may vary from the supply of pure or enhanced gases such as N_2, O_2 and He extracted from air, CO_2 and H_2S extraction from acid, separation of H_2 in the petrochemical and chemical industries and other varieties of smaller applications (Scoth 1990).

The pioneer of using membrane technology in petrochemical industry began in 1950 where the first experiment of was started (Weller and Steiner 1950). Later, in 1977, the first notable membrane unit was installed to adjust H_2/CO ratio, and in 1978, the first membrane was used for hydrogen recovery in petrochemical plants from gases elimination process. Hence, leading from the success of this application

led to another more than 200 membrane systems were built worldwide (Johnson and Schulman 1993). Recently, the use of membrane in refinery or petrochemical industry not only limited to separation and recovery of gas such as hydrogen recovery, nitrogen production, natural gas sweetening, pervaporation process but has advanced to the application of solid–liquid and liquid–liquid separation.

It was reported that in 1998, Mobil Oil Corporation used the new membrane technology where membrane of reverse osmosis (RO) was installed to separate solvent from lube oil (Baker 2004). The polymeric membrane was also the first kind of membrane made from polyamide which not only solvent resistance but also stable in high temperature up to 300 °C. Another successful application implemented such as the separation of organic liquid, aromatic compound such as benzene from paraffin and oleo chemicals. The use of ultrafiltration membrane developed by Exxon to recover and enhance their residuals feed product during thermal cracking was also considered as one of innovative ways where the application of membrane technology was used in petrochemical industry (Bernardo and Driolo 2010).

Currently, excellent progression of membrane technology has led to another application of membrane which is to be used to treat produce water (PW) in petrochemical industry. Membrane has demonstrated good application in surface water management brought the idea to use this technology in water at refinery plant not only for treating the water before being discharged to environment but also to recover any oil from the water itself. The expansion of this industry has led to the exploitation of high amount of water to be used. Thus, it is essential to review more innovations in membrane technology process to implement in petroleum industry. This chapter will describe previous studies have shown successful various works of membrane technology such as microfiltration (MF), ultrafiltration (UF), nanofiltration (NF) and reverse osmosis (RO) as well as the membrane-driven system used for this application. Therefore, this chapter will briefly contribute and assist to any further innovation for the applications of membrane technology in PW treatment especially in the petroleum or petrochemical refinery process (Alzahrani and Mohammad 2014).

2.1 Microfiltration (MF)

Microfiltration or MF is a sieving process to separate particulates or molecules by membrane where considered by the pore size of the membrane. Normally, the pore size for MF has approximately to be in the range of 0.1–50 μm. There are a lot of studies on using polymeric or ceramic MF membrane for the treatment of PW, and it is widely available. Figure 5 depicts the typical set-up of MF membrane for PW treatment. The earliest report on using MF membrane for PW treatment was by Mulder (1996) using ceramic polyacrylonitrile (PAN) to separate oil. Study by Li and Lee (2009), Cui et al. (2008) using PAN with pore size 0.8, 0.2 μm and 0.1 μm to synthetic PW indicated that by increasing the oil concentration, cross flow velocity (CFV), transmembrane pressure (TMP) and water flux were reduced whereas only slight impact on temperature (*T*) for final flux. It was discovered that there was increase in final flux where permeate containing oil concentration lower than 6 ppm lower and suspended solids of diatomaceous earth at 250 ppm. This may due to adsorption of oil by suspended solids at the membrane's surface has broken down the oil layer hence making the permeation flux to increase. In

addition, PAN membrane with big pore size ranges from 0.2 to 0.8 μm was appeared to possess both fouling on the outside and inside of the membrane whereas small pore size which is 0.1 μm and below where the fouling discovered only on the outside which is fouling on the membrane surface. Due to this oil fouling which forming an oil layer, the membrane's surface had become hydrophobic and becoming resistance to hydrodynamic shear method to be removed.

Abbasi et al. (2010) reported of using mullite and mullite–alumina ceramic MF membranes from kaolin and α-alumina extruded with water mixture dried at room temperature, followed by sintering the membrane at 1250 °C for 3h and leaching process using strong alkali for free silica removal to study on the effect of control cross flow velocity (CFV), temperature, pressure, salt and oil concentration on synthetic oil wastewater (synthetic PW) and real PW. It was found that by increasing the alumina content, temperature and flow rate of the mullite–alumina membranes, the permeation flow and the rejection had also increased. Whereas, the increase of oil concentration had decreased the permeation in the PW, although there was decreasing in fouling resistance at some elevated temperature and high CFV. It should be noted from this study that the real PW resulted in lower rejection and permeation flux compared to synthetic PW. Same goes for the result of rejection and permeation flux of total organic carbon (TOC) for real PW was less compared to synthetic PW though using the same membrane. Similar result was reported from study by Abadi et al. (2011) of using tubular ceramic α-Al$_2$O$_3$ MF for Tehran refinery PW where the PW content is less than 4 mg/L of oil after treatment which passed the allowed limit. From the study also, backwashing could prevent the flux decline significantly. In addition, study by Zhong et al. (2003) using MF membrane made up of zirconia (ZrO$_2$) suggested that, pre-treatment of PW by flocculation prior to the membrane

Fig. 5 Microfiltration pilot plant (Zsirai et al. 2018). Lic. No 4616890849059

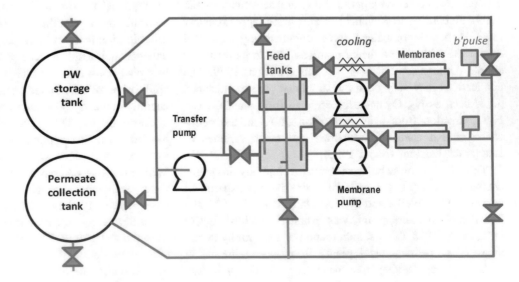

filtration, the concentration of oil in the permeate was reduced from 200 to 8.7 mg/L at 1.1 bar which has passed the Chinese standard for PW discharge.

With referring to previous MF membrane studies, there is something common which is the use of inorganic or ceramic membrane which attracted interest. Pre-treatment step by flocculation prior to filtration process improved the efficiency of the membrane and avoids the fouling formation of cake layer on the membrane surface (Alzahrani and Mohammad 2014).

2.2 Ultrafiltration (UF)

One the hand, the membrane filtration where the pore size of the membrane is between 0.01 and 0.1 μm is called ultrafiltration or UF. Initially, UF was accompanied along with MF to enhance the MF. However, in PW or oil treatment, it was found that UF membranes are prone to fouling due to its high permeation flux. Hence, in UF studies, this matter is being stress on how to reduce the fouling especially on membrane's surface also the effect of the fouling to the performance of the system. One of effective ways to reduce fouling is by reducing the roughness of the membrane surface and make it more hydrophilic.

Wandera et al. (2011) has reported that in his study of using UF membrane for synthetic PW treatment with modified hydrophilic cellulose and unmodified to enhance the surface roughness of the membrane. From the study, the unmodified membrane showed higher water flux if to be compared with modified due to grafted polymers which covered the surface of modified membrane. However, the unmodified membrane appeared to recover only 81% of permeation flux compared to modified membrane where almost 100% permeation flux recovery was recorded using simple water rinse. Further study by the same group Wandera et al. (2012), where the study on the effect of grafting density of polymer to prevent fouling indicated that there was decreasing in permeation flux as the density of the grafted polymer was increased. Further decline of permeation flux was showed as the operation was continued, and it was remarkably significant compared to unmodified membrane showing that grafted polymer thus brought effect to the pore size of the membrane and hence influences the permeation flux.

Another study by Yan et al. (2009) where the ability of modified and unmodified UF membrane for PW treatment was investigated using polyvinylidene fluoride (PVDF) mixed with nano-sized alumina particles fabricated in a tubular-shaped module. The results discovered that in term of organic pollutants removal, modified UF showed better performance compared to unmodified where the modified UF membrane exhibited more advantages such as high permeation flux more than 170 L/m^2 h bar, upgraded anti-fouling performance as well as almost 100% permeation flux was recovered after a few cleaning testing using some chemicals. It was also suggested that back-wash using OP-10 surfactant could be used for an enhanced cleaning testing to maintain high performance of the UF membrane whenever fouling was happened.

On the other hand, study by Kang et al. (2007) uses a modified UF membrane from PAN on the effect of hydrophilic modification to counter membrane's fouling in PW treatment. 20 wt% of PAN-g-PEO was added, and three different sources of PW were used, and for comparison, a commercial Serpo PAN400 was used as an unmodified membrane. From the study, the membranes showed a successful removal of grease and oil in the PW using a dead-end filtration system as shown in Fig. 6. The modified hydrophilic PAN membrane was able to recover up to 25% initial flux irreversibly compared to unmodified PAN membrane. In addition, the hydrophilic modified UF membrane showed a significant increase in rate of permeation flux compared to unmodified membrane which clearly indicated the success of hydrophilicity modification on the membrane's surface compared to commercial unmodified UF membrane where the permeation flux was declined in the PW treatment operation.

Report on using a hybrid PVDF/MWCNT nanocomposite UF membrane has been done by Moslehyani et al. (2015) used in a photocatalytic reactor together with UF filtration for petroleum refinery wastewater treatment. The membranes were fabricated using multi-wall carbon nanotubes (MWCNT) combined with 200 ppm of TiO_2 and UV light to enhance the photocatalytic reaction of the PW treatment. From the study, up to 90% of organic pollutant was removed, and more than 90% of organic matters were decomposed by the photocatalytic reaction after 6 h UV radiation process by studying the content in the permeate. Interestingly, it also discovered that the TiO_2 photocatalyst also recovered to more than 99% efficiency compared to normal catalytic reactor. Hence, the modified membrane acted as double usage simultaneously to filter the PW water and also for photocatalyst purpose.

In order to optimize an effective separation of oil in PW, other factors such as physiochemical nature of the UF membrane should be considered for a practical use in membrane technology. Seyed Shahabadi and Reyhani (2014) used commercial PAN350 UF membrane to investigate the effect of operating condition of membrane including cross flow velocity (CFV), temperature (T), transmembrane pressure (TMP), TOC rejection, permeation flux and resistance to fouling. The study reported that TMP showed significant influence on the fouling resistance and permeation flux while TOC removal greatly influenced by the CFV. The claim was supported by comparison in experimental and full factorial design methodology for data optimization and from

Fig. 6 UF dead-end system for enrichment and washout methods (Huang and Feng 2019). 4616910281625

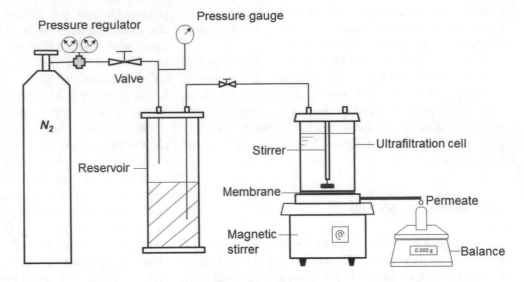

the outcome it was clearly implied that the optimization was important to achieve a best result depending on the PW quality and operating parameters unit of the membrane.

Another related study by Salahi et al. (2012) where two commercial UF membranes, i.e. PAN and PSf, were used to investigate the membrane operating parameter for operational optimization in pilot-scale PW treatment. In order to compare from the existing conventional biological treatment, refinery wastewater taken from Tehran API was used. It was discovered that except COD removal, all parameters including TMP, CFV, pH and temperature were highly removed by the UF membrane compared to existing biological treatment used in the refinery plant. In addition, the study also indicated that PSf UF membrane showed less water flux performance and low resistance to fouling compared to PAN UF membrane.

Furthermore, studies on the use of UF for inorganic or ceramic membrane were reported by Ebrahimi et al. (2009), for TiO_2/Al_2O_3 UF membrane performance in PW treatment. From the result, almost 80% of organic content was removed along with the capability of the membrane to work at low pressures, 0.5 bar for salt removal. Besides, the efficiency of the ceramic UF membranes was improvized in order to investigate the effect of pressure of the membrane by increasing the applied pressure up to 2 bar, instead of 0.5–1.5 bar at ambient pressure.

Ultimately, the main focus of UF membrane in PW treatment is more to develop an anti-fouling membrane or modification in order to reduce the membrane's fouling where fouling has been the major drawback to apply the UF membrane system in large scale of PW treatment plant. Hence, the future of anti-fouling and smart UF membrane is still in research, and the ongoing process will be better for membrane technology in PW treatment.

2.3 Nanofiltration (NF) and Reverse Osmosis (RO)

Considered as latest in membrane technology, NF and RO membranes possess the smallest pore size compared to MF and UF where the pore size ranges below 0.1 μm to 50 nm. Normally, the membranes operated at high pressure and relatively effective to remove inorganic minerals. Furthermore, the significant difference amid NF and RO membrane is their selectivity. For RO membrane, the water permeation flux is directly proportional to the operating pressure, whereas the salt permeation is independent of pressure, Baker (2004). Apparently, RO membrane can reject almost all iconic species including monovalent ions whereas NF more to selected divalent ions, and only allows several monovalent ions such as Na^+ and Cl^-. Mostly, the membranes were fabricated as thin film composite (TFC) and deposited on the surface asymmetric UF membrane (acted as a support). These membranes are considerably simply prone to fouling due to its very particular selectivity and required to operate at high pressure condition. Hence, a relative clean feed must be required such as clay, organic foulants, suspended solids, etc.

Study on the use of ultra low pressure RO (ULPRO) and NF membranes by Xu and Drewes (2006) for methane recovery in PW has discovered that, enhanced fouling resistance and high permeation flux using pure water were recorded by the membrane which possess high hydrophilic properties and low surface roughness. It was hypothesized that from the result of FESEM and FTIR taken, the decline in flux was not majorly due by organic fouling. From the study also, the NF membrane passed the standard to be used as primary drinking water, nevertheless, failed to be considered at secondary drinking water standards due to high content of chloride and TDS, whereas high permeation flux and rejection were recorded for ULPRO membranes compared to conventional RO membrane.

Study reported by Kim et al. (2011) in pre-treatment of oil sands process-affected water (OSPW), NF and RO membrane was used for desalination process. Prior to the process, the PW has undergone coagulation–flocculation–sedimentation (CFS) process followed by the pre-treatment process to compare for the PW without pre-treatment process. The results showed that by utilizing CFS the flux reduction can be reduced below 40%. Nevertheless, for COD or TDS did not show any sign of reduced from the pre-treatment of the feed and it was proposed that fouling may be occurred due to high solid content during the CFS process. Chemical cleaning was suggested to be required for foulant layers removal formed on the membrane surface either by using acid or base. Similar studied by Alpatova et al. (2014), irreversible membrane fouling might be occurred if direct filtration of PW was used without pre-treatment. Without proper CFV, fouling prone to occur even increased in permeation flux. New innovation by Miller et al. (2013) where a combination of a surface modified UF & RO membranes were used in a flowback water system treatment. In order to obtain hydrophilic and fouling resistance membranes, surface of the membranes was modified with in situ polymerization of dopamine followed by the addition PEG-NH$_2$. The modification has shown significant lower transmembrane pressure, higher permeate flux and efficient cleaning compared to the unmodified UF membrane. On the other hand, high salt rejection was recorded the modified RO membrane compared to unmodified RO membrane.

Excitingly, nowadays, ceramic NF and RO membranes used for PW treatment have risen which includes the membrane fabricated from zeolite and alumina apart from typical polymeric membrane fabricated from polyvinylidene fluoride, polyethersulfone and polyamide. Lee and Dong (2004) used RO membranes fabricated from synthetic zeolite for salinity reduction of the PW treatment. The TDS concentration was reduced more than 11% with high pressure applied. Similar to study by Liu et al. (2008) where RO membrane fabricated from MFI silicate zeolite was used for organic solvent rejection and approximately 96% was removed. For salt rejection, the membrane was discovered able to reject up to 99%. In cases of RO membrane from polymer-based, Fakhru'l-Razi et al. (2009) used RO membrane fabricated from PVDF and PES in post PW treatment water and discovered that all the membranes displayed high performance in term of flux and removal efficiencies up to 2% and 70%, respectively. Supplementary by chemical cleaning, the membranes can still further be operated, and 98% of recovery rate was recorded with no changing on the treated water quality.

On the other hand, by using commercially available ceramic NF, UF and MF membrane for real and synthetic PW treatment, Ebrahimi et al. (2009) reported that although almost 99% of oil was removed for all membranes, but not in the case of TOC where the results considered as modest. Follow up study by the same author, Ebrahimi et al. (2010) where dissolved air flotation (DAF) process was applied for pre-treatment of the water has shown better results. Nevertheless, the author has suggested that DAF process was the reason behind the improvement not by the membranes. Thus, according to the author, the efficiency of the ceramic membranes still not comparable with polymeric membranes perhaps due to the ceramic membrane containing high MWCO.

In summary, the use of NF and RO membranes in PW treatment can be very selective due to the characteristic of the membranes itself. The fouling of the membrane can be reduced by modifying the surface of the membrane and polymeric-based membranes showed better performance compared to ceramic-based membranes. Figure 7 depicts the schematic diagram of fouling at membrane's surface. In future, better understanding of the membranes should be studied for a better application of the membrane.

2.4 Forward Osmosis (FO)

The mechanism behind forward osmosis (FO) membrane is driving force was applied by the osmotic pressure where the transfer of water as feed across the membrane's wall to draw solution driven by the different chemical potential gradient (Chen et al. 2015) as shown in Fig. 8. Normally, in a semipermeable membrane, high salinity or concentration of draw solution such as NaCl was used with water (produced water) at low concentration in order to allow the permeation process to occur. Few studies have been done by few researchers to investigate the used of FO in refinery plant for hydrocarbon separation McGinnis et al. (2013), Hickenbottom et al. (2013) and Li et al. (2014) using cellulose triacetate (CTA) membrane. The idea of FO process is that the process would yield similar rejection as reverse osmosis (RO) but at the same time preventing any irreversible fouling and early membrane failure especially during pre-treatment.

On the other hand, there are previous studies of using FO in refinery plant. Study by Hickenbottom et al. (2013) used real hydrocarbon from Hydration Technology Innovations (Albany, OR (HTI)), while Yun et al. (2014) and Li et al. (2014) investigated the FO membranes in of using cellulose triacetate (CTA) membrane using synthetic solutions which has been modified chemically to tolerate the operating parameters at the plant. Hence, exploration on the use of FO membranes in pilot plant of sing real wastewater is still in search (Coday et al. 2015).

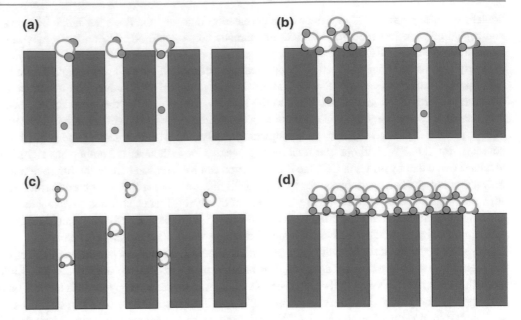

Fig. 7 Schematic diagram of fouling in membrane **a** normal blocking, **b** agglomeration, **c** big pore size **d** layer cake

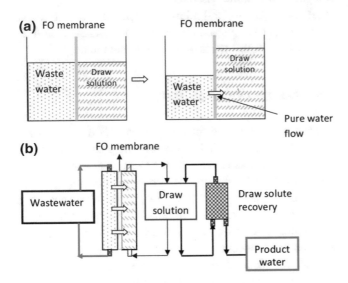

Fig. 8 Mechanism of forward osmosis (FO)

Previously, FO was designated to be an ideal technology of produced water treatment, and it was discovered that FO could be demonstrated as more environmental and economic friendly if to be compared to conventional method as such deep well injection. The use of the semipermeable membrane for FO and the implementation of difference osmotic pressure between the feed (PW) and draw solution (high salinity) eventually rejected any dissolved components in the PW. According to Bella et al. (2017), in oil and gas industry, FO was discovered could recover almost 85% of PW and operated at high concentration of total dissolved solid (TDS) which approximately 150,000 mg/L prior to be used for advanced technology such as RO and can be reused directly to the upstream. In addition, it was predicted that FO membrane eventually will minimize any fouling during

pre-treatment compared to normal pressure-driven membrane processes.

On the other hand, fouling in FO membrane still something needs to be taken seriously. Along with high content of TDS in PW which cause the formation of inorganic layer on the membrane's surface where most FO membranes were applied, another fouling also followed which are biofouling due to bacterial or microorganism activities, as well as organic fouling from the hydrocarbon itself. However, it should be noted that FO possess less fouling possibility compared to RO especially in PW treatment (Holloway et al. 2015). Despite that, due to high cost to operate the FO system in order to decrease the fouling and cleaning of the membrane still concerning which hindering the used of the FO membrane in real plant.

Chen et al. (2015) used cellulose triacetate (CTA) membranes, while Duong and Chung (2014) used polyamide thin film composite (TFC) membranes to study the tendency of fouling occurred on these membranes using PW. It was discovered that CTA membranes possess high surface roughness, high initial water flux and strong hydrogen bonding capability compared to TFC. Even surface roughness has been one of the factors to increase the tendency of fouling on membranes, but according to Coday et al. (2015), there was no solid correlation between the factors to fouling of FO membranes. Instead, it was because of high water flux during the initial process which attributed to the fouling where the PW brought along the TDS which attached at the membrane's wall. Hence, due to this attached TDS at the membrane's surface, along with the present of strong hydrogen bond from which sited on the TFC membrane and also active layer of carboxyl groups (–COOH) which display a strong attraction compared to hydroxyl groups (–OH) which enhanced the fouling.

3 Driving Force in Hydrocarbon Separation

Generally, operation of membrane separation process does not involve any heating, thus making this technology required less energy to operate compared to the thermal conventional separation such as distillation, sublimation or crystallization. Membrane technology is widely applicable in most industry including food technology, biotechnology and pharmaceutical industries. Nowadays, this technology becomes extensively important in the wastewater treatment process with the help of ultra or microfiltration process. This such development had solved most of the wastewater issues by protecting the water from the infection via the elimination of particles, colloids and also macromolecules.

Apart from removing unnecessary substance from the wastewater, separation of hydrocarbon with water became another concern that needs an urgent care. Basically, water is one of the main contaminants in a hydrocarbon fluid system and one of the most damaging agents. This problem usually occurs during the oil and gas processing and water inevitably usually being collected as part of stream. This collected water is known as a produced water (PW). PW not only needs to be removed; PW should be treated before releasing to the environment. PW treatment will differ according to its dumping purpose or reuse reason. Traditionally, PW management has been a three-step process, from water–oil separation to secondary treatment to final filtration. With the all experts and knowledge, the treatment of the PW became more advanced with the help of the membrane technology by separating the hydrocarbon from the PW.

An emerging membrane separation technology not only could treat the oily wastewater, but it also increased the production of oil, consequently. Most of the oily wastewater or PW comprise various polycyclic aromatic hydrocarbons (PAHs). The advantages of this membrane technology in separating hydrocarbon are having no addition of chemicals involved, only low energy required, easy to be handled and have well-arranged process conductions. In hydrocarbon membrane separation, various mechanism could be involved such as microfiltration, ultrafiltration, nanofiltration, reverse osmosis, vapour permeation, pervaporation, membrane distillation and membrane contactor. All of these mechanisms will operate with the help of driving force. Driving force usually comes from the system's surrounding, where the membrane technology operated. It also depends on what type of mechanism involve during the separation process. There are several driving forces involved during the hydrocarbon separation process including thermal-driven, electrical-driven and biological-driven.

3.1 Thermal-Driven Membrane

Thermal-driven in membrane separation system occurs when there are thermal difference exits between a hot feed stream and a cold permeate stream during the separation process. Both feed stream and permeate stream are separated by a hydrophobic, microporous membrane (Wang and Chung 2015). There are several processes that involved in the thermal-driven membrane separation such as membrane distillation (MD), membrane evaporation, membrane crystallization (MCr) and pervaporation. In MD process, feed salinity gives an effect on the cell performance where it permits a desalination process for high salinity of produce water (Shaffer et al. 2013). Air gap MD, direct contact MD, sweeping gas MD, AND vacuum MD are typical module configuration in MD system. Furthermore, most commonly studied is the DCMD configuration. While, flat sheet is the MD membrane configuration that most widely examined among the other membrane configurations i.e., flat sheet, hollow fibre, capillary, tubular and spiral wound (González et al. 2017). Besides, polytetrafluoroethylene (PTFE), polypropylene (PP) and polyvinylidene fluoride (PVDF) are the materials that usually used in the fabrication of the MD membrane.

MD system can operate at lower temperatures (30–90 °C) with pressures relative to conventional desalination technologies and treat wastewaters with TDS up to 350,000 mg/L. In contrast, Rao and Li (2015) indicated that MD might be a suitable method in treating FPW in range medium to high TDS. However, the high energy consumption required in the MD system making it less attractive than RO for low-TDS FPW. Li et al. (2014) reported that diluted FO draw solution can be treated by using vacuum MD, and the quality of product from this technique was comparable to bottled drinking water. However, due to the high content of inorganic or organic particles in the recovered water being a factor of the decreased in the MD performance. Moreover, Jang et al. (2017) found that MD performance more better than RO and evaporative crystallization in treating PW with the outstanding removal result of above 99% efficiencies for all tested ions (Ca^{2+}, Li^+, Mg^{2+}, K^+, Na^+, Sr^{2+}, Ba^{2+}, Cl^- and Br^-).

After all, there are two major obstacles facing in the MD applications which include membrane fouling and pore wetting conditions. In MD operation, primary fouling mechanism that usually occurred is inorganic scaling, organic fouling and biofouling (Razaei et al. 2018). Despite, limited effort has been taken in order to produce anti-fouling MD membranes. Currently, membrane modification only focuses on the optimizing surface wettability. For example, omniphobic PVDF membrane had been fabricated in order to minimize the wetting problem, and it shows that a slight flux decline where the membrane will repel both water and oil. This repletion is due to the grafting silica nano-particles coated with fluoroalkylsilane (perfluorodecyltrichlorosilane). Furthermore, Lokare et al. (2017) stated that membrane properties were having a great influence on the flux permeation process of MD. Besides, membranes with highly porous support are important for DCMD system. For all that, MD technology is hardly to be commercialized due to high energy consumption and the lack of commercially available of high-performance membranes.

Pre-treatment (e.g., MF) (Kim et al. 2018) or posttreatment (e.g., crystallization) is required for PW treatment in order to maintain the long-term operation of membranes and/or minimize membrane fouling in the MD process. In study conducted by Kim et al. (2017), it reported that total recovery increased to 37.5% from 20 to 25% after removing organic constituents like oil and grease prior to MD application. While Cho et al. (2018) had suggested that MD flux decline could be avoided by performing a pre-treatments process such as flocculation–sedimentation, vortex-based anti-fouling membrane and flocculation–sedimentation-MF. It shows that the flux reduction ratios were 3.6–6.9% and 8.9–16.2% for PP and PVDF membranes compared with 13.6% and 27.7% with raw PW wastewater, respectively, after the pre-treatments. Moreover, Li et al. (2019) confirmed that sweeping air, elevated feed solution temperature (from 59 to 65 °C), increased the VMD flux by 33%, 50% and 19% and enhanced vacuum pressure (from 40 to 58 kPa), respectively. But, there is not necessary to having a pre-treatment for FPW treatment by using DCMD, because the TDS, TSS, oil and grease, and volatile organics had minimal impact on DCMD membrane fouling and scaling.

Thermal-driven is also being a driven factor for membrane evaporation and pervaporation process. In term of heat and mass transfer, membrane evaporation is similar to MD system, but it is used for thermally sensitive solutions where the water vapour will not be recovered at the permeate side (Johnson et al. 2017). While, the pervaporation process is using thermal-driven mechanism especially in the desalination technology. This pervaporation process is involving the combination of membrane evaporation and permeation for selective separation of aqueous mixtures. Pervaporation offers the possibility of separating solutions, mixtures of components with close boiling points or azeotropes that are difficult to separate by distillation or other means. The first systematic work on pervaporation was done by Binning and co-workers at American Oil in the 1950s. The process was not commercialized until 1982 when the first commercial pervaporation plant, GFT (Gesellschaft für Trenntechnik GmbH, Germany), was installed. This plant works in separating water from alcohol solutions. Polyvinyl alcohol composite membranes were being used in this plant. Distillation and pervaporation process were combined in this plant to produce dry alcohol. Its working by removing the water as permeate and producing pure ethanol contains of 1% of water without any azeotropic distillation problem. About 50 such GFT plant had been installed at that time.

Another commercial pervaporation application is the separation of dissolved volatile organic compound (VOCs) from water, developed by Membrane Technology and Research, Inc. This system was using hydrophobic composite membranes such as silicone rubber coated on a microporous polyimide support membrane in a separation of the organics and water. The difference in polarity between organic solvents and water eases the separation process. The first pilot plant was reported by Separec in 1988 in separating methanol from methyl t-butyl ether/isobutene mixtures. More recently, Exxon started a pervaporation pilot plant for the separation of aromatic/aliphatic mixtures, using polyimide/poly urethane block co-polymer membranes. Furthermore, pervaporation process had demonstrated to be achievable procedure in separating organic micro-pollutants from water. Higher molecular size of 47–86% of aromatic micro-pollutants could be rejected by hydrophilic pervaporative tubular membrane (Sule et al. 2016). Besides, graphene oxide/polyimide hollow fibre membranes are prepared in study conducted by Huang and Feng (2018), for the desalination of seawater by pervaporation. In result, an high rate of water permeability and almost 99.8% of salt rejection had been achieved at 90 °C operating temperature.

Last but not least, membrane crystallization (MCr) being another process involves in thermal-driven concept. MCr is an addition of the MD concept where membrane technology and crystallization are working together in a single step (Drioli et al. 2012). With the help of crystallizers, the hybrid MD-crystallization technology succeeds in treating wastewater that contains high content of saline water. As compared to the single MD which having low recovery (37.5%), MD-crystallization process could recover about 62.5% of the water by reducing inorganic loading in the presence of crystallization. Additionally, Kim et al. (2018) reported that MD-crystallization system is using in the Eagle Ford PW where around 84% water success to be recovered and leads to the solid production of 2.72 kg/m^2 per at optimal operating conditions. Low energy consumed, 28.2 kWh/m^3 by using this MD-crystallization technology under optimal operating conditions. Furthermore, with the aid of osmotic pressure, MCr also has advanced capability in extracting the freshwater and valuable components from various streams without any limitations (Drioli et al. 2011). In oilfield PW treatment, 16.4 kg per m^3 of high purity (>99.9%) of crystal NaCl were recovered via MCr process by using hollow fibre PP and PVDF membranes with a 37%

of recovery factor. Both MD and MCr could be implemented in the industrial wastewater treatment consisting high Na_2SO_4. Direct treatment using unfiltered wastewater might be more suitable for these both technologies compared to NF-treated wastewater (Quist-Jensen et al. 2017).

3.2 Electrical-Driven

Most of the materials will involve with the electrical charge when dispersed in polar media. The charge properties of the ions involved could be a driven factor for the separation process. So, separation rate of the colloids or fine particles will depend on the surface charge of those particles. Therefore, in membrane technology, the separation process by the electrically-driven membrane was occurred due to the alternate ion exchange membranes. There are two main ED membranes in separation field, cation exchange membrane (CEM) and anion exchange membrane (AEM) where the positive charged ions and negative charged are transported selectively depend on the which type of the membrane used. The ED processes are potentially feasible for the (partial) desalination of high salinity waters including PW.

There are several ED-related processes are involved for treatment of FPW consisting reverse electrodeionization (REDI) (i.e., the combination of EDI and reverse electrodialysis (RED)) (Lopez et al. 2016), electrodeionization (EDI), (CDI), ion-concentration polarization desalination (ICP) (Kwak et al. 2016) and membrane capacitive deionization. A typical EDI unit has same configuration with the ED, but at least one channel is filled with ion exchange resin. Contrary to ED, which uses energy to remove salts, RED is a membrane-based process that producing electricity by mixing two solutions at different concentrations [225]. Lopez et al. (2016) state that the incorporating ion exchange wafers in each cell could enhance the REDI performance in order to shorten the passage ways for the diffusion process and limit the shadow spacer effect. As compared to a conventional ED system, a degree of enhanced ion selectivity occurred by increasing the wafer size and varying the wafer compositions.

Capacitive deionization (CDI) is a desalination process in which an electrical energy is used to high surface area electrodes in order to adsorb organic and inorganic species that have charger. The factor parameter for this technology is the capacity of the electrode adsorption. Biesheuvel and van der Wal (2010) had performed a research to examine the performance of a pilot-scale membrane CDI that involved an ion exchange membrane. The energy efficiency of CDI membrane with low salinity of feed water is higher compared to RO technology. The clear superiority of membrane CDI in current efficiency and specific energy consumption

has also been recorded in the previous study by Kim et al. (2010). Efficiency of salt removal was enhanced by 32.8–55.9% by using the CDI membrane, and it is depending on the operating condition. Zhao et al. (2013) showed that the current efficiency was up to four times reading for membrane CDI because the "co-ion" effect could be avoided by the usage of the ion exchange membranes.

ICP can be applied in treating high salinity solutions by using desalination process. Kim et al. (2016) demonstrated that brine salinity up to 100,000 mg/L of TDS could be treated by using ICP desalination and resulting in 70% of salt rejection. This happens due to the implementation of multi-stage operation with less membrane fouling/scaling than ED. Kim et al. (2010) run the ICP studies with high salinity water (TD $\sim$30,000 mg/L) and manage to get $\sim$99% rejection of TDS (at 50% recovery rate) with low power consumption below 3.5 kWh/m^3. Kwak et al. (2016) examined the behaviours of a novel ICP desalination approach by adopting unipolar ion conduction. For example, conducting only cations (or anions) with the unipolar ion exchange membrane stack and the salt removal was recorded at a given current. As compared to the convention ED, the power consumption excellently decreased by 50% with the unipolar cation conduction. Only 3.08 \$/m^3 with minimal power consumption of 5.6 kWh/m^3 are required for optimal water cost when using ICP at a salt rejection ratio of 50% (Kim et al. 2017).

3.3 Biological

Biological membrane of bio-membrane is a separating membrane that works together with a biological activity or process. The main challenge in the FPW treatment by using biological processes is the high salinity whereby the hypersaline pressure is bigger than plasmolysis and osmotic stress of bacterial cells. However, study by Akyon et al. (2015) had proved that the engineered microbial mats had the ability to treat saline FPW with TDS reading of >100,000 mg/L; 1.45 mg of chemical oxygen demand (COD). While Freedman et al. (2017) confirmed that biologically active filtration (BAF) could be adapted to treat SOG wastewater (TDS = 10.5–18.2 g/L, COD = 770–6360 mg/L), with 80% of COD removal. Besides, the FPW treatment at the Piceance and DJ Basins demonstrates that BAF had treated about 67–87% organics in DOC removal. However, tryptophan-like compounds are difficult to be removed due to complexation bound to humic/fulvic. Lester et al. (2014) reported with increasing TDS concentrations of 22,000–45,000 mg/L after 31 h, dissolved COD removal reduced from 90 to 60%. It shows the need for further research into the potential and robustness of biological treatments for PW.

In PW treatment, TDS could be removed together with the organic matter via membrane filtration process with the help of biological treatment. Study showed that, almost 99% of DOC and 94% of TDS were successfully rejected in PW treatment by coupling the ultrafiltration/nanofiltration (UF/NF) membrane technology with the bioaccumulation factor (BAF) process (Riley et al. 2016). Besides, membrane fouling also could be reduced by combination of membrane separation and biological process. Membrane bioreactor (MBR), microbial fuel cell (MFC), microbial desalination cell (MDC) are biologically active membrane processes that have been studied for PW treatment.

The MBR process was introduced by the late 1960s, as soon as commercial scale ultrafiltration (UF) and microfiltration (MF) membranes were available. MBR process is a process where the membrane process like ultrafiltration of microfiltration combines with biological wastewater treatment. In PW treatment, biological reaction works together with low-pressure membrane filtration in MBR process in order to produce high quality effluent without the aid of any chemicals. The progress in the PW treatment was extensively investigated. An integrated process comprised of reverse osmosis, ion exchange and MBR was implemented at the Pinedale Anticline FPW treatment facility. Besides, removal of 90% DOC from DJ Basin PW was achieved excellently by a hybrid reactor of (SBR)-MBR process combined with hollow fibre UF membrane (0.03 μm) (Frank et al. 2017).

MFC or known as microbial fuel cell is a bioelectrochemical system, where the electric current was driven by the usage of bacteria. MFC use biological treatment of organic pollutants in saline wastewaters to produce electricity [236]. One of the advantages of the MRC process is it can operate under an extreme salinity. Study reported that MFC successfully yield a power density up to 71 mW/m^2 with 42% coulombic efficiency under an extreme salinity of 250 g/L NaCl with the help of appropriate exoelectrogenic halophiles colonizing the anode (Monzon et al. 2015). Moreover, the addition of trace levels of exogenous quorum-sensing signals had increased the generated power density up to 30%. While the MFC-fed Barnett PW yield a power output of 47 mW/m^2 and 68% of COD removal. Since the energy generated by a hypersaline MFC can force desalination in a CDI unit, it is attainable in using a hybrid MFC + CDI system for PW treatment (Monzon et al. 2016).

MDC is a biological electrochemical system, where the electro-active bacteria is implemented to power the in-situ water desalination. In brief, MDC is a modified MFC that contains three chambers that detached by a pair of ion exchange membranes. Salt removal occurred when ion travels from the middle chamber to the cathode and anode chambers (Saeed et al. 2015). MDC process is important in the removal of contaminants and electricity generation. Desalination performance could be enhanced by coupling the MDC system with FO or RO process. Compared to the MDC system, the combination of MDC-FO system minimized the wastewater volume by 64% and improved the conductivity reduction (99.4%) in saline water two-fold. In addition, the MFC-MDC system shows high reduction of conductivity (>85%) from the salt solution containing 10–50 g/L NaCl, suggesting the potential application of this hybrid membrane technology to desalinate and treat PW (Zhang and He 2013).

Salt migration issue in the MDC system had been solved by the introduction of MCDC integrated technology. This system is operated by combining the MDC with CDI in a three-chamber configuration (i.e., anode, cathode and middle chamber). In MCDC, ions will not move to electrode chambers, because the electrodes capacitor will adsorb it before any migration occurred. Forrestal et al. (2015) proved that 2760 (mg/L)/h of TDS rate can be removed by using this system as well as removal of COD at a combined rate of 170 (mg/L)/h while treating Piceance PW. It is 18 times and five times faster than the traditional MDC. Stoll et al. (2015) demonstrated that 0.25–0.28 V of electrical energy for desalination is being generated by using SGPW contained sufficient biodegradable organic matter via MCDC system. Additionally, a 2.2-L MCDC system working continuously for nearly two years generated 89–131 W/m^3 and success in removing 75% of COD with 10.2 g/L TDS per day from actual FPW. Extra water would produce by applied in-expensive MCDC system by adopting sodium percarbonate as an electron acceptor. Shrestha et al. (2018) proved that more TDS was removed by MCDC in Bakken PW wastewater, while MFC manage to give higher COD removal and better electrical performance compared to MCDC.

3.4 Hybrid Technology in Hydrocarbon Separation

The speedy advancement in separation industry had brought many outstanding numerous studies and practices in separating one substance to another substance. Such rapid progress develops an idea to design a hybrid technology in separation system, not only for good performance result, but it will establish more sustainable and economical processes. For instance, energy consumption could be reduced, and qualities of distillation cuts can be improved by using a combination of distillation with membrane separation system in separating hydrocarbon. Besides, hybrid technology may result in significant cut in its capital and operational cost.

Hybrid distillation combined with vapour membrane separation system had been designed by using mathematical approach (Caballero et al. 2009). A two stages of path consist of distillation column and membrane separation system. This work tested on the separation of ethane and ethylene as a case study. Normally, separation of such hydrocarbons performs via cryogenic distillation. However, this technology is needed high energy consumption which required high capital. Therefore, exploration of this technology is aimed to lowering the energy and cost expenditure. This system consists of compressor, cooler and distillation column which developed via process simulator called and Unisim Design; MATLAB was prepared to analyse that adoption of pallor membrane. By using this approach, separation of ethylene from ethane is proved could be savings up to 20% of cost, and at the same time, 30% energy consumption can be reduced (Caballero et al. 2009).

Computer simulation was practised in order to test an economical method of pervaporation-distillation in separating azeotropic mixtures of alcohol-ether and consequently producing ether (González and Ortiz 2002). gPROMS software had been used in developing the modelling and simulation of such hybrid technology. This software works together with mass transfer model in order to test on hybrid system which having different formation for its feed stream. From the testing, it shows that in order to obtain 99.8 wt% of ether special grade, a membrane that having 1590 m^2 of area is needed. While 2110 m^2 of membrane area is needed for recovering 98 wt% of gasoline. Both conditions having different percent of methanol escape the reactor. Moreover, such combination of pervaporation–distillation could minimize the complexity and capital needed for the conversion plant. Besides, this hybrid process produces almost pure ether in the bottom of the stream; at the same time, it also could be applied in generating 2-methyl propane via thermal decomposition.

Membrane distillation system had been coupled together with heat pump system in a propylene/propene separation. In study conducted by Park et al. (2019), 99.6 wt% of C_3H_4 of separation recovery was targeted via membrane distillation system hybrid with recompression heat pump. Membrane characters also play an important role in this technology such as permeability, selectivity, pressure ratio and stage-cut. This alternative technology shows a significant energy saving and reducing investment cost by cutting the number of stages in the distillation part.

4 Challenges in Membrane Technology for Hydrocarbon Separation

There are a lot of challenges faced to implement membrane technology for hydrocarbon separation or refinery produced water treatment. The major challenges to achieve a sustainable operation include reducing the membrane fouling and degradation, lack of real example of pilot and full-scale plant and high capital and operation cost.

Membrane fouling has been the major challenge faced by researchers where this problem will influence the performance and operation of the membrane in long term. It is also greatly influenced by the quality of the produced water, condition and location of the membrane operation as well as the characteristic of the membrane itself. Typical solutions to membrane fouling are the use of extensive pre-treatment, development of an effective cleaning regimen, operation optimization and proper selection of membrane materials. However, cleaning by chemicals may lead to the damage of the membrane, and hence, proper handling and further study should be made for a solution.

Finally, although many innovations have been made to encounter the problem, however, no real implementation has been reported. Almost all reported study used synthetic PW and only in lab scale. The high energy consumption needed to operate the membrane system also has been one of the challenges to implement the system in real life. Proper costing and full operational analysis should be done prior to be used.

5 Conclusion and Future Development

In conclusion, excellent progression in the development of membrane technology gives huge benefits to the wastewater treatment especially in treating PW in petrochemical industry. Membrane technology not only useful for treating the water from the PW collected, but it also increases the production of oil from the PW itself. High performance of ultrafiltration, nanofiltration, microfiltration, reverse osmosis, vapour permeation, pervaporation, membrane distillation and membrane contactor technology have successfully implemented in many petrochemical industries. With the help of driven factor including thermal-driven, electrical-driven and biological-driven, membrane technology could achieve a high performance of separation system with high removal of total dissolved solids (TDS) and COD.

Recently, the developments chemical synthesis has brought to the great progress in membrane materials. The need to develop novel new membrane materials has emerged to be improvement in membrane technology especially in hydrocarbon separation. Membrane's capacity and permeability together with cost-effective and simple fabrication have been studied to overcome membrane fouling in various applications. Werber et al. (2016) have pointed out some key parameters for desalination are selectivity rather than permeability. Some modifications or improvement suggested are as below:

- New alternative materials from ceramic-based materials such as bentonite clay, kaolin which also including zeolite, aquaporin, nanocomposite nanotube and graphene-based membranes.

- Advanced membrane surface modification and also development of self-assembled materials.
- Fully evaluation on performance of the membrane in long-term operation.
- Clarification on the interactions with different membrane materials.
- New analytical techniques for a better characterization of organics in PW treatment plant.
- Organic removal efficiency by membrane technology should be analysed qualitatively and quantitatively in feed and permeate.
- Reducing the energy consumption by utilizing the salinity, organics or waste heat in the feed and to generate power.
- The use of organics and nutrients from MFC and MDC to produce electricity.
- Introducing blue energy, which is a renewable, clean and sustainable by salinity gradients from two membrane technologies (PRO and RED).
- Alternative materials to fabricate the membranes with good and sustainable chemicals properties as well as low cost and economic.

References

Abadi, S. R. H., Sebzari, M. R., Hemati, M., Rekabdar, F., & Mohammadi, T. (2011). Ceramic membrane performance in microfiltration of oily wastewater. *Desalination, 265,* 222–228.

Abbasi, M., Mirfendereski, M., Nikbakht, M., Golshenas, M., & Mohammadi, T. (2010). Performance study of mullite and mullite–alumina ceramic MF membranes for oily wastewaters treatment. *Desalination, 259,* 169–178.

Akyon, B., Stachler, E., Wei, N., & Bibby, K. (2015). Microbial mats as a biological treatment approach for saline wastewaters: The case of produced water from hydraulic fracturing. *Environmental Science & Technology, 49,* 6172–6180.

Alpatova, A., Kim, E.-S., Dong, S., Sun, N., Chelme-Ayala, P., & Gamal El-Din, M. (2014). Treatment of oil sands process-affected water with ceramic ultrafiltration membrane: Effects of operating conditions on membrane performance. *Separation and Purification Technology, 122,* 170–182.

Alzahrani, S., & Mohammad, A. W. (2014). Challenges and trends in membrane technology implementation for produced water treatment: A review. *Journal of Water Process Engineering, 4*(C), 107–133.

Baker, R. W. (2004). *Membrane technology and applications* (2nd ed.). West Sussex, England: Wiley.

Bella, E. A., Poynora, T. E., Newharta, K. B., Regnerya, J., Codaya, B. D., & Cath, T. Y. (2017). Produced water treatment using forward osmosis membranes: Evaluation of extended-time performance and fouling. *Journal of Membrane Science, 525,* 77–88.

Bernardo, P., & Drioli, E. (2010). Membrane technology: Latest applications in the refinery and petrochemical field. In D. Enrico &
G. Lidietta (Eds.), *Comprehensive membrane science and engineering* (pp. 211–239). Oxford: Elsevier.

Caballero, J. A., Grossmann, I. E., Keyvani, M., & Lenz, E. S. (2009). Design of hybrid distillation-vapor membrane separation systems. *Industrial and Engineering Chemistry Research, 48*(20), 9151–9162.

Chang, H., Li, T., Liu, B., Vidic, R. D., Elimelech, M., & Crittenden, J. C. (2019). Potential and implemented membrane-based technologies for the treatment and reuse of flowback and produced water from shale gas and oil plays: A review. *Desalination, 455,* 34–57.

Chen, G., Wang, Z., Nghiem, L. D., Li, X.-M., Xie, M., Zhao, B., et al. (2015). Treatment of shale gas drilling flowback fluids (SGDFs) by forward osmosis: Membrane fouling and mitigation. *Desalination, 366,* 113–120.

Cho, H., Choi, Y., & Lee, S. (2018). Effect of pretreatment and operating conditions on the performance of membrane distillation for the treatment of shale gas wastewater. *Desalination, 437,* 195–209.

Coday, B. D., Almaraz, N., & Cath, T. Y. (2015). Forward osmosis desalination of oil and gas wastewater: Impacts of membrane selection and operating conditions on process performance. *Journal of Membrane Science, 488,* 40–55.

Cui, J., Zhang, X., Liu, H., Liu, S., & Yeung, K. L. (2008). Preparation and application of zeolite/ceramic microfiltration membranes for treatment of oil contaminated water. *Journal of Membrane Science, 325,* 420–426.

Drioli, E., Profio, G. D., & Curcio, E. (2012). Progress in membrane crystallization. *Current Opinion in Chemical Engineering, 1,* 178–182.

Drioli, E., Stankiewiczd, A. I., & Macedonio, F. (2011). Membrane engineering in process intensification—An overview. *Journal of Membrane Science, 380,* 1–8.

Duong, P. H., & Chung, T. S. (2014). Application of thin film composite membranes with forward osmosis technology for the separation of emulsified oil–water. *Journal of Membrane Science, 452,* 117–126.

Ebrahimi, M., Ashaghi, K. S., Engel, L., Willershausen, D., Mund, P., Bolduan, P., & Czermak, P. (2009). Characterization and application of different ceramic membranes for the oil-field produced water treatment. *Desalination, 245,* 533–540.

Ebrahimi, M., Willershausen, D., Ashaghi, K. S., Engel, L., Placido, L., Mund, P., Bolduan, P., & Czermak, P. (2010). Investigations on the use of different ceramic membranes for efficient oil-field produced water treatment. *Desalination, 250*(3), 991–996.

Fakhru'l-Razi, A., Pendashteh, A., Abdullah, L. C., Biak, D. R., Madaeni, S. S., & Abidin, Z. Z. (2009). Review of technologies for oil and gas produced water treatment. *Journal of Hazardous Materials, 170,* 530–551.

Forrestal, C., Stoll, Z., Xu, P., & Ren, Z. J. (2015). Microbial capacitive desalination for integrated organic matter and salt removal and energy production from unconventional natural gas produced water. *Environmental Science: Water Research & Technology, 1,* 47–55.

Frank, V. B., Regnery, J, Chan, K. E., Ramey, D. F., Spear, J. R., & Cath, T. Y. (2017). Co-treatment of residential and oil and gas production wastewater with a hybrid sequencing batch reactor-membrane bioreactor process. *Journal of Water Process Engineering, 17,* 82–94.

Freedman, D. E., Riley, S. M., Jones, Z. L., Rosenblum, J. S., Sharp, J. O., Spear, J. R., & Cath, T. Y. (2017). Biologically active filtration for fracturing flowback and produced water treatment. *Journal of Water Process Engineering, 18,* 29–40.

González, B., & Ortiz, I. (2002). Modelling and simulation of a hybrid process (pervaporation-distillation) for the separation of azeotropic mixtures of alcohol-ether. *Journal of Chemical Technology and Biotechnology, 77*(1), 29–42.

González, D., Amigo, J., & Suárez, F. (2017). Membrane distillation: Perspectives for sustainable and improved desalination. *Renewable & Sustainable Energy Reviews, 80,* 238–259.

Hickenbottom, K. L., Hancock, N. T., Hutchings, N. R., Appleton, E. W., Beaudry, E. G., Xu, P., & Cath, T. Y. (2013). Forward osmosis treatment of drilling mud and fracturing wastewater from oil and gas operations. *Desalination, 312,* 60–66.

Holloway, R. W., Achilli, A., & Cath, T. Y. (2015). The osmotic membrane bioreactor: A critical review. *Environmental Science: Water Research & Technology, 1,* 581–605.

Huang, A., & Feng, B. (2018). Synthesis of novel graphene oxide-polyimide hollow fiber membranes for seawater desalination. *Journal of Membrane Science, 548,* 59–65.

Huang, Y., & Feng, X. (2019). Polymer-enhanced ultrafiltration: Fundamentals, applications and recent developments. *Journal of Membrane Science, 586,* 53–83.

Igunnu, E. T., & Chen, G. Z. (2012). Produced water treatment technologies. *International Journal of Low-Carbon Technologies, 9,* 157–177.

Jang, E., Jeong, S., & Chung, E. (2017). Application of three different water treatment technologies to shale gas produced water. *Geosystem Engineering, 20,* 104–110.

Johnson, D. W., Muppavarapu, N., & Shipley, H. J. (2017). Aeration waste heat for membrane evaporation of desalination brine concentrate. *Journal of Membrane Science, 539,* 1–13.

Johnson, H. E., & Schulman, B. L. (1993). *Assessment of the potential for refinery applications of inorganic membrane technology: An identification and screening analysis* (p. 217). Technical Report prepared by SFA Pacific, Inc. for U.S. Department of Energy, DOE/FE/61680-H3, Washington, DC.

Kang, S., Asatekin, A., Mayes, A. M., & Elimelech, M. (2007). Protein antifouling mechanisms of PAN UF membranes incorporating PAN-g-PEO additive. *Journal of Membrane Science, 296*(1), 42–50.

Kim, B., Kwak, R., Kwon, H. J., Pham, V. S., Kim, M., Al-Anzi, B., et al. (2016). Purification of high salinity brine by multi-stage ion concentration polarization desalination. *Scientific Reports, 6,* 31850.

Kim, E.-S., Liu, Y., & Gamal El-Din, M. (2011). The effects of pretreatment on nanofiltration and reverse osmosis membrane filtration for desalination of oil sands process-affected water. *Separation Science and Technology, 81,* 418–428.

Kim, J., Kim, J., & Hong, S. (2018). Recovery of water and minerals from shale gas produced water by membrane distillation crystallization. *Water Research, 129,* 447–459.

Kim, J., Kwon, H., Lee, S., & Hong, S. (2017). Membrane distillation (MD) integrated with crystallization (MDC) for shale gas produced water (SGPW) treatment. *Desalination, 403,* 172–178.

Kim, S. J., Ko, S. H., Kang, K. H., & Han, J. (2010). Direct seawater desalination by ion concentration polarization. *Nature Nanotechnology, 5,* 297–301.

Kwak, R., Pham, V. S., Kim, B., Chen, L., & Han, J. (2016). Enhanced salt removal by unipolar ion conduction in ion concentration polarization desalination. *Scientific Reports, 6,* 25349.

Lee, R. L., & Dong, J. (2004). *Modified reverse osmosis system for treatment of produced waters.* Technical Report prepared by New Mexico Institute of Mining and Technology for U.S. Department of Energy, 2004.

Lester, Y., Yacob, T., Morrissey, I., & Linden, K. G. (2014). Can we treat hydraulic fracturing flowback with a conventional biological process? The case of guar gum. *Environmental Science & Technology Letters, 1,* 133–136.

Li, L., & Lee, R. (2009). Purification of produced water by ceramic membranes: Material screening, process design and economics. *Separation Science and Technology, 44,* 3455–3484.

Li, Z., Rana, D., Matsuura, T., & Lan, C. Q. (2019). The performance of polyvinylidene fluoride—polytetrafluoroethylene nanocomposite distillation membranes: An experimental and numerical study. *Separation and Purification Technology, 226,* 192–208.

Li, X.-M., Zhao, B., Wang, Z., Xie, M., Song, J., Nghiem, L. D., et al. (2014). Water reclamation from shale gas drilling flow-back fluid using a novel forward osmosis-vacuum membrane distillation hybrid system. *Water Science & Technology, 69,* 1036–1044.

Liu, N., Li, L., McPherson, B., & Lee, R. (2008). Removal of organics from produced water by reverse osmosis using MFI-type zeolite membranes. *Journal of Membrane Science, 325,* 357–361.

Lokare, O. R., Tavakkoli, S., Wadekar, S., Khanna, V., & Vidic, R. D. (2017). Fouling in direct contact membrane distillation of produced water from unconventional gas extraction. *Journal of Membrane Science, 524,* 493–501.

Lopez, A. M., Dunsworth, H., & Hestekin, J. A. (2016). Reduction of the shadow spacer effect using reverse electrodeionization and its applications in water recycling for hydraulic fracturing operations. *Separation and Purification Technology, 162,* 84–90.

McGinnis, R. L., Hancock, N. T., Nowosielski-Slepowron, M. S., & McGurgan, G. D. (2013). Pilot demonstration of the NH_3/CO_2 forward osmosis desalination process on high salinity brines. *Desalination, 312,* 67–74.

Miller, D. J., Huang, X., Li, H., Kasemset, S., Lee, A., Agnihotri, D., et al. (2013). Fouling-resistant membranes for the treatment of flowback water from hydraulic shale fracturing: A pilot study. *Journal of Membrane Science, 437,* 265–275.

Monzon, O., Yang, Y., Li, Q., & Alvarez, P. J. J. (2016). Quorum sensing autoinducers enhance biofilm formation and power production in a hypersaline microbial fuel cell. *Biochemical Engineering Journal, 109,* 222–227.

Monzon, O., Yang, Y., Yu, C., Li, Q., & Alvarez, P. J. J. (2015). Microbial fuel cells under extreme salinity: Performance and microbial analysis. *Environmental Chemistry, 12,* 293–299.

Moslehyani, A., Ismail, A. F., Othman, M. H. D., & Matsuura, T. (2015). Design and performance study of hybrid photocatalytic reactor-PVDF/MWCNT nanocomposite membrane system for treatment of petroleum refinery wastewater. *Desalination, 363,* 99–111.

Mulder, M. (1996). *Basic principles of membrane technology.* Dordrecht, The Netherlands: Kluwer academic Publishers.

Munirasu, S., Haija, M. A., & Banat, F. (2016). Use of membrane technology for oil field and refinery produced water treatment—A review. *Process Safety and Environmental Protection, 100,* 183–202.

Park, J., Kim, K., Shin, J. W., & Park, Y. K. (2019). Analysis of multistage membrane and distillation hybrid processes for propylene/propane separation. *Chemical Engineering Transactions, 74,* 871–876.

Quist-Jensen, C. A., Macedonio, F., Horbez, D., & Drioli, E. (2017). Reclamation of sodium sulfate from industrial wastewater by using membrane distillation and membrane crystallization. *Desalination, 401,* 112–119.

Rao, G., & Li, Y. (2015). Feasibility study of flowback/produced water treatment using direct-contact membrane distillation. *Desalination and Water Treatment, 57,* 21314–21327.

Ravanchi, M. T., Kaghazchi, T., & Kargari, A. (2009). Application of membrane separation processes in petrochemical industry: A review. *Desalination, 235,* 199–244.

Rezaei, M., Warsinger, D. M., Lienhard, V. J. H., Duke, M. C., Matsuura, T., & Samhaber, W. M. (2018). Wetting phenomena in membrane distillation: Mechanisms, reversal, and prevention. *Water Research, 139,* 329–352.

Riley, S. M., Oliveira, J. M. S., Regnery, J., & Cath, T. Y. (2016). Hybrid membrane bio-systems for sustainable treatment of oil and gas produced water and fracturing flowback water. *Separation and Purification Technology, 171,* 297–311.

Saeed, H. M., Husseini, G. A., Yousef, S., Saif, J., Al-Asheh, S., Fara, A. A., et al. (2015). Microbial desalination cell technology: A review and a case study. *Desalination, 359,* 1–13.

Salahi, A., Mohammadi, T., Rahmat Pour, A., & Rekabdar, F. (2012). Oily wastewater treatment using ultrafiltration. *Desalination and Water Treatment, 6,* 289–298.

Scoth, K. (1999). *Handbook of industrial membranes* (2nd ed.). Amsterdam: Elsevier.

Scoth, K. (1990). *Membrane separation technology: Industrial applications and markets.* British Hydromechanics Association.

Seyed Shahabadi, S. M., & Reyhani, A. (2014). Optimization of operating conditions in ultrafiltration process for produced water treatment via the full factorial design methodology. *Separation Science and Technology, 132,* 50–61.

Shaffer, D. L., Chavez, L. H. A., Ben-Sasson, M., Castrillón, S. R.-V., Yip, N. Y., & Elimelech, M. (2013). Desalination and reuse of high-salinity shale gas produced water: Drivers, technologies, and future directions. *Environmental Science & Technology, 47,* 9569–9583.

Shrestha, N., Chilkoor, G., Wilder, J., Ren, Z. J., & Gadhamshetty, V. (2018). Comparative performances of microbial capacitive deionization cell and microbial fuel cell fed with produced water from the Bakken shale. *Bioelectrochemistry, 121,* 56–64.

Stoll, Z. A., Forrestal, C., Ren, Z. J., & Xu, P. (2015). Shale gas produced water treatment using innovative microbial capacitive desalination cell. *Journal of Hazardous Materials, 283,* 847–855.

Sule, M. N., Templeton, M. R., & Bond, T. (2016). Rejection of organic micro-pollutants from water by a tubular, hydrophilic pervaporative membrane designed for irrigation applications. *Environmental Technology, 37,* 1382–1389.

Takht Ravanchi, M., Kaghazchi, T., & Kargari, A. (2009). Application of membrane separation processes in petrochemical industry: A review. *Desalination, 235*(1–3), 199–244.

van der Biesheuvel, P. M., & Wal, A. (2010). Membrane capacitive deionization. *Journal of Membrane Science, 346,* 256–262.

Wandera, D., Wickramasinghe, S. R., & Husson, S. M. (2011). Modification and characterization of ultrafiltration membranes for treatment of produced water. *Journal of Membrane Science, 373,* 178–188.

Wandera, D., Himstedt, H. H., Marroquin, M., Wickramasinghe, S. R., & Husson, S. M. (2012). Modification of ultrafiltration membranes with block copolymer nanolayers for produced water treatment: The roles of polymer chain density and polymerization time on performance. *Journal of Membrane Science, 403–404,* 250–260.

Wang, P., & Chung, Y.-S. (2015). Recent advances in membrane distillation processes: Membrane development, configuration design and application exploring. *Journal of Membrane Science, 474,* 39–56.

Weller, S., & Steiner, W. A. (1950). Engineering aspects of separation gases: Fractional permeation through membranes. *Chemical Engineering Progress, 46,* 585–590.

Werber, J. R., Osuji, C. O., & Elimelech, M. (2016). Materials for next-generation desalination and water purification membranes. *Nature Reviews Materials, 1,* 16018.

Witze, A. (2015). Race to unravel Oklahoma's artificial quakes. *Nature, 520,* 418–419.

Xu, P., & Drewes, J. E. (2006). Viability of nanofiltration and ultra-low pressure reverse osmosis membranes for multi-beneficial use of methane produced water. *Separation Science and Technology, 52,* 67–76.

Yan, L., Hong, S., Li, M. L., & Li, Y. S. (2009). Application of the Al_2O_3–PVDF nanocomposite tubular ultrafiltration (UF) membrane for oily wastewater treatment and its antifouling research. *Separation and Purification Technology, 66*(2), 347–352.

Yun, T., Koo, J. W., Sohn, J., & Lee, S. (2014). Pressure assisted forward osmosis for shale gas wastewater treatment. *Desalination Water Treatment, 54,* 1–9.

Zhao, Y., Wang, Y., Wang, R., Wu, Y., Xu, S., & Wang, J. (2013). Performance comparison and energy consumption analysis of capacitive deionization and membrane capacitive deionization processes *Desalination, 324,* 127–133.

Zhang, Z., & He, B. (2013). Improving water desalination by hydraulically coupling an osmotic microbial fuel cell with a microbial desalination cell. *Journal of Membrane Science, 441,* 18–24.

Zhong, J., Sun, X., & Wang, C. (2003). Treatment of oily wastewater produced from refinery processes using flocculation and ceramic membrane filtration. *Separation Science and Technology, 32,* 93–98.

Zsirai, T., Qiblawey, H., Buzatu, P., Al-Marri, M., & Judd, S. J. (2018). Cleaning of ceramic membranes for produced water filtration. *Journal of Petroleum Science and Engineering, 166,* 283–289.

Advanced Membrane Technology for Textile Wastewater Treatment

Mohd Hafiz Dzarfan Othman, Mohd Ridhwan Adam,
Roziana Kamaludin, Nurul Jannah Ismail, Mukhlis A. Rahman,
and Juhana Jaafar

Abstract

The utilization of membrane technology wastewater treatment process has gained great attention in the industrial practitioners worldwide. The conventional and current treatment processes devoted in the textile wastewater treatment are discussed thoroughly throughout this chapter. Additionally, the main converge of this chapter is the application of the advanced membranes on textile industries namely reverse osmosis, nanofiltration, ultrafiltration, microfiltration, electrodialysis, membrane bioreactor and photocatalytic membrane. The advantages and restrictions of these techniques are carefully addressed in their respective subchapters. At the end of this chapter, an attempt is also made to show the future direction of the advanced membrane technology toward the advance wastewater technology processes such as membrane distillation, membrane contactor and many others.

Keywords

Advanced membrane technology • Textile wastewater • Treatment process

1 Introduction and History

Textile industries have been known as one of the largest industrials that have grown rapidly, especially in the developing countries. The word textile was originated from the Latin word 'texere' which means to weave. The gist of the

M. H. D. Othman (✉) · M. R. Adam · R. Kamaludin · N. J. Ismail · M. A. Rahman · J. Jaafar
Advanced Membrane Technology Research Centre (AMTEC), School of Chemical and Energy Engineering, Faculty of Engineering, Universiti Teknologi Malaysia, 81310 UTM Johor Bahru, Johor, Malaysia
e-mail: hafiz@petroleum.utm.my

textile industries can be classified due to the types of textile fibers it is working with. The different types of fibers that categorized the industries are protein fibers which mainly derived from animals, manmade fibers that normally synthesized, as well as cellulose-based either from natural sources or regained. The sources of these fibers can be varied such as plant sources of the cellulose-based fibers to produce rayon, ramie, cotton, linen, lyocell, viscose and hemp. On the other hand, fibers like wool, silk, cashmere, angora and mohair are among the protein-based fibers which are normally obtained from the animal sources. Meanwhile, the most used fibers due to the abundant source and mass production are the artificially synthesized manmade fibers (polymer-based) including spandex, nylon, acrylic, polypropylene, polyester, Ingeo and acetate. Figure 1 summarizes the type, source and example of the textile fibers exist in this era. It is worth to be mentioning here that most of the textiles were produced from the petrochemicals, wood pulp and cotton liners. Apart from accommodate the primary needs of humans (cloth and fashion), the textile industry normally associated by the large volume and variability of the wastewater generation.

Textile industrial wastewater has become as one of the major pollution contributors to the water stream. This can be attributed by the production process of the textiles that gets chemically intensive due to the application of dye for coloring purpose. On that note, the textile industries have generated massive quantities of chemical including dyes as the form of wastewater during the manufacturing of the textiles. The usage of the dyes can be varied depending on the types of fibers used. For instance, the direct dye, reactive dye, indigo dye and naphthol dye are more attracted to the cellulose-based fibers. Meanwhile, the protein-based fibers tend to react with the acid dye. The polymer-based of the manmade fibers in other way have high affinity to basic, direct and disperse dyes. Despite having an attractive feature due to its pleasant appearance to the human eyes, these highly colored materials could be highly disturbing the water

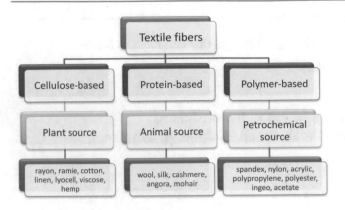

Fig. 1 Classification of textile fibers

properties when it discharged with the wastewater. These components may disrupt the reoxygenation capacity of the water besides cutting-off the sunlight receival to the water bodies, thus distressed the aquatic lives' biological activities (Nassar and Magdy 1997). The situation is more go downhill upon the introduction of different dyestuffs with varying chemical properties used in these industries. This occurrence has further complicated the treatment process of the textile wastewater.

The textile wastewater can be of harmful toward human health and the environment. The wastewater normally related to the high concentration of the chemical oxygen demand (COD), high temperature and pH value, strong color and low biodegradability (Gao et al. 2007). Table 1 shows the regulation of the permitted content of textile wastewater in India. The stringent regulation of the wastewater content has become the major concern of the industrial players since the textile wastewater contained of extremely high chemical contents.

The main problem faced by the environment authorities or the industrial players is the removal of the color due to the remaining recalcitrant dyes. The elimination of the color from the wastewater has become the great challenge for many decades. Figure 2 summarizes the environmental burden caused by the textile industries. To date, many great efforts have been devoted, and yet, there is no single and

economical process of decolorization has come to totally succeed in treating the textile wastewater. The utilization of the conventional approaches such as adsorption, coagulation/flocculation, chemical oxidation, so on and so forth have gained so much attempts and still have been tested with some new substances such as new coagulant and adsorbent. Recently, the appliance of new single or combination processes such as ultrasonic, electrochemical as well as photochemical processes has attracted much attention for the textile wastewater treatment. Of all the methods, membrane separation process has been rapidly evolved in its technology, and yet, resulting in the promising outcomes.

This chapter is to bring an overview of the treatment process of textile wastewater that is considered as one of the most threaten pollutions in humankind. Although many treatment processes of the real and synthetic textile wastewaters are available and practiced, only few processes have been implemented using the newest technologies such as advanced membranes approaches. Therefore, this chapter will be discussing the textile wastewater treatment using the cutting-edge membrane technologies. Besides, the brief discussion on the conventional treatment processes has also included.

2 Conventional Textile Wastewater Treatment Process

Many great efforts and techniques have been explored in removing dyes and other contaminants from the textile wastewater for the past few decades. These processes include coagulation/flocculation, adsorption, ion exchange, oxidation as well as biological treatment process. These techniques nevertheless may have advantages and disadvantages.

2.1 Coagulation/flocculation

Coagulation/flocculation (Fig. 3) is a common and essential physicochemical treatment employed in textile wastewater

	Serial No	Parameters	Standards
Table 1 Indian industry standards for pollutant content	1	pH	6.9
	2	COD	250 mg/L
	3	BOD	30 mg/L
	4	TDS	2000 mg/L
	5	Chloride	500 mg/L
	6	Sulfide	2 mg/L
	7	Magnesium	50 mg/L
	8	Calcium	75 mg/L

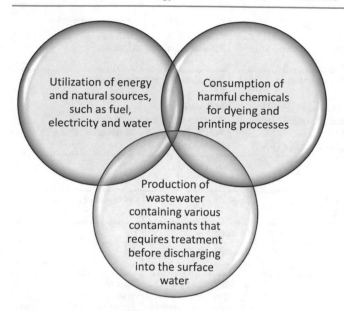

Fig. 2 Environmental burden as implication from the textile industries

treatment. Coagulation/flocculation efficiently decolorizes effluent completely and reduces the total load of pollutants. Coagulation/flocculation process has been found to be cost effective, easy to operate and energy-saving treatment alternatives. However, the major limitations of coagulationflocculation process are the sludge production and ineffective color degradation of decolorization of some soluble dyes (Sabur et al. 2012).

Chemical coagulants (Table 2) can be categorized based on the effectiveness to decolorize the dye, however, the efficiency depends on the characteristics of raw wastewater, pH and temperature of the solution, the type and dosage of coagulants, and the intensity and duration of mixing.

Among the three categorized coagulants, pre-hydrolyzed metallic salts are more effective than the hydrolyzing metallic salts due to efficient dye decolorization and low sludge production. In textile industries, most of the dyes used are of negatively charged. Therefore, cationic polymer coagulants are often chosen for efficient dye removal and degradation. Table 3 summarized some of the chemical coagulants and their efficiency in textile wastewater treatment.

Coagulation/flocculation offers complete decomposition of dye from textile wastewater without intermediate by-products, therefore, this process is totally harmless and non-toxic. Despite its effectiveness in decolorization of waste stream, the coagulation process also has some limitations. Some highly soluble and low molecular weight cationic dyes might be unaffected by coagulation/flocculation process in addition to the need for disposal of the sludge. Furthermore, the sludge production only can be reduced by the elimination of low volume of the highly colored dye bath by direct chemical treatment after the dyeing process (Golob et al. 2005). In addition, coagulation/flocculation is a complicated process. It involves a series of physical–chemical interactions which are electrostatic attraction, sorption, bridging and inclusion in metal precipitates. Although coagulation/flocculation has been widely employed in textile wastewater treatment, to maximize the dye removal and reduce the chemical consumptions, the operating conditions depend on the particular characteristics of the textile effluents which need to be carefully examine.

2.2 Adsorption

Adsorption is a phenomenon on the surface where a fluid of multiple component is attached to the surface of a solid adsorbent to form attachments via physical or chemical bond (Sasaki et al. 2014). The substance that remains on the solid surface is known as adsorbent, whereas the removed material is termed as adsorbate. It also employed as a stage of integrated of various wastewater treatment processes which are physical, chemical and biological method (Geenens et al. 2001).

In general, there are four consecutive steps during the adsorption process of dye molecules (Seow and Lim 2016). The first step starts with the dispersion or convection of dye molecules through the bulk of solution. Next, the dye pigments diffuse into the boundary layer or known as film diffusion and followed by the diffusion from the surface layer onto the interior of adsorbent materials. The dye pigments then attach to the material surface due to the interaction between molecules. There are several factors that

Fig. 3 Overview of coagulation/flocculation process

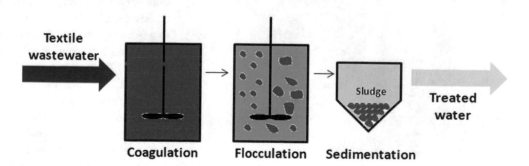

Table 2 Common chemical coagulants

Chemical coagulants		
Hydrolyzing metallic salts	Pre-hydrolizing metallic salts	Synthetic cationic polymers
Ferric chloride Ferric sulfate Magnesium chloride Aluminum sulfate	Polyaluminum chloride (PACI) Polyferric chloride (PFCI) Polyaluminum ferric chloride (PAFCI) Polyaluminum sulfate (PAS)	Aminomethyl polyacrylamide Polyalkylene Polyamine Polyethylenimine Polydiallyldimethyl ammonium chloride (Poly-DADMAC)

Table 3 Examples of chemical coagulants and their optimum conditions

Coagulant	Optimized dose (mg/L)	Type of dyes	Optimum pH	Removal (%)
Polyaluminum chloride (PACl)	10	Disperse	7.2	99.9
	0.1	Reactive, acid and direct	8.5	80
Alum	20	Reactive, acid and direct	Near to neutral	98
Ferric chloride	400	Sulfur	8.3	100
Magnesium chloride	120	Reactive and disperse	8.5	98

Adapted from Verma et al. (2012)

affect this adsorption process such as agitation and concentration of dye. Besides, the rate of adsorption of dye pigments can be determined based on time taken for the diffusion on the interior adsorbent materials. While step 4 is frequently depend on the nature of the dye whether anionic or cationic substances (Fig. 4).

On the other hand, there are two kinds of adsorption that are chemical sorption and physical sorption. Chemisorption or named as chemical sorption can be elucidated as strong chemical association formed between adsorbate molecule as a result of electron exchange and this process is irreversible (Anusha 2013). As for physisorption or physical sorption, it is reversible process in which formation of weak van der Waals bonds between adsorbent and adsorbate is produced (Allen and Koumanova 2005).

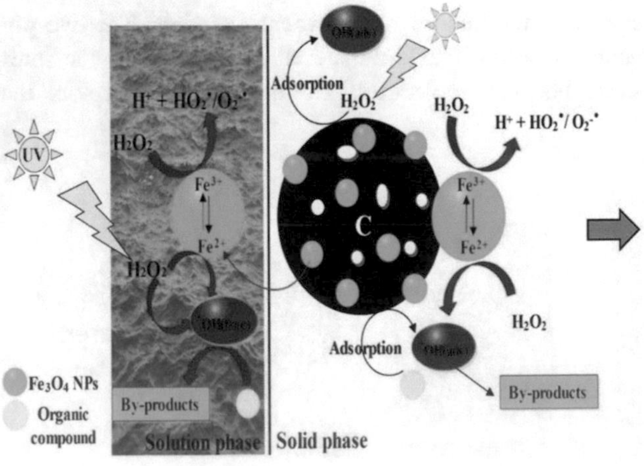

Adsorption methods have drawn considerable attention because of their higher effectiveness in decoloration of wastewater containing a range of colors. The primary features to be regarded during the choice of an adsorbent for color removal are high affinity, compound capacity and adsorbent regeneration capacity. Adsorption methods are commonly used to extract certain pollutant classes from water, particularly, those that are not readily biodegradable. Adsorption materials are available from various sources such as natural sources, agricultural and industrial wastes. Although the adsorption process using commercially available materials are preferred due to their effectiveness, their widespread use is restricted due to relatively high cost which led the researches to find an alternative using non-commercial low-cost adsorbents (Rafatullah et al. 2010). The usage of these kinds of materials is attractive as they can contribute to the cost reduction for waste disposal thus have a hand to the environmental protection.

Several adsorbents were studied to determine their ability to adsorb colors from aqueous effluents, but activated carbon is the most widely used and readily available coloring adsorbent from commercial sources (Liu et al. 2013). Activated carbon is the oldest known adsorbent and generally yielded from physical or chemical activation. There are several resources of commercial activated carbon including coconut shell, lignite, coal, fly ash and woods. It is known that almost all carbonaceous material can be utilized as starting material for carbon adsorbent. Even though activated

Fig. 4 Overview of adsorption mechanism (Seow and Lim 2016)

carbon is a preferred means for adsorption process, the disadvantages of this material are due to high cost and complex preparation process as they are derived from natural resources could contribute to raise the total cost of manufacturing. Furthermore, according to Singh and Arora (2011), activated carbons are hard to distinguish from the solution and were discarded with the process sludge after use in water and wastewater treatment, leading in secondary pollution (Singh and Arora 2011).

Many studies explored the potential of using a broad variety of alternative adsorbents to adsorb textile wastewater from industrial by-products, mineral deposits and agricultural waste as a substitute for commercially costly activated carbon. The search for an alternative adsorbent is driven by its abundance, low-cost, low handling requirements and highly efficient treatment of textile wastewater. Researchers have, therefore, made countless efforts to develop a fresh and novel adsorbent with superior efficiency in terms of greater adsorption capability, greater surface area and enhanced mechanical stability of the adsorbent. Table 4 summarizes some of the major works pertaining to the preparation and performance of alternative adsorbent in textile wastewater treatment.

2.3 Ion-Exchange

The ion-exchange process has also been widely applied in the textile wastewater treatment. This process is normally used for the elimination of the inorganic matters such as salts and some other organic compounds such as anionic components like phenol. This process involved the reversible interchange of the ions between the ion-exchange materials (normally in solid form) and the liquid containing the ion of interest (waste product). It is to be noted that there should not be permanent change in the solid structure of the ion-exchanger. In terms of the industrial wastewater treatment, the wastewater is contained of the unwanted ions that should be eliminated. In most cases, the ion-exchanger is of the complex ions with functionalized porous or gel polymer acts as the ion source. This material is used to be exchanged with the unwanted ions of the contaminant presence in wastewater. Theoretically, the anion-exchangers such as weak base resins exchange the negatively charged ions while the cation-exchangers namely the weak acid cation-exchange the positively charged ions contained in the solution (wastewater). Figure 5 depicted the mechanism of a simple ion-exchange process via ion-exchanger. As for the cationic exchange resin, the weak hydrogen ion bound to the negatively charged resin. Upon the treatment, the hydrogen ions are given into the solution containing sodium and chloride ions and replaced by the sodium ions. The same thing happened in the anionic-exchange resin where the hydroxide ion is replaced by the chloride ions that contained in the solution.

The application of the ion-exchange process in the textile industrial wastewater treatment normally recombinant with other main processes such as electrochemical or biological treatment processes. Usually, the biological treatment cannot fully recover the contaminants presence in the textile wastewater. The treated wastewater normally consists of residual dissolved organic carbon (DOM) including dye which is non-biodegradable and possess extreme toxicity. The dye in textile industries is normally made up of auxochromes and chromophores that are defined as typical anionic. Therefore, the utilization of the anionic-exchange resins is of interesting approach to be employed in treating the textile wastewater containing DOM. Few study has been

Table 4 Alternative adsorbents used for textile wastewater treatment

Starting material	Dye	Removal rate (%)	References
Bentonite	Acid green	N/A	Koswojo et al. (2010)
Bottom ash	Light green SF (Yellowish)	88	Mittal et al. (2010)
Fly ash (coal)	Fabric color	55–83	Zaharia and Suteu (2013)
Kaolin	Methylene blue Malachite green Basic yellow	65–99	El Mouzdahir et al. (2010) Tehrani-Bagha et al. (2011)
Kenaf fiber char	Methylene blue	95	Mahmoud et al. (2012)
Natural clay	Acid Red 88 Methylene blue	98 90–99	Akar and Uysal (2010) Elass et al. (2010)
Pinecone	Acid Black 26 Acid Green 25 Acid Blue 7	93 97 94	Mahmoodi et al. (2011)
Pistachio hull	Methylene blue	94	Moussavi and Khosravi (2011)
Sawdust	Methylene blue Methyl green	N/A	Djilali et al. (2016)

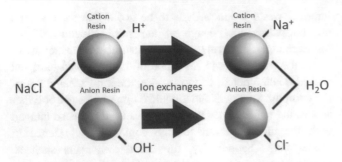

Fig. 5 Mechanism of ion-exchange process

reported on the use of magnetic anion-exchange resin in eliminating the non-biodegradable DOM consisted in the textile wastewater (Fan et al. 2014). However, the application of the ion-exchange resin is restricted due to the immobility, expensive and low flux challenges.

2.4 Oxidation

Oxidation process is one of basic process in fundamental of chemistry. It defined as a process of electron/s loss by an element in a chemical reaction. The element that responsible for the loss of electron/s called oxidizing agent or oxidant. This process of reaction happened between a compound and oxygen gas which is the quintessential oxidizer where the oxygen is reduced in the reaction, but it causes oxidation for the chemical substances with high oxidation states (Bajpai 2018). For example, the oxidation that occurs in H_2O chemical reaction, and the hydrogen is oxidized by oxygen which acts as oxidizing agent.

$$\text{Full reaction}: 2H_2 + O_2 \rightarrow 2H_2O$$
$$\text{Half-reaction}: H \rightarrow H^+ + e^-$$
$$O^2 + 2e^- \rightarrow O^{2-}$$

Oxidation process is widely applied in textile wastewater treatment for dyes degradation. The main concern of oxidation process is the decolorization of textile effluent by various chemical method. Table 5 shows the relative oxidation potentials of several chemical oxidizers.

Among the oxidizing agents, hydrogen peroxide (H_2O_2) is among the strongest existing oxidizing agents. Variously activated to form hydroxyl radicals (OH), H_2O_2 are able to decolorize a wide range of dyes effluents Fenton reaction is a first method to activate OH radical formation from H_2O_2, where in this reaction, hydrogen peroxide is added to an acidic solution (pH = 2–3) containing Fe^{2+} ions. The Fenton reaction is exothermic, however, in large scale plant, the reaction is commonly carried out at ambient temperature and large excess of iron and H_2O_2. Although widely applied for dyes degradation, the significant addition of acid and alkali

to reach the optimum pH, the need to reduce the residual iron concentration, and high sludge production becomes the major limitations of oxidation by hydrogen peroxides (Lin and Chen 1997) (Fig. 6).

In the meantime, ozone (O_3) is a powerful disinfection and a strong oxidant agent to remove color and odor, eliminating trace toxic synthetic organic compounds and assisting in coagulation. Fine bubble contactor is the most widely used conventional ozone generator used due to its high performance and 90% of ozone transfer (Zhou and Smith 2002). However, the concentration and types of dye will affect the increase in biodegradability index of textile wastewater. O_3 at 300 mg/dm^3 increased the biodegradability index by 1.6 times meanwhile biodegradability index was increased 11–66 times for azo dye wastewater. Furthermore, the biodegradability index increased to 80 times for simulated reactive dye and reactive yellow 84 textile wastewater (Koch et al. 2002). O_3 decolorize all dyes, except non-soluble disperse and reactive dyes which react slowly and take longer time. However, the color removal from textile wastewater is depended on dye concentration. In this regard, higher initial dye concentration of textile wastewater causes more ozone consumption which enhances mass transfer that causes an increase in ozone concentration in liquid phase, which increase color removal. In addition, increasing the temperature from 25 to 50 °C and increasing pH solution color removal efficiency increased with dye concentration (Al-Kdasi et al. 2004).

Ozone oxidation is very effective in the removal of most of the dyes with the double bond. Ozonation can easily break the double bond, hence, offers fast decolorizing process for textile wastewater treatment. In addition, ozonation can inhibit the foaming properties of residual surfactants and oxidize a significant portion of COD without increases the volume of wastewater or the sludge production. The ozone oxidation is applied in various applications, especially in final polishing treatment. However, up-stream treatments such as filtration are required to reduce the suspended oxidizing agents and increase the decolorization performance (Wang et al. 2011).

2.5 Biological Treatment

2.5.1 Aerobic Process

Despite the fact that many physical and chemical treatment resulted in incomplete degradation of dyes and number of researches conducted on exploring the ability of bacteria, fungi and algae to treat dye wastewater, many findings are the capability of pure culture to decolorize the dye. Typically, the color decoloration by bacterial strains is triggered by azoreductase-catalyzed anaerobic decrease or cleavage of dye bonds accompanied by aerobic or anaerobic degradation

Table 5 Oxidizing potential for conventional oxidizing agents (Al-Kdasi et al. 2004)

Oxidizing agent	Electrochemical oxidation potential (EOP), V	EOP relative to chlorine
Fluorine	3.06	2.25
Hydroxyl radical	2.80	2.05
Oxygen (atomic)	2.42	1.78
Ozone	2.08	1.52
Hydrogen peroxide	1.78	1.30
Hypochlorite	1.49	1.10
Chlorine	1.36	1.00
Chloride dioxide	1.27	0.93
Oxygen (molecular)	1.23	0.90

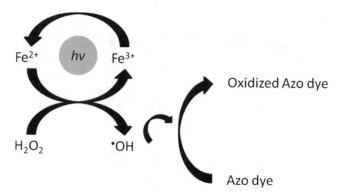

Fig. 6 Oxidation by hydrogen peroxide as oxidizing agent

by a blended bacterial community of resulting aromatic amines. Biological techniques for the complete degradation of textile wastewater have advantages such as: (a) environmental-friendly, (b) cost-competitive, (c) less sludge manufacturing, (d) less water consumption of non-hazardous metabolites or full mineralization and (e) higher concentration or less dilution requirement compared to physical/oxidation technique.

Commonly, only dissolved material in the textile wastewater can be eradicated and by employed aforementioned method, there are several aspects that affect the removal efficiency which are the types of organisms and their loading, ratio of organic load and dye, oxygen concentration and temperature of the system. From most of these factors, chosen microorganisms seemed to be the primary criteria to determine the efficiency as different types of microbes possessed different levels of enzymatic activities.

An aerobic method uses microbes for the treatment of the textile wastewater in the presence of oxygen. Selected microbes are then been isolated and cultured prior to test on the degradation of several types of dyes. Some of the aerobic bacteria decolorizing dyes are listed in Table 6.

Most of aerobic microorganisms that capable to decolorize the dyes require organic carbon to growth due to inability to use the dye pigments as growth substrate. For

instance, *P. aeruginosa* only able to degrade textile dye named fast blue whenever glucose and oxygen are existed. Similar conditions are applied to *Bacillus sp.* and *A. hydrophila* which unable to react without oxygen. A study conducted by Kolekar et al. revealed that *B. fusiformis* able to tolerate and degrade Disperse Blue 79 and Acid Orange 10 under anoxic condition at a concentration of 1.5 g/l of dye within 48 h (Kolekar et al. 2008). There are very few bacteria known that can grow on azo dye compounds utilizing them as sole carbon source and others reductively cleave azo bonds and utilize amines as sole source of carbon and energy for growth.

2.5.2 Anaerobic Process

The anaerobic process is a branch of biological treatment that utilize the anaerobic bacteria to decompose the organic matter in the absence or limited access of oxygen. In the past, this method was used for the digestion of sludge. Over the years, this technique has been gradually used for the treatment of various concentrations of organic wastewater treatment. Meanwhile, the textile wastewater contained of many high concentrations of organic waste namely dyeing waste, textile printing waste and wool washing sewage. The organic waste content which is as high as 1000 mg/L or more can be harmful toward human and aquatic organism if it is discharged into the surface water resources without further treatment. Unlike the aerobic biological treatment which aims to treat low concentration wastewater, the anaerobic treatment process is usually adopted for the high concentration wastewater. To date, the main anaerobic biological treatment normally used in the industrial practices is the hydrolysis acidification. This technique can eventually increase the sewage biodegradability that consequently facilitates the subsequent biological treatment process. Figure 7 shows the membrane bioreactor system that utilizing the anaerobic treatment step of hydrolysis acidification process in polymer-flooding wastewater treatment plant.

The hydrolysis acidification process is the two primary stages of the anaerobic biological treatment. During this

Table 6 Decolorization of textile wastewater using aerobic microorganisms

Microorganisms	Dyes	References
Bacillus fusiformis	Acid orange	Kolekar et al. (2008)
Enterococcus faecalis	Methyl red	Feng et al. (2012)
Halomonas sp.	Remazol black	Asad et al. (2007)
Klebsiella sp.	Reactive Black 5	Elizalde-González et al. (2009)
Pseudomonas fluorescens	Acid yellow	Pandey et al. (2007)
Shewanella putrefaciens	Reactive Black 5	Khalid et al. (2008)

(continued)

Table 6 (continued)

Microorganisms	Dyes	References
Staphylococcus aureus	Congo red	Xing et al. (2012)
Vibrio logei *Pseudomonas nitroreductase*	Methyl red	Adedayo et al. (2004)

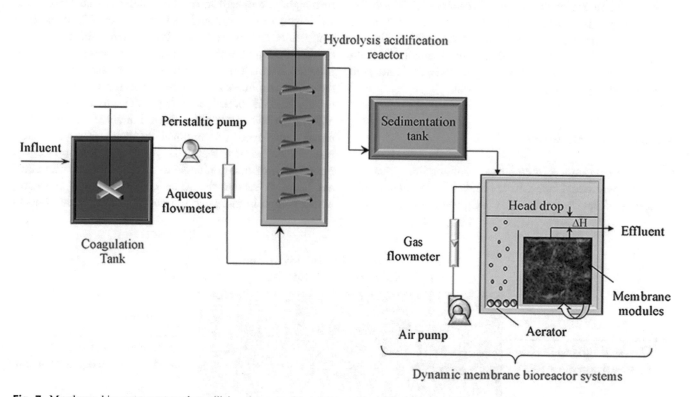

Fig. 7 Membrane bioreactor system that utilizing the anaerobic treatment step of hydrolysis acidification process (Shen et al. 2018)

process, the presence of the facultative and anaerobic bacteria has facilitated the decomposition of the heterocyclic organic matter, macromolecules and other recalcitrant biodegradable organic matters into smaller organic matter molecules. This has subsequently increasing the biodegradability of the matter and the wastewater as well as destructing the colored group of the dye, thus, degrading the color of the wastewater. In addition, the assistance of the anaerobic bacteria in decomposing the molecular structure of the colored material and the organic matter has eased the decomposition and decolorization of the wastewater in the aerobic condition. On the other hand, the anaerobic treatment process has significantly changed the pH of the wastewater. The treated wastewater in the hydrolysis tank usually decreased its pH value up to 1.5 unit, thus, neutralized the alkalinity of the sewage to about pH 8. This condition is subsequently providing a good condition for the aerobic treatment process to take place. Hence, the anaerobic digestion process is known to plays a vital role in determining the efficiency of the biological treatment of the textile wastewater treatment. Up to the present time, there are many anaerobic processes used in the dyeing and textile wastewater treatment such as up-flow anaerobic fluidized bed (UABF), anaerobic biological filter, up-flow anaerobic sludge bed (UASB) and anaerobic baffled reactor (ABR).

3 Advanced Membrane Separation for Textile Wastewater Treatment Process

Membrane technology has gained a great attention and frequently utilized in the chemical processing and technologies. The usage of this technology is mainly due to the capacity of this technique to meet the separation requirement in various applications including wastewater treatment. Additionally, this technique is capable of producing stable water in the absence of chemicals consumption as well as relatively low energy requirement. This technique can be subdivided into several types of operation depending on the membrane pore size and the application of the treatment.

3.1 Reverse Osmosis

One of the common membrane configurations that are intensely researched upon in the field of membrane technology is reverse osmosis (RO). RO is defined as a water purification process which employs a partially permeable membrane for the removal of ions, unwanted molecules and larger particles from water bodies (Jiang et al. 2017). Unlike osmosis, which is a natural process where a water molecule moves from a solution of lower solute concentration to a solution on higher solute concentration across a semipermeable membrane, reverse osmosis requires pressure to be exerted on the salt side and forces the water to permeate through the semi-permeable membrane, leaving only unwanted salt and particulates behind (Shenvi et al. 2015). As per current literature and technology, RO is heavily used for desalination to produce clean water from seawater. Figure 8 shows the overview of a continuous RO system.

In general, RO systems consist of two water streams, which is called the low solute concentration (permeate), while the second water stream is a higher solute concentration compared to the feed (reject/concentrate). As the high solute concentration enters into the RO system under pressure higher compared to osmotic pressure, water molecules are forced to pass through the semi-permeable membrane while salts and other impurities are left behind and discharged through the reject (reject/concentrate) stream (Kang and Cao 2012). This stream will either be fed back into the feed tank or removed completely from the system. Usually, up to 90–95% of salt and impurities are removed from effluent via RO mechanism. Even though RO is very commonly employed for removal of salts from seawater, research has shown that it is versatile to be applied for remediation of other effluents too, particularly, textile effluent, since it retains the same characteristic of seawater, which is high in inorganic salt concentration. In an effort to

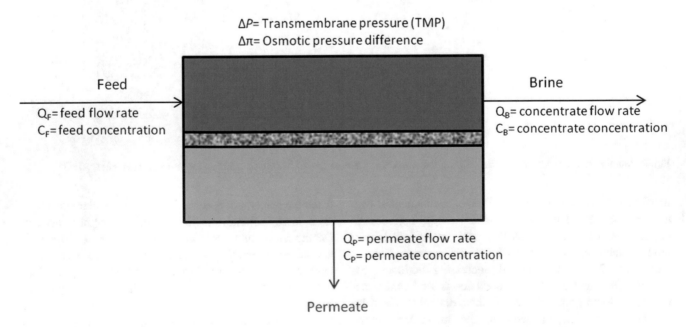

ΔP = Transmembrane pressure (TMP)
$\Delta\pi$ = Osmotic pressure difference

Feed

Q_F = feed flow rate
C_F = feed concentration

Brine

Q_B = concentrate flow rate
C_B = concentrate concentration

Q_P = permeate flow rate
C_P = permeate concentration

Permeate

Fig. 8 Schematic of a continuous RO system

reuse water that is utilized in the water intense textile industry, researchers have turned to RO as a viable option to realize this goal. Textile effluents are generally high with acids, alkalis, dyes, hydrogen peroxide, starch, surfactants dispersing agents and metals in the wastewater, not to mention the high BOD and COD exhibited by these effluents (Sahinkaya et al. 2018). There are plenty of evidence where RO works effectively to recover water from textile effluents. A study conducted by Cinperi et al., where RO system was coupled with membrane bioreactor (MBR) and ultraviolet (UV) process, where a turbidity and color value were reduced by 97% and 73.7%, respectively, while exhibiting salt removal of 97% (Cinperi et al. 2019). The water recovered from this effluent was used for subsequent textile dying process and did not pose any change in quality of textile color. Another study conducted by Sahinkaya et al., where a dual RO a membrane integrated pellet reactor was used to precipitate scaling cations showed promising results too. Removal of Ca^{2+} and Mg^{2+} was more than 95%, while almost 85% of water from textile effluent was recovered. Wang et al. coupled RO system with electro-oxidation process with controlled oxidation–reduction potential (ORP) for treatment of textile effluent (Wang et al. 2018). This robust system was able to exhibit high removal efficiencies of COD (72%), total nitrogen (TN) (18%) and chroma (99%). Both electro-oxidation process and ORP assists in mineralization of macromolecules into smaller molecules, which further minimize the production of brine waste after RO treatment. The works that have been done shows that RO may not be able to work by itself to treat textile effluents, due to several factors, including high scaling/fouling of membrane by various salts and high pressure required to manipulate osmotic pressure. Hence, RO system would work efficiently when coupled with other treatment processes as described earlier.

3.2 Nanofiltration

Nanofiltration (NF) has become a widely accepted process not only for producing drinking water but also for recovering wastewater in industrial processes or removing pollutants from industrial wastewater effluent. It is a membrane-based method that maintains comparatively small molecular weight organic compounds and divalent ions or big molecular ions such as 700–1000 molecular weight hydrolyzed reactive dyes as well as auxiliary dyeing. It has the benefit of recovering valuable materials and should be preferred to reverse osmosis with higher maintenance and operating costs (Chakrabarty et al. 2008). Nanofiltration membranes are also currently produced of ceramic materials that can resist elevated temperatures. Preparation flexibility and the range of raw materials for nanofiltration training will improve and

spread its implementation in various procedures. With such versatile choice of raw materials and ease of modification for various applications, nanofiltration is quickly become the largest and most widely used membrane filtration technology requiring a more focused research community on nanofiltration development.

Nanofiltration has been increasingly being used by the textile industry to manage color discharges. Nanofiltration membranes contain organic complexes of low molecular weight, divalent ions, large monovalent ions, hydrolyzed reactive colors and dyeing auxiliary. The quantity of mineral salts does not exceed 20 g/L in most accessible research on dye house discharges, and the quantity of dyestuff does not exceed 1.5 g/L. There are three main aspects that controlling the mass transfer of nanofiltration. Diffusion and convection mechanisms are frequently observed in the nanofiltration process as they possess small pore size. Since the active membrane layer of nanofiltration usually consists of negatively charged group, thus, the migration of ions on an electrical field need to be considered.

Nanofiltration can play a significant role in separating precious chemicals or removal from liquid streams of a dangerous or unwanted substance that can save costs and enhance industry's environmental impact. Dye production is a series source of various pollutants before the effluent from the discharge dye sector should be handled to decrease the negative impact on human and aquatic life. In the textile industry, nanofiltration can be used to separate distinct substances; however, further study is required to enhance effectiveness and overcome expected issues. Table 7 displayed several experiments conducted using nanofiltration system to treat textile wastewater.

3.3 Ultrafiltration

The ultrafiltration (UF) is a type of membrane with porous membrane structure and medium separation performance. The pore width of 2–200 nm has categorized this membrane in the middle between large microfiltration and small nanofiltration membranes counterparts. Similarly to that of other membranes, the UF membrane is used for the separation purpose due to the permeability for specific substances that contained in the solution medium which to be removed or to be concentrated. To be more specific, the UF membrane is normally used to concentrate, fractionate and to treat the macromolecules in fluid systems. Figure 9 shows the selectivity of the UF membrane separation for different types of materials.

The implementation of the UF membrane in the textile wastewater treatment has been reported in the literature. For instance, there was an investigation on the textile wastewater treatment using the polysulfone UF membrane that was

Table 7 Nanofiltration method on textile wastewater treatment

Dyes	Treatment condition	Removal rate (%)
Everzol black Everzol blue Everzol red	Initial concentration: 600 ppm Pressure: 3–12 bars	>90
Eriochrome black	Initial concentration: 1 ppm Pressure: 4 bars	>99
Sunset yellow	Initial concentration: 100 ppm Pressure: 6.2–6.9 bars pH: 6.8	82.2
Reactive black	Concentration: 0.4–2 ppm Pressure: 0.3–1.7 bars	60–97
Safranine orange	Initial concentration: 50 ppm Pressure: 5 bars	86

Adapted from Lau and Ismail (2009)

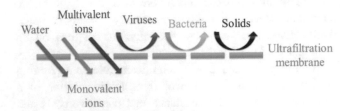

Fig. 9 Separation selectivity of the ultrafiltration membrane

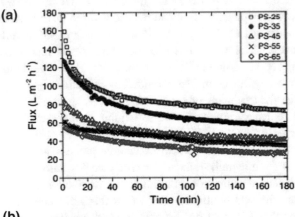

fabricated using phase inversion technique at a different temperature (Koseoglu-Imer 2013). In the study, the membrane performances were evaluated using a real textile wastewater consisting of 3094 mg/L of COD, suspended solids of 33 mg/L, color of 1.47 with absorbance wavelength of 530 nm, pH 9.0 and 5370 μS/cm of conductivity value. Upon the separation process, the highest removal efficiency was obtained by membrane produced at 65 °C with 99% of both color and COD removal. Figure 10 depicts the polysulfone UF membrane performance in the real textile wastewater treatment. Although the performance could be higher and seen effective, the UF is only limited to the removal of macromolecules of the contaminant presence in the wastewater. Therefore, this limitation can only be encountered when the smaller pore membrane is used and combined to this type of membranes.

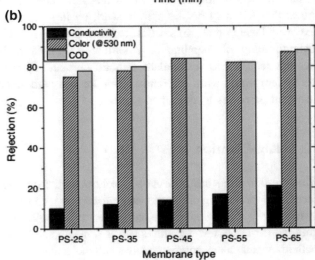

Fig. 10 Separation efficacy of the polysulfone ultrafiltration membrane of the dye-bath filtration experiment, **a** flux graphs of the membranes and **b** the rejection graphs of the membranes (Koseoglu-Imer 2013)

3.4 Microfiltration

The basis of membrane filtration is via size exclusion. Microfiltration (MF) membranes are classified as membranes which have the filterability of between 0.1 and 10 μm and a molecular weight cut off (MWCO) between less than 100,000 g/mol. MF membranes are commonly used for the treatment of large sized molecules, including oil effluent, milk/whey processing, bacteria in food industry and

production of paints and adhesives (Kumar and Ismail 2015). However, reports in employment of MF membrane for remediation of textile effluent is scarce, due to the fact that MF membranes are not able to remove dye molecules via size exclusion because size of dye molecules are smaller compared to the pore size of average MF membranes. This has propelled researchers to focus on employment of nanofiltration (NF) and RO membranes for remediation of textile effluent via size exclusion due to the significantly smaller pore size and diameter (Karisma et al. 2017). However, some research has been carried out using MF membranes for remediation of these effluents via MF membrane modification, including surface modification and incorporation of novel nanomaterials into membrane matrix. Jedidi et al. successfully produced a MF membrane using mineral coal fly ash via slip-casting method for removal of dyes (Jedidi et al. 2011). The fabricated membranes were able to exhibit high flux values 475 L/hm^2 bar while retaining respectable dye rejection (75–90%) using raw textile effluent. The ability of fly ash to produce intrinsic macropores throughout the membrane matrix when sintered at 800 °C is cited as a probable factor toward high permeation and respectable rejection performance. In line with inorganic membranes, Tahri et al. developed an inorganic carbon-based MF membranes for remediation of textile effluent, where asymmetric tubular MF membranes were prepared using mineral coal and phenolic resin (Tahri et al. 2013). The MF membranes exhibited a flux value of 280 L/h m^2 bar with rejection of dyes of more than 80%. Beqquor et al. prepared a ceramic-based MF membrane which is modified with micronized phosphate in means to reduce membrane pore size during fabrication process (Beqquor et al. 2019). The modification exhibited the desired effect, as the developed MF membrane showed flux values between 220 and 240 L/h m^2 bar and dye rejection of more than 99%. Addition of micronized phosphate significantly reduced membrane pore size, which instantaneously made the MF membrane competitive in terms of flux and rejection capabilities compared to UF or NF membranes. On the other hand, Homem et al. attempted to deposit graphene oxide (GO) on the surface of PES-PEI MF membranes via layer-by-layer self-assembly method utilizing electrostatic interaction (Homem et al. 2019). The surface modification was able to reduce the size of pores on membrane selective layer, where a pure water permeability of 99.4 L/hm^2 bar and a dye rejection up to 90% were achieved, while a flux recovery ratio up to 80% after fouling was observed.

3.5 Electrodialysis

Electrodialysis (ED) is a latest sophisticated approach of membrane separation used in the commercial application of industrial effluents. This method consists of an ion-exchange membrane and electrical potential are the driving force needed to apply the method. After passing through the ion-selective membrane barrier, electrical potential ions from one solution are transmitted to another solution. The efficiency of an ED method hinges on the current density, pH, flow rate, ED cell structure, water ionic concentration and ion-exchange membrane attributes. Membrane fouling is a significant factor that improves power usage and reduces membrane flux.

This process is not only part of the electrochemistry applied, but also part of the field of separation techniques. It utilizes electrical current rather than pressure to permit ions to move through the membrane compared to other types of membrane methods. As well, a small electrical present amount can be used, which helps to decrease the price of ED. Furthermore, ED has extra appealing features in which it is highly selective, able to operate continually, does not involve chemicals for therapy, green technology and cost-effectiveness (Li et al. 2017).

For ED application, there are at least five supplementary components that involves in this technology which are (a) direct present supply to strengthen ion migration, (b) electrodes where oxidation/reduction responses happened, (c) membrane for ion exchange, (d) solvents and (e) electrolytes (current carrier between anode and cathode) to transform ionic conduction into electron conduction and thus provide the initial driving force for ion migration. Illustration on mechanism of electrodialysis is as per Fig. 11.

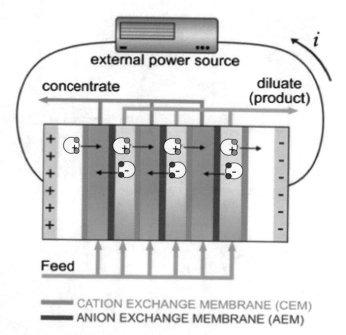

Fig. 11 Illustration on electrodialysis mechanism

3.6 Membrane Bioreactor (MBR)

The limitation of a single technology usually encountered by the combination of several treatment processes. One of its kind is the membrane bioreactor (MBR) systems. As be seen from its name, this treatment process has combining the membrane filtration and the biological wastewater treatment processes. Nowadays, this treatment process has gained great attention and is widely used for the industrial and municipal wastewater treatment. Unlike the conventional biological treatment process, the MBR is differ in the separation of the treated wastewater and the activated sludge. In this process, the common sedimentation in the secondary clarifier found in the normal biological treatment process is replaced by the membrane filtration technique.

MBR has become a famous choice in the wastewater treatment technology due to its ultimate benefit which produces high quality of treated water. In addition, this technique possesses several other advantages namely low maintenance, consistent high quality of treated water, small

footprint, excellent contaminants removal, low sludge production and high concentration of mixed liquid suspended solids (Chang et al. 2002). Although this process is limited by the drawbacks of the membrane fouling derived from the composition of the wastewater and the grown biomass, the fouling control strategies during the treatment has combat these weaknesses. On top of that, the area required for the MBR system installation as compared to that of conventional system is much smaller attributable to membrane direct installation inside the biological reactor. Figure 12 illustrates the MBR system installed in the laboratory-scale unit for the textile wastewater treatment process with the enhanced fouling control approaches. The combination of the membrane filtration system has eventually enhanced the performance of the wastewater treatment system, thus, increasing the water recovery as the treated water.

In term of the process efficacy, the MBR system has been evaluated in treating the textile wastewater as influent using the aerobic submerged MBR system (Friha et al. 2015). The long-term experiment run for 6 months using a flat sheet

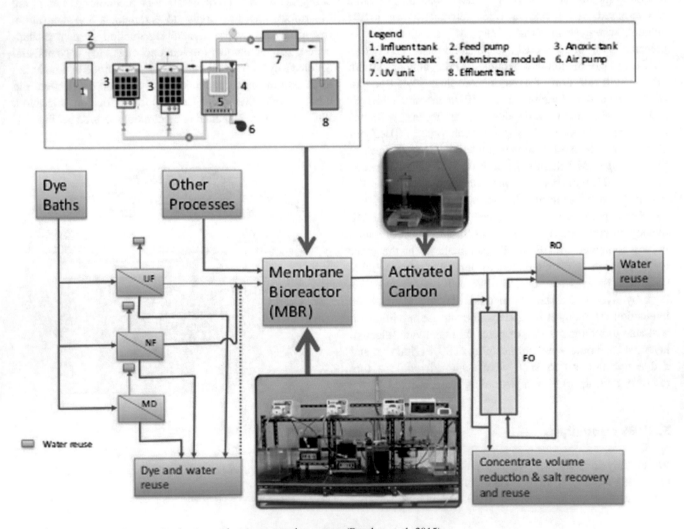

Fig. 12 Process flow diagram for the dye and water recoveries system (Rondon et al. 2015)

membrane module with transmembrane pressure (TMP) of 70–350 mbar has shown a stable treatment result toward the end of the operation. The outstanding of the pollution parameters removal efficiencies with 100% removal of color and suspended solids, 98% removal of COD and 96% of biochemical oxygen demand (BOD_5) elimination has indicated that the process can be effectively operated for the treatment of textile wastewater. Most importantly, this process has successfully decreased the toxicity of the wastewater.

3.7 Photocatalytic Membrane

In recent years, the interest for advanced oxidation processes (AOPs) has increased, especially heterogeneous photocatalytic processes. The heterogeneous photocatalytic process starts with the irradiation of heterogenous semiconductors by light irradiation as a source of energy. The electron excitation from the valence band to the conduction band upon light irradiation will generate electron–hole pairs that can react with water and dissolved oxygen to form various oxidizing species such as hydroxyl superoxide and perhydroxyl radicals. These highly reactive oxidizing species are able to oxidize and mineralize completely various organic contaminants (Kamat 1993).

Typically, semiconductors such as titanium dioxide (TiO_2), zinc oxide (ZnO), cadmium sulfide (CdS) and zinc sulfide (ZnS) are employed in photocatalysis due to their electronic structure (Saggioro et al. 2011). Interestingly, TiO_2 is the most common heterogenous photocatalyst with good photocatalytic activity due to its ideal properties such as nontoxic, chemical inertness and low-cost operation. Despite the excellent performance, the main limitation of the photocatalytic process is the recovery of the photocatalyst from the solution. To overcome this problem, heterogeneous photocatalytic oxidation may be combined with different membrane processes, such as microfiltration (MF),

ultrafiltration (UF), nanofiltration (NF) and direct contact membrane distillation (DCMD, MD) (Buscio et al. 2015). In the meantime, photocatalytic membrane reactors (PMRs) divided into reactors with suspended photocatalyst and photocatalysts suspended in a feed solution and (II) reactors with immobilized photocatalysts (Mendret et al. 2013) (Fig. 13).

The MPR offers several advantages for textile wastewater treatment such as continuous treatment process with simultaneous separation of photocatalyst from the reaction environment and constantly confined the photocatalyst in the reaction surrounding. Additionally, MPR also able to control the molecules retention time in a reactor. These remarkable improvements provided by MPR over the conventional treatments will ensure efficient treatment to decolorize, detoxify and treat industrial dye wastewater. Table 8 shows various photocatalytic MPR used for the degradation of textile wastewater.

Several issues have to be improved in order to increase the feasibility of photocatalysis process for textile wastewater treatment in the future. The most important is the development of a photocatalyst with high photo-efficiency under the irradiation of a wider solar spectra. Besides, catalyst immobilization strategies need to be addressed properly to provide a cost effective for the separation of photocatalyst from the reaction environment. There is also a need for the improvement in the photocatalytic reaction for wider pH range as well as to minimize the addition of oxidizing agents. Finally, good photocatalytic reactor with an effective design would significantly reduce the electricity costs without jeopardizing its performances.

4 Conclusions and Recommendation

Textile effluents in the organisms cause bio-toxicity, subsequent in growth inhibition and low plant chlorophyll content. Alternative treatments need to be found that are efficient

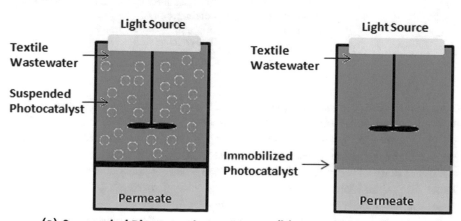

Fig. 13 Photocatalytic membrane photoreactor (MPR) for **a** Suspension photocatalyst and **b** Immobilized photocatalyst

(a) Suspended Photocatalyst MPR   **(b) Immobilized Photocatalyst MPR**

Table 8 Degradation of textile wastewater by photocatalytic MPR

Photocatalytic MPR	Reaction condition	Type of dyes	Findings	References
Photocatalytic membrane reactor (PMR) with TiO2 nanotubes (TNTs)	The optimal pH value and catalyst loading of the reaction system with application of TNTs were 4.5 and 0.5 g/L, respectively	Reactive brilliant red X-3B dye (X-3B)	The decolorization rate of X-3B with application of TNTs was up to 94.6% after 75 min	Wang et al. (2016)
MPR with Suspended TiO_2-P25 and cellulose triacetate membranes with entrapped TiO_2 (CTA–TiO_2–Al-10)	Oxygen concentration 20 mg/L with catalyst amount 1 g/L and initial pH 6.42. The lamp was switched on after 30 min of mixing in the dark	Congo red ($C_{32}H_{22}N_6Na_2O_6S_2$) Patent blue ($C_{27}H_{31}N_2NaO_6S_2$)	The reactor containing the suspended photocatalyst was significantly more efficient than the reactor containing the catalyst entrapped into the membrane. It was possible to treat successfully highly concentrated solutions (500 mg/L) of both dyes	Molinari et al. (2004)
MPR in the presence of zinc oxide capped with polyethylene glycol (ZnO-PEG)	Initial pH 11, 0.10 g/L of ZnO-PEG nanoparticles, and 75% dilution of textile wastewater	Textile wastewater (SDWW)-dark blue color in room temperature	MPR with ZnO-PEG resulted in maximum photocatalytic degradation efficiency and minimum membrane fouling with complete removal of the color and turbidity of SDWW. The COD and electrical conductivity degradation of the permeate were lower compared to the original SDWW	Desa et al. (2019)
MPR with TiO_2 Aeroxide P25 hollow fiber	pH 4, an initial dye concentration of 50 mg L^{-1}, and a TiO_2 loading of 2 g L^{-1}	C.I. Disperse Red 73	60 and 90% of dye degradation and up to 98% COD removal	Buscio et al. (2015)

in removing colors from big effluent quantities and are cost effective, such as chemical, biological or combined systems. However, in order to guarantee the creation of an eco-friendly technology, further research works are needed to explore the toxicity of color degradation metabolites, and the probable fate of the biomass used.

The goal of wastewater treatment plants in the textile industry is to introduce techniques that provide minimum or zero pollution of water. These wastewater treatment plants in the textile industry are the most widely accepted approaches to environmental safety. However, for all types of textile effluents, no specific treatment procedures are suitable or widely acceptable. Thus, the textile wastewater treatment is performed by a combination of several techniques that include physical, chemical and biological techniques depending on the type and quantum of pollution load. This review addressed several techniques for treating the dye in textile wastewater and reducing the pollution load. Physical and oxidation techniques are only efficient when the quantity of textile effluent is low for the degradation of dye in textile wastewater.

References

Adedayo, O., Javadpour, S., Taylor, C., Anderson, W., & Moo-Young, M. (2004). Decolourization and detoxification of methyl red by aerobic bacteria from a wastewater treatment plant. *World Journal of Microbiology and Biotechnology, 20*, 545–550.

Akar, S. T., & Uysal, R. (2010). Untreated clay with high adsorption capacity for effective removal of CI Acid Red 88 from aqueous solutions: Batch and dynamic flow mode studies. *Chemical Engineering Journal, 162*, 591–598.

Al-Kdasi, A., Idris, A., Saed, K., & Guan, C. T. (2004). Treatment of textile wastewater by advanced oxidation processes—A review. *Global NEST Journal, 6*, 222–230.

Allen, S., & Koumanova, B. (2005). Decolourisation of water/wastewater using adsorption. *Journal of the University of Chemical Technology and Metallurgy, 40*, 175–192.

Anusha, G. (2013). Feasibility studies on the removal of iron and fluoride from aqueous solution by adsorption using agro based waste materials.

Asad, S., Amoozegar, M., Pourbabaee, A. A., Sarbolouki, M., & Dastgheib, S. (2007). Decolorization of textile azo dyes by newly isolated halophilic and halotolerant bacteria. *Bioresource Technology, 98*, 2082–2088.

Bajpai, P. (2018). Introductory chemistry reviews (Chap. 20). In P. Bajpai (Ed.), *Biermann's handbook of pulp and paper* (3rd ed., pp. 401–426). Amsterdam: Elsevier.

Beqqour, D., Achiou, B., Bouazizi, A., Ouaddari, H., Elomari, H., Ouammou, M., et al. (2019). Enhancement of microfiltration performances of pozzolan membrane by incorporation of micronized phosphate and its application for industrial wastewater treatment. *Journal of Environmental Chemical Engineering, 7*, 102981.

Buscio, V., Brosillon, S., Mendret, J., Crespi, M., & Gutiérrez-Bouzán, C. (2015). Photocatalytic membrane reactor for the removal of CI Disperse Red 73. *Materials, 8*, 3633–3647.

Chakrabarty, B., Ghoshal, A., & Purkait, M. (2008). Ultrafiltration of stable oil-in-water emulsion by polysulfone membrane. *Journal of Membrane Science, 325*, 427–437.

Chang, I.-S., Clech, P. L., Jefferson, B., & Judd, S. (2002). Membrane fouling in membrane bioreactors for wastewater treatment. *Journal of Environmental Engineering, 128*, 1018–1029.

Cinperi, N. C., Ozturk, E., Yigit, N. O., Kitis, M. (2019). Treatment of woolen textile wastewater using membrane bioreactor, nanofiltration and reverse osmosis for reuse in production processes. *Journal of Cleaner Production, 223*, 837–848.

Desa, A. L., Hairom, N. H. H., Ng, L. Y., Ng, C. Y., Ahmad, M. K., & Mohammad, A. W. (2019). Industrial textile wastewater treatment via membrane photocatalytic reactor (MPR) in the presence of ZnO-PEG nanoparticles and tight ultrafiltration. *Journal of Water Process Engineering, 31*, 100872.

Djilali, Y., Elandaloussi, E. H., Aziz, A., & De Menorval, L.-C. (2016). Alkaline treatment of timber sawdust: A straightforward route toward effective low-cost adsorbent for the enhanced removal of basic dyes from aqueous solutions. *Journal of Saudi Chemical Society, 20*, S241–S249.

El Mouzdahir, Y., Elmchaouri, A., Mahboub, R., Gil, A., & Korili, S. (2010). Equilibrium modeling for the adsorption of methylene blue from aqueous solutions on activated clay minerals. *Desalination, 250*, 335–338.

Elass, K., Laachach, A., Alaoui, A., & Azzi, M. (2010). Removal of methylene blue from aqueous solution using ghassoul a low-cost adsorbent. *Applied Ecology and Environmental Research, 8*, 153–163.

Elizalde-González, M., Fuentes-Ramirez, L., & Guevara-Villa, M. (2009). Degradation of immobilized azo dyes by *Klebsiella* sp. UAP-b5 isolated from maize bioadsorbent. *Journal of Hazardous Materials, 161*, 769–774.

Fan, J., Li, H., Shuang, C., Li, W., & Li, A. (2014). Dissolved organic matter removal using magnetic anion exchange resin treatment on biological effluent of textile dyeing wastewater. *Journal of Environmental Sciences, 26*, 1567–1574.

Feng, J., Kweon, O., Xu, H., Cerniglia, C. E., & Chen, H. (2012). Probing the NADH-and Methyl Red-binding site of a FMN-dependent azoreductase (AzoA) from Enterococcus faecalis. *Archives of Biochemistry and Biophysics, 520*, 99–107.

Friha, I., Bradai, M., Johnson, D., Hilal, N., Loukil, S., Ben Amor, F., et al. (2015). Treatment of textile wastewater by submerged membrane bioreactor: In vitro bioassays for the assessment of stress response elicited by raw and reclaimed wastewater. *Journal of Environmental Management, 160*, 184–192.

Gao, B.-Y., Wang, Y., Yue, Q.-Y., Wei, J.-C., & Li, Q. (2007). Color removal from simulated dye water and actual textile wastewater using a composite coagulant prepared by ployferric chloride and polydimethyldiallylammonium chloride. *Separation and Purification Technology, 54*, 157–163.

Geenens, D., Bixio, B., & Thoeye, C. (2001). Combined ozone-activated sludge treatment of landfill leachate. *Water Science and Technology, 44*, 359–365.

Golob, V., Vinder, A., & Simonič, M. (2005). Efficiency of the coagulation/flocculation method for the treatment of dyebath effluents. *Dyes and Pigments, 67*, 93–97.

Homem, N. C., Beluci, N. D. C. L., Amorim, S., Reis, R., Vieira, A. M. S., Vieira, M. F., et al. (2019). Surface modification of a polyethersulfone microfiltration membrane with graphene oxide for reactive dyes removal. *Applied Surface Science, 486*, 499–507.

Jedidi, I., Khemakhem, S., Saïdi, S., Larbot, A., Elloumi-Ammar, N., Fourati, A., et al. (2011). Preparation of a new ceramic microfiltration membrane from mineral coal fly ash: Application to the treatment of the textile dying effluents. *Powder Technology, 208*, 427–432.

Jiang, S., Li, Y., & Ladewig, B. P. (2017). A review of reverse osmosis membrane fouling and control strategies. *Science of the Total Environment, 595*, 567–583.

Kamat, P. V. (1993). Photochemistry on nonreactive and reactive (semiconductor) surfaces. *Chemical Reviews, 93*, 267–300.

Kang, G.-D., & Cao, Y.-M. (2012). Development of antifouling reverse osmosis membranes for water treatment: A review. *Water Research, 46*, 584–600.

Karisma, D., Febrianto, G., & Mangindaan, D. (2017). Removal of dyes from textile wastewater by using nanofiltration polyetherimide membrane. In *IOP Conference Series: Earth and Environmental Science* (p. 012012). IOP Publishing.

Khalid, A., Arshad, M., & Crowley, D. E. (2008). Decolorization of azo dyes by *Shewanella* sp. under saline conditions. *Applied Microbiology and Biotechnology, 79*, 1053–1059.

Koch, M., Yediler, A., Lienert, D., Insel, G., & Kettrup, A. (2002). Ozonation of hydrolyzed azo dye reactive yellow 84 (CI). *Chemosphere, 46*, 109–113.

Kolekar, Y. M., Pawar, S. P., Gawai, K. R., Lokhande, P. D., Shouche, Y. S., & Kodam, K. M. (2008). Decolorization and degradation of

Disperse Blue 79 and Acid Orange 10, by *Bacillus fusiformis* KMK5 isolated from the textile dye contaminated soil. *Bioresource Technology, 99,* 8999–9003.

Koseoglu-Imer, D. Y. (2013). The determination of performances of polysulfone (PS) ultrafiltration membranes fabricated at different evaporation temperatures for the pretreatment of textile wastewater. *Desalination, 316,* 110–119.

Koswojo, R., Utomo, R. P., Ju, Y.-H., Ayucitra, A., Soetaredjo, F. E., Sunarso, J., & Ismadji, S. (2010). Acid Green 25 removal from wastewater by organo-bentonite from Pacitan. *Applied Clay Science, 48,* 81–86.

Kumar, R., & Ismail, A. (2015). Fouling control on microfiltration/ultrafiltration membranes: Effects of morphology, hydrophilicity, and charge. *Journal of Applied Polymer Science, 132.*

Lau, W.-J., & Ismail, A. (2009). Polymeric nanofiltration membranes for textile dye wastewater treatment: Preparation, performance evaluation, transport modelling, and fouling control—A review. *Desalination, 245,* 321–348.

Li, X., Jin, X., Zhao, N., Angelidaki, I., & Zhang, Y. (2017). Novel bio-electro-Fenton technology for azo dye wastewater treatment using microbial reverse-electrodialysis electrolysis cell. *Bioresource Technology, 228,* 322–329.

Lin, S. H., & Chen, M. L. (1997). Treatment of textile wastewater by chemical methods for reuse. *Water Research, 31,* 868–876.

Liu, W., Yao, C., Wang, M., Ji, J., Ying, L., & Fu, C. (2013). Kinetics and thermodynamics characteristics of cationic yellow X-GL adsorption on attapulgite/rice hull-based activated carbon nanocomposites. *Environmental Progress & Sustainable Energy, 32,* 655–662.

Mahmoodi, N. M., Hayati, B., Arami, M., & Lan, C. (2011). Adsorption of textile dyes on pine cone from colored wastewater: Kinetic, equilibrium and thermodynamic studies. *Desalination, 268,* 117–125.

Mahmoud, D. K., Salleh, M. A. M., Karim, W. A. W. A., Idris, A., & Abidin, Z. Z. (2012). Batch adsorption of basic dye using acid treated kenaf fibre char: Equilibrium, kinetic and thermodynamic studies. *Chemical Engineering Journal, 181,* 449–457.

Mendret, J., Hatat-Fraile, M., Rivallin, M., & Brosillon, S. (2013). Hydrophilic composite membranes for simultaneous separation and photocatalytic degradation of organic pollutants. *Separation and Purification Technology, 111,* 9–19.

Mittal, A., Mittal, J., Malviya, A., Kaur, D., & Gupta, V. (2010). Decoloration treatment of a hazardous triarylmethane dye, Light Green SF (Yellowish) by waste material adsorbents. *Journal of Colloid and Interface Science, 342,* 518–527.

Molinari, R., Pirillo, F., Falco, M., Loddo, V., & Palmisano, L. (2004). Photocatalytic degradation of dyes by using a membrane reactor. *Chemical Engineering and Processing: Process Intensification, 43,* 1103–1114.

Moussavi, G., & Khosravi, R. (2011). The removal of cationic dyes from aqueous solutions by adsorption onto pistachio hull waste. *Chemical Engineering Research and Design, 89,* 2182–2189.

Nassar, M. M., & Magdy, Y. H. (1997). Removal of different basic dyes from aqueous solutions by adsorption on palm-fruit bunch particles. *Chemical Engineering Journal, 66,* 223–226.

Pandey, A., Singh, P., & Iyengar, L. (2007). Bacterial decolorization and degradation of azo dyes. *International Biodeterioration & Biodegradation, 59,* 73–84.

Rafatullah, M., Sulaiman, O., Hashim, R., & Ahmad, A. (2010). Adsorption of methylene blue on low-cost adsorbents: A review. *Journal of Hazardous Materials, 177,* 70–80.

Rondon, H., El-Cheikh, W., Boluarte, I. A. R., Chang, C.-Y., Bagshaw, S., Farago, L., et al. (2015). Application of enhanced membrane bioreactor (eMBR) to treat dye wastewater. *Bioresource Technology, 183,* 78–85.

Sabur, M., Khan, A., & Safiullah, S. (2012). Treatment of textile wastewater by coagulation precipitation method. *Journal of Scientific Research, 4,* 623–633.

Saggioro, E. M., Oliveira, A. S., Pavesi, T., Maia, C. G., Ferreira, L. F. V., & Moreira, J. C. (2011). Use of titanium dioxide photocatalysis on the remediation of model textile wastewaters containing azo dyes. *Molecules, 16,* 10370–10386.

Sahinkaya, E., Sahin, A., Yurtsever, A., & Kitis, M. (2018). Concentrate minimization and water recovery enhancement using pellet precipitator in a reverse osmosis process treating textile wastewater. *Journal of Environmental Management, 222,* 420–427.

Sasaki, T., Iizuka, A., Watanabe, M., Hongo, T., & Yamasaki, A. (2014). Preparation and performance of arsenate (V) adsorbents derived from concrete wastes. *Waste Management, 34,* 1829–1835.

Seow, T. W., & Lim, C. K. (2016). Removal of dye by adsorption: A review. *International Journal of Applied Engineering Research, 11,* 2675–2679.

Shen, X., Lu, L., Gao, B., Xu, X., & Yue, Q. (2018). Development of combined coagulation-hydrolysis acidification-dynamic membrane bioreactor system for treatment of oilfield polymer-flooding wastewater. *Frontiers of Environmental Science & Engineering, 13,* 9.

Shenvi, S. S., Isloor, A. M., & Ismail, A. (2015). A review on RO membrane technology: Developments and challenges. *Desalination, 368,* 10–26.

Singh, K., & Arora, S. (2011). Removal of synthetic textile dyes from wastewaters: A critical review on present treatment technologies. *Critical Reviews in Environmental Science and Technology, 41,* 807–878.

Tahri, N., Jedidi, I., Cerneaux, S., Cretin, M., & Amar, R. B. (2013). Development of an asymmetric carbon microfiltration membrane: Application to the treatment of industrial textile wastewater. *Separation and Purification Technology, 118,* 179–187.

Tehrani-Bagha, A., Nikkar, H., Mahmoodi, N., Markazi, M., & Menger, F. (2011). The sorption of cationic dyes onto kaolin: Kinetic, isotherm and thermodynamic studies. *Desalination, 266,* 274–280.

Verma, A. K., Dash, R. R., & Bhunia, P. (2012). A review on chemical coagulation/flocculation technologies for removal of colour from textile wastewaters. *Journal of Environmental Management, 93,* 154–168.

Wang, J., Zhang, T., Mei, Y., & Pan, B. (2018). Treatment of reverse-osmosis concentrate of printing and dyeing wastewater by electro-oxidation process with controlled oxidation-reduction potential (ORP). *Chemosphere, 201,* 621–626.

Wang, L., Xiong, W., Yao, L., & Wang, Z. (2016). Novel photocatalytic membrane reactor with TiO_2 nanotubes for azo dye wastewater treatment. In *MATEC Web of Conferences* (p. 06020). EDP Sciences.

Wang, Z., Xue, M., Huang, K., & Liu, Z. (2011). Textile dyeing wastewater treatment. In *Advances in treating textile effluent*. IntechOpen.

Xing, M., Shen, F., Liu, L., Chen, Z., Guo, N., Wang, X., et al. (2012). Antimicrobial efficacy of the alkaloid harmaline alone and in combination with chlorhexidine digluconate against clinical isolates of Staphylococcus aureus grown in planktonic and biofilm cultures. *Letters in Applied Microbiology, 54,* 475–482.

Zaharia, C., & Suteu, D. (2013). Coal fly ash as adsorptive material for treatment of a real textile effluent: Operating parameters and treatment efficiency. *Environmental Science and Pollution Research, 20,* 2226–2235.

Zhou, H., & Smith, D. W. (2002). Advanced technologies in water and wastewater treatment. *Journal of Environmental Engineering and Science, 1,* 247–264.

Solid Electrolyte Membranes for Low- and High-Temperature Fuel Cells

Siti Munira Jamil, Mazlinda Abd Rahman, Hazrul Adzfar Shabri, and Mohd Hafiz Dzarfan Othman

Abstract

In a fuel cell technology, the membrane will function as an ion-exchange membrane where it will act as a semipermeable membrane that enables the passage of selectively dissolved ions while blocking the others. This chapter describes the recent progress on the research of solid electrolyte membranes in fuel cells. It focuses on the utilization of ceramic and polymeric membranes in solid oxide fuel cell, proton exchange membrane fuel cell, and direct methanol fuel cell, respectively. The chapter then discusses the membrane configurations that have been widely used in fuel cell as well as their fabrication technique. Recent performance evaluation was also discussed for each type of membrane fuel cell. A brief discussion on the applications, potential, and future direction of this membrane fuel cell is also included. Finally, this chapter concludes the challenges of membrane electrolyte utilization in fuel cell.

Keywords

Membranes • Fuel cell • Electrolytes • Ceramic • Polymeric

S. M. Jamil · M. A. Rahman · H. A. Shabri · M. H. D. Othman (✉)
Advanced Membrane Technology Research Centre, Universiti Teknologi Malaysia (UTM), 81310 Johor Bahru, Johor, Malaysia
e-mail: hafiz@petroleum.utm.my

M. A. Rahman · H. A. Shabri · M. H. D. Othman
Faculty of Engineering, School of Chemical and Energy Engineering, Universiti Teknologi Malaysia (UTM), 81310 Johor Bahru, Johor, Malaysia

S. M. Jamil
School of Professional and Continuing Education (SPACE), Centre for Degree and Foundation Programmes, Universiti Teknologi Malaysia (UTM), 81310 Johor Bahru, Johor, Malaysia

1 Introduction

Membrane is a selective barrier that consists of thin sheets of layer of the cell, where it allows selective things to pass through but stop the others. Basically, membrane depends on its materials sources, either ceramic membrane or polymeric membrane. Ceramic membrane is an artificial membrane that made from inorganic materials like alumina, zirconia oxide, silicon carbide, or some glassy materials. While polymeric membrane is made of different kinds of polymer that usually soluble in organic solvents, membrane technology has been explored widely which covers all engineering approaches for substance transportation from one fraction to another fraction such as water treatment industry, food technology industry, and pharmaceutical industry. Fuel cell technology is also not excluded from using membrane as its main component's base.

In a fuel cell technology, the membrane will act as ion-exchange membrane where it will behave as semipermeable membrane that allows selectively dissolved ions to pass through, while blocking the others. This ions exchange membrane will serve as electrolyte that lies between two electrodes: anode and cathode. There are several main types of fuel cell ion-exchange membrane consisting of cation-exchange membrane (CEM), anion-exchange membrane (AEM), and bipolar membrane. CEM, known as proton-exchange membrane, usually will transport H^+ while AEM usually used in certain alkaline fuel cells to transport OH^- anions from one electrode to another electrode, whereas a bipolar membrane is a combination of anionic and cationic membrane that laminated together (Tanaka 2007).

Commonly, fuel cell technology uses a membrane that allows the transportation of ions from one electrode to another electrode where ions will meet at one point and generate the electrons through an electrochemical reaction. Ions exchange membrane is made up from organic or inorganic materials with charger. For instance, CEM contains fixed anionic groups with abundantly mobile cationic groups; thus, most of the conductivity comes from cation transport. The basic operations for AEM are vice versa with CEM. Membrane structure could be divided into two main structures of heterogeneous and homogeneous. This classification depends on the degree of heterogeneity of the membrane (Ariono et al. 2017). Homogeneous membrane formed from a polymer, while heterogeneous membrane formed from two different polymers. Heterogeneous membranes are less expensive compared to homogeneous membrane, but the composition of these membranes is thicker with rough surface. Thus, heterogeneous membrane is having higher resistance than homogeneous membrane.

Essentially, for example, proton-exchange membrane fuel cell (PEMFC) consists of porous composite of polymer electrolyte binder and supported nanoparticle catalyst on carbon particles (Litster and McLean 2004). Function of polymer electrolyte binder is to provide ionic conductivity, whereas electrical conductivity maintains by the carbon support catalyst. Its electrode consists of carbon support which acts as an electrical conductor; Pt particles as the reaction site; Nafion ionomer which provides pathway for proton conduction and Teflon binder which increases hydrophobicity of the cell. Apart from that, gas diffusion layer also important for PEM cell where it provides electrical connection between the current collector and catalyst. This layer must be thin and porous as well as electrically conductive.

Direct methanol fuel cell (DMFC) usually will have a thin membrane that is covered with sparse layer of platinum-base catalyst on its both sides, which sandwiched between two electrodes. A methanol solution introduces to the electrode with negative charges. Typically, anode structure of DMFC membrane composed of supported/unsupported catalyst layers bonded with Nafion resin, Teflon-bonded carbon black diffusion layer (GDL), and a carbon cloth or paper diffusion layer (Allen et al. 2005). This type of fuel cell will setup according to its Standard Newcastle flow bed-design (Allen et al. 2005).

In contrast, solid oxide fuel cell (SOFC) membrane usually comprises of thin and dense electrolyte, porous asymmetric anode, and porous cathode. Thin electrolyte is significant to transport the oxide ions from cathode to anode while the dense structure is a must in order to ensure there is no gas leaking or crossover between the fuel and the oxidizing agent. A porous asymmetric anode is meant by two different structures that composed of anode layer that consists of finger-like void and sponge-like void. Finger-like void is essential in providing a pathway for fuel to enter the cell; sponge-like void gives a support to the whole cell and also being sites for chemical reaction to take place. Porous cathode will allow the oxidizing agent like oxygen or air to pass through before entering the electrolyte.

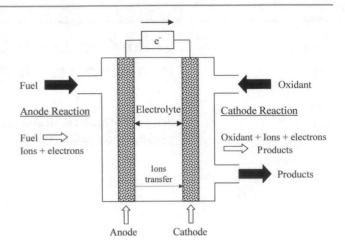

Fig. 1 Schematic diagram of fuel cell

2 Membranes Applications in Fuel Cells

Since all fuel cells involve the transfer or movement of ion (O^{2-} in SOFC, H^+ in PEM and DMFC) in electrolyte between anode and cathode; it is thus important to make the electrolyte layer to be thin to reduce the distance needed for the ion to travel but thick enough to separate the anode and cathode to prevent spillage. A general schematic of a fuel cell is provided in Fig. 1. Fabrication of the electrode and electrolyte layer in the form of thin membrane has become one of the challenges in fuel cell technology nowadays.

2.1 Solid Oxide Fuel Cell

Solid oxide fuel cell (SOFC) involves solid electrode and electrolyte, whereby the oxide ion (O^{2-}) moves from anode to cathode through the electrolyte layer. The oxygen from atmosphere is reduced at the cathode to form O^{2-} where it travels to anode through the electrolyte layer. The fuel, for example, H_2 gas will combine with the O^{2-} ion at the anode releasing the electron which will flow through external circuit to the cathode layer. The electrolyte layer consists of ceramic metal oxide material, usually having lattice structure of fluorite or perovskite and doped with metal with different valencies or atomic radii to introduce defect into the lattice. This defect will create oxygen vacancy in the lattice structure that will allow O^{2-} ion to hop from atom to atom when the material is heated to a certain temperature (operational temperature). Examples of the material are zirconia doped with 8 mol% yttria called yttria-stabilized zirconia (YSZ) and ceria doped with 10 mol% gadolinium called gadolinium-doped ceria (GDC).

Fabrication of SOFC using membrane generally involves the mixture of the ceramic electrolyte or electrode material with polymer to form the structure of the initial thin layer membrane followed by the removal of the polymer membrane leaving the ceramic material in the desired structure. This removal can be achieved by heating the membrane at high temperature to burn off the polymer material, or simply by heating to the sintering temperature of the ceramic material. The resulting material will then further heat to sinter temperature to strengthen the structure of the ceramic.

In terms of configuration, the membrane is being produced either in planar or tubular or micro-tubular design as shown in Fig. 2. Screen printing and tape casting techniques are common fabrication techniques in producing a planar membrane. Besides, compressing or dried pressing technique also being a most famous technique in fabricating a membrane with button cell design or disk design (Horri et al. 2012; Yoo and Lim 2013). In contrast, extrusion technique usually used to fabricate membrane with tubular configuration (Jamil et al. 2015). Conventionally, dry-jet wet and plastic mass ram are methods that employed to produce a tubular membrane via the extrusion technique. Both techniques are used to fabricate a single support layer of the membrane either anode-supported, electrolyte-supported, or cathode-supported. Until then, an advanced dry-jet wet extrusion technique known as phase inversion-based extrusion technique had been introduced which successfully produce a smaller size of tubular membrane which called as micro-tubular solid oxide fuel cell.

2.1.1 Planar SOFC

Planar form of SOFC (P-SOFC) involves layering flat sheet of electrode and electrode material. Electrolyte in planar membrane will lie and sandwich between anode and cathode. One of the advantages of the planar membrane is its configuration shorter the path lengths between anode to cathode for the electron movement which offers a production of high power output. However, the larger area of this configuration is being a factor to the gas sealing problem during the high-temperature operation, making the thermal stability of the cell reduces. Conventional technique of fabricating planar SOFC usually involves pressing the ceramic material at high temperature followed by sintering.

However, fabrication of planar SOFC via membrane route is also widely found in literatures. Example of this is by using polymer material like polyvinyl butyral as the binder. Through this method, the ceramic material is mixed into polymer polyvinyl butyral that was dissolved in methyl ethyl ketone solvent together with other additives such as pore former, plasticizer, and dispersant to form suspension. The suspension then can be transformed into thin layer by tape casting using doctor blade and consequently dried to remove the solvent. Thickness of the membrane layer can be controlled by the application of several layers. Two layers of electrolyte and electrode can be fabricated by tape casting the anode onto the electrolyte layer. The membrane then heated to remove the other material leaving the ceramic material and then to sinter temperature and to sinter the ceramic anode and electrolyte layer. Finally, the cathode layer is then deposited onto the electrolyte layer using painting and heated to sinter temperature (Kaur and Basu 2015).

2.1.2 Tubular and Micro-tubular SOFC

Tubular configuration that having a cylindrical shape offers more advantages and ultimately solving the problem facing by the planar configuration. Same like planar membrane, tubular membrane configuration can be differentiated according to its support. For example, anode-supported tubular membrane will produce a thicker anode layer among the electrolyte and cathode layer; it applied to the cathode- and electrolyte-supported tubular membranes. Anyhow, tubular configuration often increasing the ohmic loss of the cell due to its longer current pathways. The enhancement of this configuration had led to the introduction of micro-tubular membrane. This micro-tubular membrane means by the tubular membrane is fabricated in a hollow fiber design with a smaller diameter of 2–3 mm. The reduction of the diameter possesses various potential benefits including higher volumetric output, quicker start-up capability, good thermal cycling as well as portable characteristics (Meng et al. 2013).

Fabrication of SOFC offers additional advantages which is the ability for two or three layers of electrode or

Fig. 2 Membrane configurations   ◼ Anode   ◼ Electrolyte   ◼ Cathode

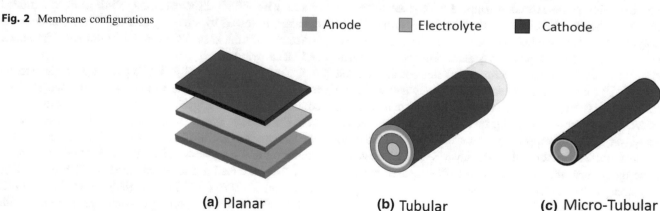

(a) Planar   **(b) Tubular**   **(c) Micro-Tubular**

electrolyte to be co-fabricated together, reducing the time and step of fabrications. This is usually can be achieved by fabrication of anode and electrolyte layer together, or the recent technique: cathode–electrolyte–anode triple layer. Currently SOFC is investigated to be used not only with H_2 fuel, but also the hydrocarbons such as methane, butane, and alcohol such as ethanol and butanol as well as biogas. The variation of the fuel source is made possible due to the fact that high operating temperature of SOFC actually encompasses the working temperature of steam reforming process together with some modification in anode layer. The modification can be either (1) allowed anode to directly catalyze oxidation of fuel or (2) by adding material that performs reforming process in the anode, or (3) by joining the reforming layer onto the anode. This modification was achieved by using metal or metal layer alloy or usage of other ceramic material with fluorite and perovskite structure.

A tubular SOFC had been developed since the late 1950s by the Westinghouse Electric Corporation (Stambouli and Traversa 2002). Commonly, plastic mass-ram extrusion technique via a die with desired dimensions will be applied in the fabrication of tubular SOFC. This technique will involve the mixing of support materials with the binder and solvent in order to form a viscous paste. The paste is then extruded through the die forming with the support tube, and it then followed up by drying and firing that tubular membrane. As in 2004, a tubular anode-supported SOFC manages to be produced via the plastic mass extrusion method assisted with vacuum dip-coating and painting (Du and Sammes 2004). 300 MPa of Ni-YSZ tube with 10–50 μm gastight YSZ layer was obtained in this work. Apart of that, thermal spraying technique is also being employed in the development of tubular SOFC. For example, Ni–Al_2O_3 cermet-supported tubular SOFC had been produced by using this technique in the study conducted by Li et al. (2006). 800 μm of porous Ni–Al_2O_3 supporting tube was successfully developed together with 25 μm of NiO-4.5YSZ, anode layer deposited on supporting tube via an atmospheric plasma spraying method.

Micro-tubular SOFC (MT-SOFC) can be fabricated using ram extrusion or co-extrusion coupled with phase inversion process, whereby the electrode or electrolyte is mixed with polymer material together with its solvent to form suspension. For co-extrusion phase inversion process, the suspension then can be extruded through tubular opening (called spinnerets) into non-solvent at which the polymer will form hollow fiber (HF) with the electrode or electrolyte material in it. MT-SOFC fabricated using this method usually requires one layer to act of support at which the layer will hold the highest mechanical strength or the layer will be fabricated to have the highest thickness to provide the required mechanical strength. The layer may consist of cathode-, electrolyte-, or anode-supported solid oxide fuel cell. The support layer

may be extruded first followed by sintering at which other layer will be deposited onto the support, or the support layer will be extruded together to form dual-layer or even triple-layer HF.

Example of anode-supported MT-SOFC was done by Azzolini et al. (2015) using ram extrusion process. The initial anode materials of GDC, CuO or Cu_2O, and $LiNO_3$ were mixed with hydroxypropyl methylcellulose and water to form paste which were later extruded and dried. The electrolyte GDC later was deposited onto the electrolyte using dip-coating method followed by sintering and deposition of cathode LSCF using the dip-coating method as well. However, there is no mention of thickness of the anode layer obtained in this study. Sumi et al. (2015, 2017) also investigated anode-supported MT-SOFC by fabricating the anode layer using ram extrusion. Mixture of 60% NiO and 40% GDC with binder acrylic resin, water, and cellulose was extruded using piston cylinder to form the micro-tube followed by air drying. The YSZ electrolyte layer was later dip-coated onto the anode, followed by brush painting of LSCF onto the electrolyte layer. Using this method, anode, electrolyte, and cathode thickness of 640, 10, and 20 μm were each obtained.

Recent research on extrusion of electrolyte layer for electrolyte-supported SOFC using phase inversion was carried by Rabuni et al. (2018). The initial YSZ material was mixed with N-methyl-2-pyrrolidone (NMP), dispersant and polyethersulfone (PESf), and extruded into microtubular form. The process produces electrolyte with two different layers, dense thick outer layer of approximately 10 μm and micro-channel/porous layer. The anodic material of Cu and CeO_2 was deposited onto the porous layer using wet impregnation technique where precursor aqueous metal nitrate was deposited followed by sintering at 1450 °C and reduction by H_2. Cathode layer of LSM was brush painted onto the electrolyte using mixture of LSM and ethylene glycol followed by sintering at 1200 °C to form complete cell. In other study, Meng et al. (2014) fabricated dual-layer HF to produce NiO and YSZ layer for electrolyte-supported SOFC. The precursor ceramic layer of NiO and YSZ was mixed with PESf, dispersant and NMP and co-extruded together, followed by sintering at 1450 °C to obtain dense structure with thickness of 32 and 210 μm each for anode and electrolyte layer.

Cathode-supported MT-SOFC is being a preference to the researchers in this area due to the reason of good stability control during redox cycles. This reliable stability comes from the cathode-supported configuration itself where thinner anode would reduce the detrimental effect due to the expansion and contraction of Ni particles within the anode layer that may lead to the re-oxidation of Ni back to NiO. Dual-layer YSZ/YSZ-LSM of cathode-supported MT-SOFC well-developed via co-spinning/so-sintering technique with

5-μm-dip-coated Ni-YSZ layer (Meng et al. 2013). This asymmetric dual-layer hollow fiber is comprised of dense YSZ electrolyte layer supported on the porous cathode layer of YSZ-LSM. Besides, an effort in broadening the three-phase boundary length had been achieved through the dual-layer MT-SOFC fabrication of cathode functional layer LSM-YSZ sandwiched with LSM cathode layer (Meng et al. 2015). The main porous cathode layer usually will act as current-collecting layer as well as site for oxygen reduction process. Afterward, Panthi et al. (2017a, b) had carried out an investigation with the purpose of lowering down the co-sintering temperature as an attempt to avoid the chemical reaction between cathode and electrolyte layer during the co-sintering process. Co-sintering temperature can be lowering down till 1250–1300 °C with the help of sintering additive, NiO and Fe_2O_3, and microcrystalline cellulose pore former, adding into electrolyte and cathode, respectively (Panthi et al. 2017a).

In many studies, the anode layer is made as the support layer due to the fact that anode layer is the layer where the oxidation of fuel takes place; thus, thicker anode layer may be anticipated to give more area for the reaction to take place yielding better performance. However due to the limitation of material thermal expansion coefficient (TEC), the anode layer is usually made with composition consisting of mixture of electrolyte material and the anode material to ensure stability of the layers during co-sintering of the dual layer. This made the TEC of the anode and electrolyte layer to be closed to each other, resulting in the opportunity of the two layers to be fabricated together. Thus in co-extrusion of dual-layer MT-SOFC, the electrolyte layer and anode layer are usually co-extruded, co-sintered and coated with the cathode layer. Omar et al. (2018) and Jamil et al. (2019) fabricated dual-layer HF using almost similar method. Using anode/electrolyte material of NiO-GDC/GDC, the ceramic material was initially mixed with PESf polymer, dispersant, and DMSO or NMP as solvent and were extruded through spinnerets with water as the non-solvent phase to produce dual-layer HF. The anode/electrolyte layer obtained was found to range from 160 to 200 μm for anode and 30 to 60 μm for electrolyte.

Co-extrusion of triple-layer SOFC consisting of cathode, electrolyte, and anode has garnered interest in recent years. This method will further ease and reduce the fabrication step compared to the co-extrusion of two layers. Jamil et al. (2018) fabricated triple-layer HF in one single step consisting of anode, electrolyte, and cathode layers. The anode/electrolyte/cathode that were made up of NiO-GDC/GDC/LSCF-GDC employed similar phase inversion technique where the ceramic materials were mixed with solvent NMP, polymer PESf, and dispersant to form initial suspension. The triple-layer HF was later heat-treated at 400 °C, 800 °C, and later sintered at 1450 °C for 8 h. By varying the extrusion rate of anode/electrolyte/cathode to each 7/2/2 ml min^{-1} is producing thickness of 234/13.5/40 μm after co-sintering.

2.2 Proton Exchange Membrane Fuel Cells

Fuel cells generally create their names by electrolyte type and responding substances. PEMFC is a type of fuel cell that uses hydrogen as the fuel, oxygen as the oxidant and a type of membrane that is only permissible to hydrogen ion or proton. During PEMFC operation, the H_2 gas flows into the fuel cell through the anode and is oxidized by the reaction at catalyst site to form hydrogen ion. The hydrogen ion will move from anode to the cathode layer through the electrolyte membrane that is only permeable to hydrogen ion but not to electron. Upon reaching the cathode, the hydrogen ion will react with oxygen from air and electron to form water. The electron flow externally from anode to cathode layer through external circuit attached to the cell to generate electricity. Chemical equation for the reaction is shown in equation below.

$$Anode: H_2 \rightarrow 2H^+ + 2e^-$$
$$Cathode: 2H^+ + \tfrac{1}{2}O_2 + 2e^- \rightarrow H_2O$$

The electrodes layer in PEMFC (anode and cathode) consists of catalyst layer (CL) that catalyzes the oxidation of hydrogen or reduction of oxygen to produce water and gas diffusion layer (GDL) that permit the diffusion of hydrogen or oxygen into or water out of the CL and conduct electron to complete the circuit. Illustration of PEMFC is shown in Fig. 3 (Mehta and Cooper 2003; Wang et al. 2011).

The electrolyte membrane in the center of PEMFC is considered to be the most significant element. Generally, the electrolyte membrane should have strong affinity for proton while being insulator to electron in order to be regarded as material for membrane. Other criteria include durability, resistance to chemical attack, and the state which needs to be solid. The range of operating temperatures is an important factor to consider when selecting membrane materials, where common operation temperature of PEMFC ranges from 30 to 200 °C.

Many distinct membranes exist and that are made from distinct material types. The selection of materials used as membrane depends on the physical and chemical properties required to ensure effective membrane efficiency (Awang et al. 2015; Mehta and Cooper 2003; Omar et al. 2018; Wang et al. 2011). PEMFC membranes are categorized into three primary classifications. These membranes are perfluorinated compound, partially fluorinated compound, and non-fluorinated. Besides these, however, we have other membranes obtained from these main classifications or using

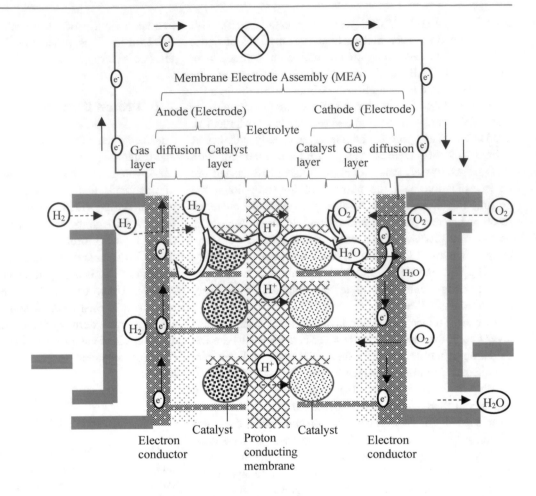

Fig. 3 Schematic diagram of polymer electrolyte fuel cell system. Adapted from Majlan et al. (2018)

extra materials such as acid–base mixtures and composite membrane supported (Wang et al. 2011). The material used for the membrane therefore accounts for the majority of cost-effective components and components would go a long way in decreasing the overall manufacturing cost. Nafion produced by DuPont USA, Aciplex and Flemion produced by Asahi Japan, and the composites are currently the raw materials used. These above components are costly, and it is becoming very crucial to need a material that is inexpensive with better results. Composite membranes were produced from, among others, materials such as hydrocarbons, ceramics, and graphene. Current study attempts are aimed at producing fully composite catalytic membranes that can replace Nafion (Mehta and Cooper 2003).

The membrane can be prepared using various methods. There are five popular techniques that have been widely researched such as polymerization technique for irradiation grafting, crosslinking technique, polymerization process for plasma grafting, sol–gel technique, and direct monomer polymerization (Ogungbemi et al. 2019). Usually, the technique used relies on the type of membrane to be produced and the materials and facilities available. In order to choose the right preparation technique, a thorough understanding of

materials and characteristics is required. On the other side, the preparation technique determines the final product and the quality thereof. PEM fuel cell technology is indeed the future of the renewable power industry, but less expensive and efficient material is needed to decrease the general price of fuel cells without limiting their efficiency.

2.3 Direct Methanol Fuel Cell

Direct methanol fuel cells (DMFC) operate on similar basis as PEMFC but with methanol as fuel to supply the hydrogen ion. The oxidation of methanol by reaction with water at the catalyst site produces carbon dioxide, electron, and hydrogen as shown in equation below.

$$\text{Anode} : CH_3OH + H_2O \rightarrow CO_2 + 6H^+ + 6e^-$$
$$\text{Cathode} : 6H^+ + 3/2O_2 + 6e^- \rightarrow 3H_2O$$

Due to their reduced weight and quantity compared to indirect fuel cells, DMFC is appealing for several applications. Solid polymers were shown as an appealing alternative to traditional liquid electrolytes in this type of fuel cells.

Polymers of Nafion perfluorosulfonic acid are the most frequently used in membranes fuel cells. The DMFC system is illustrated in Fig. 4. The DMFC has the ability to substitute rechargeable lithium-ion batteries in mobile electronic systems, but is presently experiencing important power density and effectiveness losses owing to elevated methanol crossover via polymer electrolyte membranes (PEMs) (Heinzel and Barragán 1999). Although spontaneously oxidizing methanol at the cathode would be appropriate, a transportation of methanol across the membrane has been noted. It creates losses of depolarization in terms of lost energy in the cathode and conversion losses. To enhance the DMFC's efficiency, it is essential to eliminate or, at least reducing fuel loss across the cell which generally referred to as "methanol crossover." The membrane technology is one of the options in this sense to try to fix this issue (Heinzel and Barragán 1999).

Significant progress in the growth of polymer electrolyte membranes for DMFCs has been produced in the latest years in terms of cost reduction and functionality enhancement along with other related technological advances. Common requirements for a polymer electrolyte membrane in DMFC application include: (1) elevated heat operation; (2) low methanol crossover; (3) high ionic conductivity; (4) high

chemical and mechanical stability; (5) low ruthenium crossover; and (6) low-cost operation. There are currently four major membrane types used in DMFCs. These included membranes of nafion and non-nafion, flouronated composite membranes and non-flouronated composite membranes. Among these, composite fluorinated and non-fluorinated (hydrocarbon) membranes with low cost, methanol and ruthenium crossover (for Pt–Ru anodes), wider temperature range (80–180 °C), and higher ionic conductivity compared to Nafion® membranes have been recorded (Teresa and Gámez 2007).

Not all DMFC requirements are met by the traditional Nafion® membranes for DMFC applications. Unlike DMFCs, thinner membrane materials are preferred in hydrogen PEMFC application because they decreased ionic strength and enhanced MEA efficiency. However, in DMFCs, thin Nafion®112 membranes lead to a strong crossover of methanol. These disadvantages exceed the advantage of low ionic resistance, and therefore, thicker membranes such as Nafion®117 are typically used. However, this membrane give a very small cell voltage in a DMFC (Thomas et al. 2002).

The main candidates for replacing the expensive Nafion® membranes are hydrocarbon membranes. Improving

Fig. 4 Schematic diagram of direct methanol fuel cell system. Adapted from Radenahmad et al. (2016)

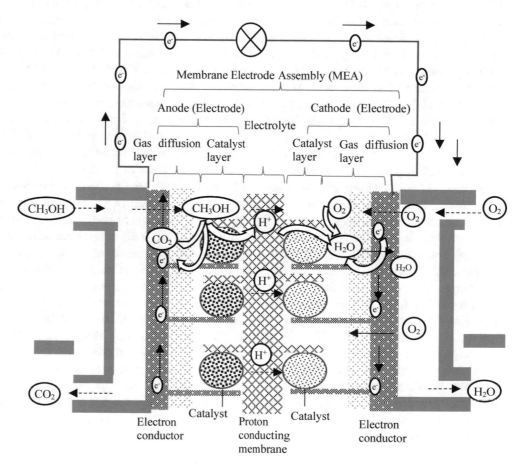

Nafion®-based membranes by adding inorganic compounds (SiO$_2$, silans, Zr, MoPh-a, etc.) and acid-based composites (polyaryl) reduces methanol crossover but does not reduce costs. Hydrocarbon membranes are cheaper for DMFC than Nafion® membranes and are more technically efficient. They have reduced methanol crossover and greater conductivity and stability. Hydrocarbon and composite fluorinated membranes presently exhibit the greatest potential for low-cost membranes with low permeability of methanol and high durability (Jörissen et al. 2002). Some of these membranes are already starting to impact the market for mobile fuel cells.

3 Performance Evaluation of Fuel Cells Involving Membrane Applications

3.1 Performance of Solid Oxide Fuel Cell

Performance evaluation of fuel cells is done in-situ under operating condition or ex-situ which is not under operating condition. Common in-situ measurement involves measuring the open-circuit voltage (OCV) and power density under operating condition and ex-situ by measuring the cell impedance using electrochemical impedance spectroscopy (EIS). Measurement of OCV for SOFC is usually accompanied by the measurement of current generated shown in I–V or I–P diagram. Since the recent interest of utilizing SOFC with hydrocarbon fueled has gained interest, performance of SOFC operating under hydrocarbon fueled is usually done in comparison with performance using H$_2$ fuel. However, the performance of SOFC running on H$_2$ fuel is still generally better compared to cell running on hydrocarbon fuel such as CH$_4$ as given in Table 1.

From the latest literature finding, there are not much differences in cell performance between planar and tubular membrane SOFC. However as described earlier, micro-tubular form offers advantages in terms of the fabrication step and the void of need to design for interconnect. The electrolyte layer which is the most important layer functions optimally at different temperature depending on the material. Current use of YSZ which is considered as high-temperature SOFC (HT-SOFC) with operating temperature ranging from, 800 to 1000 °C received the same interest as GDC electrolyte, which considered as intermediate temperature SOFC (IT-SOFC) with operating temperature ranging from 500 to 800 °C. IT-SOFC hs higher ionic conductivity at lower working temperature compared to HT-SOFC but the tendency for cerium oxide to form Ce^{4+}/Ce^{3+} species in reducing environment resulted in undesired electronic conductivity that cause current leakage, hence lowering the performance of the cell.

Electrochemical impedance spectroscopy (EIS) has become major tool to characterize and test fuel cells. Electrical impedance (usually denoted by Z) similar to electrical resistance (R) is a measurement of the circuit resistance to electrical flow when electrical potential is applied across the circuit. However unlike resistance, impedance measurement also concerns the phase shift of the current flowing when AC potential is applied. This phase shift is the characteristics of resistor–capacitor circuit (RC circuit) which is be used to model the layer of the fuel cell in the form of equivalent electrical circuit. Electrochemical impedance is measure by applying AC potential across the membrane at various frequencies, and the resultant impedance is measured and expressed in complex form consisting of real and imaginary part called Nyquist plot as illustrated in Fig. 5. The data from Nyquist plot then will be fitted into mathematical model to find equivalent electrical circuit consisting of electrical elements of resistor, inductor, and capacitor. EIS interpretation correlates to the cell characterization due to the fact that certain process works like electrical components (Pivac and Barbir 2016).

From literatures, most of studies on SOFC performance also include EIS as part of result reporting to not only show the resistance value of the cell, but also investigated the effect of modification of electrode or electrolyte on the EIS and the postulated microscopic change associated with the change. Effect of metal and metal alloy layer on the ohmic and polarization resistance change has been investigated by Meng et al. (2014), Yan et al. (2016), Wu et al. (2016), Jamil et al. (2019), Lee et al. (2016) and Harris et al. (2017). EIS is also used to investigate degradation at the anode by carbon deposition associated with the utilization of hydrocarbon fuel by Sarruf et al. (2017), Panthi et al. (2017b), Omar et al. (2018) and Akdeniz et al. (2016) where the utilization of methane fuel was shown to cause increase in ohmic resistance of the cell.

Modification of traditional anode layer of Ni-YSZ to enable SOFC to be able to be used with varieties of fuel seems to be the current direction of recent research. Since Ni-YSZ suffers from performance degradation due to carbon deposition when used with hydrocarbon fuel, anode modification through the insertion of ceria, metal alloy or used of fluorite or perovskite material to act as oxidation catalyst has been extensively studied. Insertion of ceria and copper metal in anode-supported MT-SOFC using YSZ electrolyte was studied by Meng et al. (2014) and Rabuni et al. (2018). The former anode composition containing YSZ, Ni, CeO$_2$, and Cu obtained increased the performance from H$_2$ and CH$_4$ fuel (0 0.15 and 0.25 W cm^{-2}, respectively) while later study, lacking Ni managed to obtain 0.55 and 0.16 W cm^{-2} each for H$_2$ and CH$_4$ fueled. The results contradicted each other but may also indicate the importance of Ni even in

Table 1 Recent literature on SOFC tested with H_2 and CH_4 fuel

Source	Support	Fabrication technique	Type	Power density (W/cm^2)	
				H_2	CH_4
Akdeniz et al. (2016)	Electrolyte	Tape casting	P-SOFC	0.23	0.15
Bochentyn et al. (2019)	Electrolyte	Commercial (pressed)[a]	P-SOFC	0.24	0.21
Harris et al. (2017)	Electrolyte	Commercial (pressed)[a]	P-SOFC	0.323	0.245
Hua et al. (2018)	Electrolyte	Pressed	P-SOFC	1.06	0.94
Lee et al. (2016)	Electrolyte	Pressed	P-SOFC	1.7	1.4
Li et al. (2019)	Electrolyte	Commercial (pressed)[a]	P-SOFC	0.6	0.6
Meng et al. (2014)	Anode	Extrusion/phase inversion	MT-SOFC	0.15	0.24
Omar et al. (2018)	Anode	Extrusion/phase inversion	MT-SOFC	0.67	0.22
Panthi et al. (2017b)	Electrolyte	Ram extrusion	MT-SOFC	0.75	0.55
Rabuni et al. (2018)	Anode	Extrusion/phase inversion	MT-SOFC	0.55	0.16
Sarruf et al. (2017)	Electrolyte	Commercial (pressed)[a]	P-SOFC	0.383	0.088
Sarruf et al. (2018)	Electrolyte	Commercial (pressed)[a]	P-SOFC	0.411	0.082
Sumi et al. (2015)	Anode	Ram extrusion	MT-SOFC	0.14	0.14
Wu et al. (2016)	Anode	Tape casting	P-SOFC	0.463	0.251
Yan et al. (2016)	Electrolyte	Tape	P-SOFC	0.25	0.25

[a]Fuel cell which are usually available commercially in form of YSZ disk is included for reference

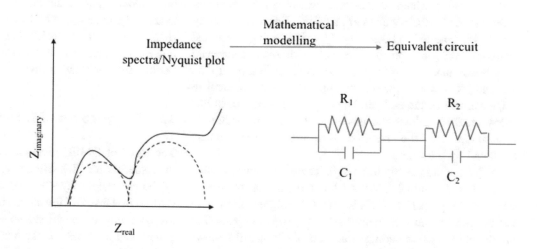

Fig. 5 General flow of interpretation of impedance spectra

hydrocarbon fueled SOFC. This is supported by Azzolini et al. (2015) wherein tubular SOFC with anode contains copper-GDC and GDC electrolyte system, maximum of 0.008 W cm^{-2} power density was achieved which is far lower than common value. This is further clarified by study from Sumi et al. (2015) with anode containing Nickel-GDC and GDC electrolyte where maximum power density 0.25 W cm^{-2} was obtained using H_2 fuel.

Co-extrusion of dual-layer hollow fiber to produce SOFC by Omar et al. (2018) and Jamil et al. (2019) shown that even though the SOFC was produced with the same composition of anode/electrolyte/cathode of Ni-GDC/GDC/LSCF-GDC, different performance SOFC can be obtained when tested on the same fuel H_2. The difference in result of

0.67 and 0.29 W cm^{-2} by Omar et al. (2018) and Jamil et al. (2019) may be attributed to the higher porosity (between 31.5 and 52.5%) in later compared to the former (25.6%). Omar et al. (2018) also tested the SOFC on CH_4 gas, obtaining a lower performance of 0.22 W cm^{-2} compared to running on H_2. This suggests that not only the material is important but also the microstructure is similar importance for the performance of SOFC.

Co-extrusion of triple-layer HF containing complete anode/electrolyte/cathode layer in single step by Jamil et al. (2018) produced roughly the same power density as the SOFC from dual-layer HF. With anode/electrolyte/cathode consisting of Ni-GDC/GDC/LSCF-GDC each, maximum power density was obtained at 0.48 W cm^{-2} running on H_2, similar to the average result of the same setting. Thus, it is evident that there are significant challenges remaining in the development of SOFC in a single step. The increase in TPB density by controlling the anode morphology (e.g. by adjusting the length of finger-like voids and enhancing the porosity) might lead to improved MT-SOFC performance. Under those circumstances, a comprehensive study on the fabrication of SOFC is important in order to produce defect-free hollow fibers with the desired morphologies that maintain high mechanical strength, decent tightness properties, and sufficient anode porosity to produce high-performance MT-SOFCs.

3.2 Performance of Proton Exchange Membrane Fuel Cell

Improvement of PEMFC performance is mainly targeted to be achieved by solving the main issues faced by the development of PEMFC which are (1) lowering cost, (2) water management system, (3) proton conductivity at high temperature, (4) lower gas crossover, and (5) improved thermal stability and mechanical and chemical strength (Peighambardoust et al. 2010; Majlan et al. 2018). Effort on lowering the cost is targeted to be achieved by developing lower cost material compared to the current traditional fluorinated membrane and by developing lower cost catalyst other than Pt and Rh while other developments are targeted at improvement of the cathode layer. Table 2 summarizes few literature that address the development of new electrolyte membrane and the resulted maximum power density achieved by the studies.

All of the studies highlighted in Table 2 involved doping or adding functional layer to the existing membrane to enhance the properties. This in effect highlighted the rise in the usage of composite membrane in PEMFC as opposed to single membrane. In addition, analysis of EIS in PEM allows the identification of (1) layers in fuel cell that occurs due to the different processes such as mass transport process and polarization effect, (2) effect of different layer to the total resistance and impedance of the cell, and (3) microscopic detail such as change in granule size that affects the impedance and performance of the cell (Pivac and Barbir 2016). Zhiani et al. (2016) studied the effect of thermal and pressure stress on PEM and found that PEM running with membrane electrode assembly (MEA) conditioned under low stress resulted in higher performance, peaking at 1.6 W cm^{-2} compared to MEA conditioned under high stress. From the EIS equivalent circuit, it was postulated that the higher performance may be the result of extension of the triple-phase boundary in MEA that increases the reactive area for reaction.

For PEM, inductive phenomenon in EIS at high frequency was known to be caused simply by the effect of the wire and cable setup of the cell (Pivac and Barbir 2016). However, inductive effect at low frequency might suggest the possibilities of side reactions occurring in the cell between intermediate species, carbon monoxide poisoning of the cell or water movement across the layer. From equivalent circuit of EIS in PEM, the total resistance, R_T of the cell can be determined together with the contribution resistance by charge transfer, R_{CT} and ionic resistance R_{ion} inside the cell (Moghaddam and Easton 2018). Low R_{CT} and high R_{ion} may significantly impact performance of MEA and R_T value may be used as basic reference to assess the performance of Nafion-based PEMFC.

EIS has also been investigated to function as online measurement tool to study and monitor the effect of load current, air humidification rate, and hydrogen flow rate effect on the online performance of PEM to be used in electric vehicle (Depernet et al. 2016). In that study, the DC bus from the vehicle is connected to inverter and to DC/AC/DC power converter and the EIS spectra was obtained at discrete and few frequency (as opposed to cycling from low- to high-frequency AC) and the resulting equivalent circuit was generated and compared with ex-situ equivalent circuit. The study found that online EIS is comparable with ex-situ offline EIS and hence can be used as reliable tool to monitor the performance of PEM online.

3.3 Performance of Direct Methanol Fuel Cell

The focus of DMFC researches is mainly targeted to solve the issue in Sect. 2.3. Similar to PEMFC, the improvement of the electrolyte layer is also one of the main focuses of studies in DMFC. However, since DMFC also faced the problem with methanol crossover, this issue added another complexity in DMFC research as compared to PEMFC. Few research that are presented with maximum power density obtained in the studies are given in Table 3.

The studies are mainly done to resolve the issues by focusing on high temperature use of DMFC, methanol crossover and by managing water and gas flow in the cell and the catalyst to improve the performance of the DMFC (Kim et al. 2015; Radenahmad et al. 2016; Zainoodin et al. 2010; Li et al. 2013; Ong et al. 2017). Overall output of DMFCs relies on several variables, the most significant of which are: (i) anode's electrocatalytic activity, (ii) ionic conductivity and methanol crossover strength of the proton conductive membrane, and (iii) water management on the cell's cathode side (flow-field and back design function). As mentioned earlier, optimization of various DMFC parts can lead to a significant increase in power density and fuel utilization.

Table 2 Recent literature on development of electrolyte membrane for PEMFC

Membrane	Max power density, mW cm^{-2}	Source
PVA/PWA composite	60	Thanganathan and Nogami (2014)
sPAES composite	360	Sung et al. (2013)
sPAES composite	800	Yu et al. (2013)
sPAES composite	540	Park et al. (2011)
sPAES composite	360	Chun et al. (2013)
sPEEK composite	170	Lee et al. (2014)
sPEEK composite	56	Cho et al. (2010)
sPEEK composite	54.93	Ilbeygi et al. (2013)
sPS composite	180	Devrim et al. (2009)
sPAES composite	600	Amirinejad et al. (2012)
PBI composite	1000	Lobato et al. (2011a)
PBI composite	450	Jin et al. (2011)
PBI composite	800	Lobato et al. (2011b)
PBI composite	170	Jheng et al. (2013)
PBI composite	600	Kannan et al. (2010)
Nafion composite	120	Amjadi et al. (2010)
Sulfonated styrene	1400	Fatyeyeva et al. (2011)
Nafion composite	800	Lee et al. (2014)

PVA/PWA—poly(vinylalcohol)/phosphotungstic acid, sPAES—sulfonated poly(arylene ethers), PBI—polybenzimidazole, sPEEK—sulfonated poly ethyl ether ketone

Table 3 Recent literature on DMFC

Membrane base	Focus of studies	Max power density, mW cm^{-2}	Source
Nafion	Catalyst	23	Ahmad et al. (2010)
Nafion	Methanol crossover	22.7	Li et al. (2010)
Nafion	Water management	19.2	Wu et al. (2010)
Nafion	Methanol crossover	33	Xu and Faghri (2010)
PBI	High temperature	12–16	Mamlouk et al. (2011)
Nafion	Water management/methanol crossover	51.6	Xu et al. (2011a, b)
Nafion	Water management/methanol crossover	34	Xu et al. (2011a, b)
Nafion	Water management	17.1	Wu et al. (2011)
Nafion	Water management	40	Wu et al. (2011)
Nafion	Water management	46	Wu et al. (2011)
Nafion	Water management	109.8	Li and Faghri (2011)
PBI	High temperature	15	Brandão et al. (2012)
Nafion	Water management	18.4	He et al. (2012)
Nafion	Catalyst	299	Santasalo-Aarnio et al. (2012)
Nafion	Water management	69.5	Wu et al. (2013)
Nafion	Water management	29.4	Zhang et al. (2014)
Nafion	Water management/methanol crossover	32.6	Yan et al. (2015)
PBI	High temperature	37.2	Zhao et al. (2015)
Nafion	Water management	21	Zhang et al. (2015)

4 Fuel Cell Applications and Technology

SOFC covers wide range of application from small portable system until power generation system. Normally, portable SOFC systems are needed primarily for emergency, transportation utilization and even it is used for certain military appliance. Seeing that SOFC system having a superior fuel flexibility characteristic; fuel others than hydrogen could be feed into the system like JP-8 military fuel, ethanol, gasoline, kerosene, and others hydrocarbon fuels (Irvine and Connor 2013). Such portable SOFC system generally consumes about milliwatts to hundred watts of power. USA company such as Adaptive Materials Inc. had been developed such portable system called Amie25 and Amie150 (Narayan and Valdez 2008). This technology base on a lightweight micro-tubular SOFC and propane utilized as a fuel. 1700 Wh of energy can be generated from this portable system that weighed about 1.5 kg, which required 1 kg of propane. Besides, Protonex Technology Corporation also offers a portable generator-based SOFC, Valta P75 with an output of 75 W which fuelled by propane.

Japan is one of the countries that lead in the application of small SOFC systems for residential combined heat and power (CHP) units. Generally, the implementation of this system for residential uses required 1–5 kW power to supply. A company from Japan Kyocera had introduced anode-supported flat tubular cells. This technology was coupled together with the hot water tank. In 2005–2006, a 1 kW of this system was operated for the first time and successfully sustain for 2000 h with 44.1% of electric efficiency and 34% hot water heat recovery efficiency (Irvine and Connor 2013). In 2017, the same company had launched the first 3 kW of SOFC cogeneration system for institutional use. The overall electrical efficiency of this system achieves nearly 90% with exhaust heat recovery. Other company such as Toto Ltd. (Japan) also had introduced such SOFC-CHP units which using cathode-supported tubular cell. After all, numerous organizations such as Hexis (Switzerland), Ceres Power (UK), Ceramic Fuel Cell Ltd. (Australia), and Siemen (USA) had commercialized this SOFC-CHP technology practical for residential use either in planar- or tubular-based SOFC configurations.

SOFC technology proved its high level of energy efficiency and acquisition when it could be used as a power generation system. SOFC power generation system firstly designs and builts in early 1984 (Singhal 2000). By referring to Singhal (2000), this system had been installed and worked at the Southern California Edison Company's Highgrove Generating Station in Grand Terrace (near San Bernardino), California, which successfully run for almost 5582 with five times thermal cycles endurance. Apart from that, Mitsubishi Hitachi Power system had brought out a hybrid SOFC-micro

gas turbine system. For the record, in September 2013, model 10 hybrid power generation system can operate for 4100 h during tested at Tokyo Gas Co., Ltd. Senju Techno Station; while model 15 hybrid power generation system can achieve 10,000 h operation when demonstrate at the Kyushu University Ito Campus on October 2016.

On the other hands, PEMFC system usually applied to the vehicle application, low power generation of the portable unit along with combined heat and power unit. Mostly, PEMFC technology would enforce on the transportation field or fuel cell vehicle (FVC). However, there are still having extensive focus on stationary and portable applications. PEMFCs promise divers advantages, including low operating temperature, long stack life, quick start-up, and compactness which lead to the launched on numerous applications, technologies along with prototype by different companies with different backgrounds such as fuel cell technology (Ballard, Plug Power, Smart fuel cell, Toshiba, NovArs, and Hydrogenics), (Ford, Nissan, Hyundai, BMW, Toyota, and, Renault), and electricity-based company (IBM, Samsung, and NTT) (Wee 2007).

In Taiwan and China, electric-powered bicycle is being the daily basic transportation. 20–250 kW is needed to power up the electric car, electric bus as well as utility vehicle, while 1–50 mW are required for stationary application and 100–1 kW generally for small-scale application (Shamim et al. 2015). According to Wee (2007), stack composed of 40 units of PEMFC could generate 378 W of power for electric bicycle with 35% of fuel cell efficiency. Mercedes-Benz Citaro fuel cell buses had been launched in Stockholm city which operate for about 200 km of the journey which produces the total power of 25 kW. By referring to Clean Urban Transportation for Europe (CUTE) program, half of 33 unit buses were run in Europe, 15% in North America and quarter in Asia regions.

In the meantime, PEMFC portable application being another focus in fulfilling the high power demands. Instead of indoor applications like mobile phone, computers, and digital camcorder; PEMFC portable is also widely utilized for outdoor conditions. It can be categorized into three PEMFC stack which are for medium power applications (forerunner stacks Ballard Mk5); automotive usage (Ballard Mk902); and OutdoorFC stack (Oszcipok et al. 2006). In addition, PEMFC could be seen as a good candidate for stationary power plant applications. PEMFC also being a great alternative technology in coping with environmental pollution and energy crises came from the combustion of fossil fuel for energy sources (Shamim et al. 2015). Altergy (USA), Ebara Ballard (Japan), P21 (Germany), and ClearEdge (USA) are among of the companies that contribute to the development and commercialization of small stationary PEMFC power plants.

Meanwhile, DMFC applications are mostly focused on the small vehicles system since it can only generate a small amount of power which make it not ideal for powering large scale of applications. Anyhow, this DMFC system could produce power for long period of time, which can supply up to 25–5 W for 100 h of operations as long as having the fuel supply. Like PEMFC, DMFC also mainly assists in the transportation or vehicle industry technology. This technology has huge potential in competing with internal combustion engine vehicles (ICEVs) fuelled by fossil fuel in terms of cost and performance as well as reducing global warming issues (Shukla et al. 1998). Besides, it also can be implemented in military applications, man-portable tactical equipment and battery chargers.

A company from Germany, Smart Fuel Cell Inc. had developed a small unit of the DMFC system with power supply of 15–150 W. Furthermore, electronic companies like Toshiba Corporation, Sony, Samsung, and MTI had stated their small development of DMFC system that applied in particular portable electronics application and some exhibits as prototypes (Narayan and Valdez 2008). In the same papers, it also stated that 1 kW of the DMFC power source had been designed and manufactured by Oorja Protonics Inc. called ad OorjaPacTM. This system works as a battery charger for vehicles and grants about 20 kWh of energy per six gallons of methanol, which provides sufficient energy for a day operation. For all that, it yields about 20% of fuel to electric efficiency. In fact, various of DMFC system had been introduced such as SFC Jenny (480 Wh), EFOY 1600-M5 (4500 Wh), and EFOY 1600-M28 (25,200 Wh). Although the energy efficiency of the DMFC system is relatively low, progress in the development of the mobile DMFC at either the research or commercialization scale has been continued to this day.

5 Potential and Future Direction

Due to the high operating temperature of SOFC that is in the region of hydrocarbon reforming, common direction of research now is aimed at utilizing readily available hydrocarbon fuel such as methane in natural gas and biogas, propane, butane from butane canister and heavier hydrocarbon such as octane and kerosene from vehicle fuel. The effort is made in finding reforming catalyst and fabricating reforming layer on top of the anode to be used as internal reformer or fixing external reformer to preprocess the fuel (Sengodan et al. 2018). Metal oxide catalyst such as $Ni-Al_2O_3$ and $Ru-CeO_2$ with various doping is actively research as internal reformer for SOFC. Other than that metal–cermet also shows promising catalytic behavior for direct oxidation of hydrocarbon fuel in the anode layer. This enables the hydrocarbon to be used directly in the anode layer with reforming layer (Mahato et al. 2015).

Another direction for SOFC research is into lowering the operating temperature of the SOFC. As the function of SOFC is mainly dictated by the ceramic electrolyte membrane that control the passing of O_2-ion, finding O_2-ceramic material that operates at low temperature is actively done. The current GDC operating at 500–800 °C classed as intermediate temperature SOFC (IT-SOFC) was used electrolyte compared to YSZ operating at 800–1000 °C. Nonetheless, GDC is also known to undergo reduction reaction under reducing hydrogen environment, causing the Ce(IV) to be reduced to Ce(III) which creates electron hole in the GDC structure due to the mixture of Ce(IV)/Ce(III) species. This made the GDC to obtain electronic conductivities, turning the material to MIEC which have electronic current leakage in the electrolyte reducing the OCV and power densities. This problem is currently researched to be resolved by adding intermediate layer between the electrolyte and cathode to avoid contact with reducing gas and is subject to under research. Apart from that other material such from perovskite class such as $La_{0.8}Sr_{0.2}Ga_{0.8}Mg_{0.2}O_{3-\delta}$ (LSGM), and bismuth vanadate, $BiVO_3$ is also actively research for application as IT-SOFC (Mahato et al. 2015). The prospect of lowering the temperature of SOFC to 500 °C is attractive due to the main challenge of finding material compatible with high temperature. Lowering the temperature of SOFC will allow better mechanical strength and compatibility among material such as anode, electrolyte, cathode, and the interconnect.

Meanwhile, the development of electrolyte membrane layer in PEMFC has managed to show remarkable improvement over recent years. While the membrane layer in PEMFC originally used commercial and high-cost Nafion membrane, the development of other membrane as alternative had produced similar or better result. The practice of doping of low-cost membrane such as PVA, sPAES, sPEEK, and PBI with acid such as phosphoric acid has shown increase in proton conductivities as well as hydration of the membrane (Kim et al. 2015; Lade et al. 2017). Composite membrane consisting of the mixture of the said polymer has also been the further researched. Compositing membrane will enable the combination of the best properties such as mechanical and chemical strength, thermal stability and gas tightness as well as lowering the cost. To enable this compositing, new fabrication method such as sputtering and electrospunning should be further researched as well as this enables the creation of layered electrolyte membrane.

High-temperature PEMFC (HT-PEMFC) that enables the operation of PEMFC at higher than 120 °C has received attention due to the fact the higher temperature will favor better kinetics for the hydrogenation oxidation reaction, allowing for lesser use of the expensive Pt catalyst, thus reducing the cost (Rosli et al. 2017). Lower thermal resistance Nafion membrane (<100 °C) can be avoided by using acid-doped PBI membrane or composite membrane consisting of organic/inorganic component such as fluorinated polymer/SiO_2, PWA/PVA, and polyalkoxysilane/phosphotungstic acid composite. The composite membrane by far is the most promising candidate for the HT-PEMFC.

Generally, the DMFC may use the same membrane electrolyte as PEMFC however with added catalyst layer at the electrode. Since the inception DMFC traditionally uses diluted methanol at concentration <1 M, effort now has focused on using higher concentration methanol with concentration up to 4 M. The higher concentration methanol enables higher energy density fuel storage, thus increasing the prospect of DMFC as mobile power source. The challenge associated with the effort such as methanol crossover is being researched actively. Methanol crossover from the anode to the cathode reduces fuel efficiency and caused mixed potential at the cathode (Li and Faghri 2013). This problem may be controlled by using hydrophobic PTFE coating at the electrode side, by employing gas diffusion layer or by using vapor fed methanol fuel system. Vapor fed DMFC has gained interest in recent years. In vapor fed DMFC, the methanol fuel is vaporized first by external heater or by pervaporation membrane and fed into the anode of the SOFC in vapor phase. At the anode, the principle operation of oxidation is the same as the liquid DMFC (Mallick et al. 2016).

In both PEMFC and DMFC, control of the carbon dioxide and water produced from the operation has been identified to play crucial role in controlling the performance of the fuel cell. Thus, research now targeted at improving the management of the water at the electrode by improving the GDL and the back layer (Majlan et al. 2018).

6 Conclusion

The implementation of membrane electrolyte in the fuel cell has been extensively researched. Membrane technology places an enormous benefit over prevalent products in terms of efficiency. The focus has already been well-founded on producing PEM for PEMFC and DMFC as well as ceramic membrane for SOFC. All the accomplishments reported in the fuel cell technology so far have been due to ongoing studies on the membrane base fuel cell and generally renewable energy. Improvement on catalysts, MEA elements, and bipolar plates is notably essential for PEMFCs and DMFCs to overcome the two significant obstacles to marketing (i.e., durability and price). While it is crucial for

SOFCs to improve the anode side in order to overcome the phenomenon of carbon deposition before using hydrocarbon such as methane in MT-SOFC, thus, further study on the utilization of membrane electrolyte in fuel cell is still needed to improve the overall fuel cell efficiency.

References

Ahmad, M. M., Kamarudin, S. K., Daud, W. R. W., & Yaakub, Z. (2010). High power passive μDMFC with low catalyst loading for small power generation. *Energy Conversion and Management, 51,* 821–825. https://doi.org/10.1016/j.enconman.2009.11.017.

Akdeniz, Y., Timurkutluk, B., & Timurkutluk, C. (2016). Development of anodes for direct oxidation of methane fuel in solid oxide fuel cells. *International Journal of Hydrogen Energy, 41,* 10021–10029. https://doi.org/10.1016/j.ijhydene.2016.03.169.

Allen, R. G., Lim, C., Yang, L. X., Scott, K., & Roy, S. (2005). Novel anode structure for the direct methanol fuel cell. *Journal of Power Sources, 143,* 142–149. https://doi.org/10.1016/j.jpowsour.2004.11.038.

Amirinejad, M., Madaeni, S. S., Lee, K.-S., Ko, U., Rafiee, E., & Lee, J.-S. (2012). Sulfonated poly(arylene ether)/heteropolyacids nanocomposite membranes for proton exchange membrane fuel cells. *Electrochimica Acta, 62,* 227–233. https://doi.org/10.1016/j.electacta.2011.12.025.

Amjadi, M., Rowshanzamir, S., Peighambardoust, S. J., Hosseini, M. G., & Eikani, M. H. (2010). Investigation of physical properties and cell performance of Nafion/TiO_2 nanocomposite membranes for high temperature PEM fuel cells. *International Journal of Hydrogen Energy, 35,* 9252–9260. https://doi.org/10.1016/j.ijhydene.2010.01.005.

Ariono, D., Khoiruddin, S., & Wenten, I. G. (2017). Heterogeneous structure and its effect on properties and electrochemical behavior of ion-exchange membrane. *Materials Research Express, 4,* 024006. https://doi.org/10.1088/2053-1591/aa5cd4.

Awang, N., Ismail, A. F., Jaafar, J., Matsuura, T., Junoh, H., Othman, M. H. D., & Rahman, M. A. (2015). Functionalization of polymeric materials as a high performance membrane for direct methanol fuel cell: A review. *Reactive & Functional Polymers, 86,* 248–258. https://doi.org/10.1016/j.reactfunctpolym.2014.09.019.

Azzolini, A., Sglavo, V. M., & Downs, J. A. (2015). Production and performance of copper-based anode-supported SOFCs. *Journal of the Electrochemical Society, 68,* 2583–2596.

Bochentyn, B., Chlipała, M., Gazda, M., Wang, S. F., & Jasiński, P. (2019). Copper and cobalt co-doped ceria as an anode catalyst for DIR-SOFCs fueled by biogas. *Solid State Ionics, 330,* 47–53. https://doi.org/10.1016/j.ssi.2018.12.007.

Brandão, L., Boaventura, M., & Ribeirinha, P. (2012). Single wall nanohorns as electrocatalyst support for vapour phase high temperature DMFC. *International Journal of Hydrogen Energy, 37,* 19073–19081. https://doi.org/10.1016/j.ijhydene.2012.09.133.

Cho, E.-B., Luu, D. X., & Kim, D. (2010). Enhanced transport performance of sulfonated mesoporous benzene-silica incorporated poly(ether ether ketone) composite membranes for fuel cell application. *Journal of Membrane Science, 351,* 58–64. https://doi.org/10.1016/j.memsci.2010.01.028.

Chun, J. H., Kim, S. G., Lee, J. Y., Hyeon, D. H., Chun, B.-H., Kim, S. H., & Park, K. T. (2013). Crosslinked sulfonated poly(arylene ether sulfone)/silica hybrid membranes for high temperature proton exchange membrane fuel cells. *Renewable Energy, 51,* 22–28. https://doi.org/10.1016/j.renene.2012.09.005.

Depernet, D., Narjiss, A., Gustin, F., Hissel, D., & Péra, M. C. (2016). Integration of electrochemical impedance spectroscopy functionality in proton exchange membrane fuel cell power converter. *International Journal of Hydrogen Energy, 41,* 5378–5388. https://doi.org/10.1016/j.ijhydene.2016.02.010.

Devrim, Y., Erkan, S., Baç, N., & Eroğlu, I. (2009). Preparation and characterization of sulfonated polysulfone/titanium dioxide composite membranes for proton exchange membrane fuel cells. *International Journal of Hydrogen Energy, 34,* 3467–3475. https://doi.org/10.1016/j.ijhydene.2009.02.019.

Du, Y., & Sammes, N. M. (2004). Fabrication and properties of anode-supported tubular solid oxide fuel cells. *Journal of Power Sources, 136,* 66–71. https://doi.org/10.1016/j.jpowsour.2004.05.028.

Fatyeyeva, K., Bigarré, J., Blondel, B., Galiano, H., Gaud, D., Lecardeur, M., & Poncin-Epaillard, F. (2011). Grafting of p-styrene sulfonate and 1,3-propane sultone onto Laponite for proton exchange membrane fuel cell application. *Journal of Membrane Science, 366,* 33–42. https://doi.org/10.1016/j.memsci.2010.09.023.

Harris, J., Lay-Grindler, E., Metcalfe, C., & Kesler, O. (2017). Degradation of metal-supported cells with Ni-YSZ or Ni-Ni$_3$Sn-YSZ anodes operated with methane-based fuels. *ECS Transactions, 78,* 1293–1304. https://doi.org/10.1149/07801.1293ecst.

He, Y.-L., Miao, Z., & Yang, W.-W. (2012). Characteristics of heat and mass transport in a passive direct methanol fuel cell operated with concentrated methanol. *Journal of Power Sources, 208,* 180–186. https://doi.org/10.1016/j.jpowsour.2012.02.033.

Heinzel, A., & Barragán, V. M. (1999). A review of the state-of-the-art of the methanol crossover in direct methanol fuel cells. *Journal of Power Sources, 84,* 70–74. https://doi.org/10.1016/S0378-7753(99)00302-X.

Horri, B. A., Selomulya, C., & Wang, H. (2012). Characteristics of Ni/YSZ ceramic anode prepared using carbon microspheres as a pore former. *International Journal of Hydrogen Energy, 37,* 15311–15319. https://doi.org/10.1016/j.ijhydene.2012.07.108.

Hua, D., Li, G., Lu, H., Zhang, X., & Fan, P. (2018). Investigation of carbon formation on Ni/YSZ anode of solid oxide fuel cell from CO disproportionation reaction. *International Communications in Heat and Mass Transfer, 91,* 23–29. https://doi.org/10.1016/j.icheatmasstransfer.2017.11.014.

Ilbeygi, H., Ismail, A. F., Mayahi, A., Nasef, M. M., Jaafar, J., & Jalalvandi, E. (2013). Transport properties and direct methanol fuel cell performance of sulfonated poly (ether ether ketone)/Cloisite/triaminopyrimidine nanocomposite polymer electrolyte membrane at moderate temperature. *Separation and Purification Technology, 118,* 567–575. https://doi.org/10.1016/j.seppur.2013.07.044.

Irvine, J. T. S., & Connor, P. (2013). *Solid oxide fuels cells: Facts and figures.* London: Springer.

Jamil, S. M., Othman, M. H. D., Mohamed, M. H., Adam, M. R., Rahman, M. A., Jaafar, J., & Ismail, A. F. (2018). A novel single-step fabrication anode/electrolyte/cathode triple-layer hollow fiber micro-tubular SOFC. *International Journal of Hydrogen Energy, 43,* 18509–18515. https://doi.org/10.1016/j.ijhydene.2018.08.010.

Jamil, S. M., Othman, M. H. D., Rahman, M. A., Jaafar, J., Ismail, A. F., Honda, S., & Iwamoto, Y. (2019). Properties and performance evaluation of dual-layer ceramic hollow fiber with modified electrolyte for MT-SOFC. *Renew Energy, 134,* 1423–1433. https://doi.org/10.1016/j.renene.2018.09.071.

Jamil, S. M., Othman, M. H. D., Rahman, M. A., Jaafar, J., Ismail, A. F., & Li, K. (2015). Recent fabrication techniques for micro-tubular solid oxide fuel cell support: A review. *Journal of the European Ceramic Society, 35,* 1–22. https://doi.org/10.1016/j.jeurceramsoc.2014.08.034.

Jheng, L., Huang, C., & Hsu, S. L. (2013). Sulfonated MWNT and imidazole functionalized MWNT/polybenzimidazole composite membranes for high-temperature proton exchange membrane fuel cells. *International Journal of Hydrogen Energy, 38,* 1524–1534. https://doi.org/10.1016/j.ijhydene.2012.10.111.

Jin, Y. C., Nishida, M., Kanematsu, W., & Hibino, T. (2011). An H$_3$PO$_4$-doped polybenzimidazole/Sn$_{0.95}$Al$_{0.05}$P$_2$O$_7$ composite membrane for high-temperature proton exchange membrane fuel cells. *Journal of Power Sources, 196,* 6042–6047. https://doi.org/10.1016/j.jpowsour.2011.03.094.

Jörissen, L., Gogel, V., Kerres, J., & Garche, J. (2002). New membranes for direct methanol fuel cells. *Journal of Power Sources, 105,* 267–273. https://doi.org/10.1016/S0378-7753(01)00952-1.

Kannan, R., Aher, P. P., Palaniselvam, T., Kurungot, S., Kharul, U. K., & Pillai, V. K. (2010). Artificially designed membranes using phosphonated multiwall carbon nanotube−polybenzimidazole composites for polymer electrolyte fuel cells. *The Journal of Physical Chemistry Letters, 1,* 2109–2113. https://doi.org/10.1021/jz1007005.

Kaur, G., & Basu, S. (2015). Physical characterization and electrochemical performance of copper-iron-ceria-YSZ anode-based SOFCs in H$_2$ and methane fuels. *International Journal of Energy Research, 39,* 1345–1354. https://doi.org/10.1002/er.3332.

Kim, D. J., Jo, M. J., & Nam, S. Y. (2015). A review of polymer–nanocomposite electrolyte membranes for fuel cell application. *Journal of Industrial and Engineering Chemistry, 21,* 36–52. https://doi.org/10.1016/j.jiec.2014.04.030.

Lade, H., Kumar, V., Arthanareeswaran, G., & Ismail, A. F. (2017). Sulfonated poly(arylene ether sulfone) nanocomposite electrolyte membrane for fuel cell applications: A review. *International Journal of Hydrogen Energy, 42,* 1063–1074. https://doi.org/10.1016/j.ijhydene.2016.10.038.

Lee, D. C., Yang, H. N., Park, S. H., & Kim, W. J. (2014). Nafion/graphene oxide composite membranes for low humidifying polymer electrolyte membrane fuel cell. *Journal of Membrane Science, 452,* 20–28. https://doi.org/10.1016/j.memsci.2013.10.018.

Lee, J. G., Jeon, O. S., Hwang, H. J., Jang, J., Lee, Y., Hyun, S. H., & Shul, Y. G. (2016). Durable and high-performance direct-methane fuel cells with coke-tolerant ceria-coated Ni catalysts at reduced temperatures. *Electrochimica Acta, 191,* 677–686. https://doi.org/10.1016/j.electacta.2016.01.091.

Li, C., Li, C., Xing, Y., Gao, M., & Yang, G. (2006). Influence of YSZ electrolyte thickness on the characteristics of plasma-sprayed cermet supported tubular SOFC. *Solid State Ionics, 177,* 2065–2069. https://doi.org/10.1016/j.ssi.2006.03.004.

Li, P., Wang, Z., Yao, X., Hou, N., Fan, L., Gan, T., et al. (2019). Effect of Sn addition on improving the stability of Ni-Ce$_{0.8}$Sm$_{0.2}$O$_{1.9}$ anode material for solid oxide fuel cells fed with dry CH$_4$. *Catalysis Today, 330,* 209–216. https://doi.org/10.1016/j.cattod.2018.04.030.

Li, X., & Faghri, A. (2011). Effect of the cathode open ratios on the water management of a passive vapor-feed direct methanol fuel cell fed with neat methanol. *Journal of Power Sources, 196,* 6318–6324. https://doi.org/10.1016/j.jpowsour.2011.03.047.

Li, X., & Faghri, A. (2013). Review and advances of direct methanol fuel cells (DMFCs) part I: Design, fabrication, and testing with high concentration methanol solutions. *Journal of Power Sources, 226,* 223–240. https://doi.org/10.1016/j.jpowsour.2012.10.061.

Li, X., Faghri, A., & Xu, C. (2010). Water management of the DMFC passively fed with a high-concentration methanol solution. *International Journal of Hydrogen Energy, 35,* 8690–8698. https://doi.org/10.1016/j.ijhydene.2010.05.033.

Litster, S., & McLean, G. (2004). PEM fuel cell electrodes. *Journal of Power Sources, 130,* 61–76. https://doi.org/10.1016/j.jpowsour.2003.12.055.

Lobato, J., Cañizares, P., Rodrigo, M. A., Úbeda, D., & Pinar, F. J. (2011a). A novel titanium PBI-based composite membrane for high temperature PEMFCs. *Journal of Membrane Science, 369,* 105–111. https://doi.org/10.1016/j.memsci.2010.11.051.

Lobato, J., Cañizares, P., Rodrigo, M. A., Úbeda, D., & Pinar, F. J. (2011b). Enhancement of the fuel cell performance of a high temperature proton exchange membrane fuel cell running with titanium composite polybenzimidazole-based membranes. *Journal of Power Sources, 196,* 8265–8271. https://doi.org/10.1016/j.jpowsour.2011.06.011.

Mahato, N., Banerjee, A., Gupta, A., Omar, S., & Balani, K. (2015). Progress in material selection for solid oxide fuel cell technology: A review. *Progress in Materials Science, 72,* 141–337. https://doi.org/10.1016/j.pmatsci.2015.01.001.

Majlan, E. H., Rohendi, D., Daud, W. R. W., Husaini, T., & Haque, M. A. (2018). Electrode for proton exchange membrane fuel cells: A review. *Renewable and Sustainable Energy Reviews, 89,* 117–134. https://doi.org/10.1016/j.rser.2018.03.007.

Mallick, R. K., Thombre, S. B., & Shrivastava, N. K. (2016). Vapor feed direct methanol fuel cells (DMFCs): A review. *Renewable and Sustainable Energy Reviews, 56,* 51–74. https://doi.org/10.1016/j.rser.2015.11.039.

Mamlouk, M., Scott, K., & Hidayati, N. (2011). High temperature direct methanol fuel cell based on phosphoric acid PBI membrane. *Journal of Fuel Cell Science and Technology, 8,* 061009. https://doi.org/10.1115/1.4004557.

Mehta, V., & Cooper, J. S. (2003). Review and analysis of PEM fuel cell design and manufacturing. *Journal of Power Sources, 114,* 32–53. https://doi.org/10.1016/S0378-7753(02)00542-6.

Meng, X., Gong, X., Yang, N., Tan, X., Yin, Y., & Ma, Z. F. (2013). Fabrication of Y_2O_3-stabilized-ZrO_2(YSZ)/$La_{0.8}Sr_{0.2}MnO_3$-α-YSZ dual-layer hollow fibers for the cathode-supported micro-tubular solid oxide fuel cells by a co-spinning/co-sintering technique. *Journal of Power Sources, 237,* 277–284. https://doi.org/10.1016/j.jpowsour.2013.03.026.

Meng, X., Gong, X., Yang, N., Yin, Y., & Tan, X. (2014). Carbon-resistant Ni-YSZ/Cu–CeO_2-YSZ dual-layer hollow fiber anode for micro tubular solid oxide fuel cell. *International Journal of Hydrogen Energy, 39,* 3879–3886. https://doi.org/10.1016/j.ijhydene.2013.12.168.

Meng, X., Yang, N., Gong, X., Yin, Y., Ma, Z.-F.F., Tan, X., et al. (2015). Novel cathode-supported hollow fibers for light weight micro-tubular solid oxide fuel cells with an active cathode functional layer. *Journal of Materials Chemistry A, 3,* 1017–1022. https://doi.org/10.1039/C4TA04635H.

Moghaddam, R. B., & Easton, E. B. (2018). Impedance spectroscopy assessment of catalyst coated Nafion assemblies for proton exchange membrane fuel cells. *Electrochimica Acta, 292,* 292–298. https://doi.org/10.1016/j.electacta.2018.09.163.

Narayan, S. R., & Valdez, T. I. (2008). High-energy portable fuel cell power sources. *Electrochemical Society Interface, 17,* 40–45.

Ogungbemi, E., Ijaodola, O., Khatib, F. N., Wilberforce, T., El Hassan, Z., Thompson, J., et al. (2019). Fuel cell membranes—Pros and cons. *Energy, 172,* 155–172. https://doi.org/10.1016/j.energy.2019.01.034.

Omar, A. F., Othman, M. H. D., Gunaedi, C. N., Jamil, S. M., Mohamed, M. H., Jaafar, J., et al. (2018). Performance analysis of hollow fibre-based micro-tubular solid oxide fuel cell utilising methane fuel. *International Journal of Hydrogen Energy.* https://doi.org/10.1016/j.ijhydene.2018.03.107.

Ong, B. C., Kamarudin, S. K., & Basri, S. (2017). Direct liquid fuel cells: A review. *International Journal of Hydrogen Energy, 42,* 10142–10157. https://doi.org/10.1016/j.ijhydene.2017.01.117.

Oszcipok, M., Zedda, M., Hesselmann, J., Huppmann, M., Wodrich, M., Junghardt, M., & Hebling, C. (2006). Portable proton exchange membrane fuel-cell systems for outdoor applications. *Journal of Power Sources, 157,* 666–673. https://doi.org/10.1016/j.jpowsour.2006.01.005.

Panthi, D., Choi, B., Du, Y., & Tsutsumi, A. (2017a). Lowering the co-sintering temperature of cathode–electrolyte bilayers for micro-tubular solid oxide fuel cells. *Ceramics International, 43,* 10698–10707. https://doi.org/10.1016/j.ceramint.2017.05.003.

Panthi, D., Choi, B., & Tsutsumi, A. (2017b). Direct methane operation of a micro-tubular solid oxide fuel cell with a porous zirconia support. *Journal of Solid State Electrochemistry, 21*(1), 255–262. https://doi.org/10.1007/s10008-016-3366-5.

Park, K. T., Kim, S. G., Chun, J. H., Jo, D. H., Chun, B.-H., Jang, W. I., et al. (2011). Composite membranes based on a sulfonated poly (arylene ether sulfone) and proton-conducting hybrid silica particles for high temperature PEMFCs. *International Journal of Hydrogen Energy, 36,* 10891–10900. https://doi.org/10.1016/j.ijhydene.2011.05.151.

Peighambardoust, S. J., Rowshanzamir, S., & Amjadi, M. (2010). Review of the proton exchange membranes for fuel cell applications. *International Journal of Hydrogen Energy, 35,* 9349–9384. https://doi.org/10.1016/j.ijhydene.2010.05.017.

Pivac, I., & Barbir, F. (2016). Inductive phenomena at low frequencies in impedance spectra of proton exchange membrane fuel cells—A review. *Journal of Power Sources, 326,* 112–119. https://doi.org/10.1016/j.jpowsour.2016.06.119.

Rabuni, M. F., Li, T., Punmeechao, P., & Li, K. (2018). Electrode design for direct-methane micro-tubular solid oxide fuel cell (MT-SOFC). *Journal of Power Sources, 384,* 287–294. https://doi.org/10.1016/j.jpowsour.2018.03.002.

Radenahmad, N., Afif, A., Petra, P. I., Rahman, S. M. H., Eriksson, S.-G., & Azad, A. K. (2016). Proton-conducting electrolytes for direct methanol and direct urea fuel cells—A state-of-the-art review. *Renewable and Sustainable Energy Reviews, 57,* 1347–1358. https://doi.org/10.1016/j.rser.2015.12.103.

Rosli, R. E., Sulong, A. B., Daud, W. R. W., Zulkifley, M. A., Husaini, T., Rosli, M. I., et al. (2017). A review of high-temperature proton exchange membrane fuel cell (HT-PEMFC) system. *International Journal of Hydrogen Energy, 42,* 9293–9314. https://doi.org/10.1016/j.ijhydene.2016.06.211.

Santasalo-Aarnio, A., Borghei, M., Anoshkin, I. V., Nasibulin, A. G., Kauppinen, E. I., Ruiz, V., & Kallio, T. (2012). Durability of different carbon nanomaterial supports with PtRu catalyst in a direct methanol fuel cell. *International Journal of Hydrogen Energy, 37,* 3415–3424. https://doi.org/10.1016/j.ijhydene.2011.11.009.

Sarruf, B. J. M., Hong, J.-E., Steinberger-Wilckens, R., & de Miranda, P. E. V. (2017). Double layered CeO_2-Co_3O_4-CuO based anode for direct utilisation of methane or ethanol in SOFC. *ECS Transactions, 78,* 1343–1351. https://doi.org/10.1149/07801.1343ecst.

Sarruf, B. J. M., Hong, J. E., Steinberger-Wilckens, R., & de Miranda, P. E. V. (2018). CeO_2–Co_3O_4–CuO anode for direct utilisation of methane or ethanol in solid oxide fuel cells. *International Journal of Hydrogen Energy, 43,* 6340–6351. https://doi.org/10.1016/j.ijhydene.2018.01.192.

Sengodan, S., Lan, R., Humphreys, J., Du, D., Xu, W., Wang, H., & Tao, S. (2018). Advances in reforming and partial oxidation of hydrocarbons for hydrogen production and fuel cell applications. *Renewable and Sustainable Energy Reviews, 82,* 761–780. https://doi.org/10.1016/j.rser.2017.09.071.

Shamim, S., Sudhakar, K., Choudhary, B., & Anwar, J. (2015). A review on recent advances in proton exchange membrane fuel cells: Materials, technology and applications. *Advances in Applied Science Research, 6,* 89–100.

Shukla, A. K., Ravikumar, M. K., & Gandhi, K. S. (1998). Direct methanol fuel cells for vehicular applications. *Journal of Solid State Electrochemistry, 2,* 117–122. https://doi.org/10.1007/s100080050075.

Singhal, S. (2000). Advances in solid oxide fuel cell technology. *Solid State Ionics, 135,* 305–313. https://doi.org/10.1016/S0167-2738(00)00452-5.

Stambouli, A. B., & Traversa, E. (2002). Solid oxide fuel cells (SOFCs): A review of an environmentally clean and efficient source of energy. *Renewable and Sustainable Energy Reviews, 6,* 433–455. https://doi.org/10.1016/S1364-0321(02)00014-X.

Sumi, H., Yamaguchi, T., Shimada, H., Hamamoto, K., Suzuki, T., & Barnett, S. A. (2017). Direct butane utilization on Ni-$(Y_2O_3)_{0.08}(ZrO_2)_{0.92}$-$(Ce_{0.9}Gd_{0.1})O_{1.95}$ composite anode-supported microtubular solid oxide fuel cells. *Electrocatalysis, 8,* 288–293. https://doi.org/10.1007/s12678-017-0369-7.

Sumi, H., Yamaguchi, T., Suzuki, T., & Shimada, H. (2015). Direct hydrocarbon utilization in microtubular solid oxide fuel cells. *Journal of the Ceramic Society of Japan, 08,* 213–216. https://doi.org/10.2109/jcersj2.123.213.

Sung, I. H., Yu, D. M., Yoon, Y. J., Kim, T.-H., Lee, J. Y., Hong, S. K., & Hong, Y. T. (2013). Preparation and properties of sulfonated poly(arylene ether sulfone)/hydrophilic oligomer-g-CNT composite membranes for PEMFC. *Macromolecular Research, 21,* 1138–1144. https://doi.org/10.1007/s13233-013-1136-0.

Tanaka, Y. (2007). Bipolar membrane electrodialysis. In *Membrane science and technology* (Chap. 3, pp. 405–436).

Teresa, L., & Gámez, M. (2007). Polymer electrolyte membranes for the direct methanol fuel cell: A review. *Journal of Polymer Science, 45,* 2007–2009. https://doi.org/10.1002/POLB.

Thanganathan, U., & Nogami, M. (2014). Proton conductivity and structural properties of precursors mixed PVA/PWA-based hybrid composite membranes. *Journal of Solid State Electrochemistry, 18,* 97–104. https://doi.org/10.1007/s10008-013-2235-8.

Thomas, S. C., Ren, X., Gottesfeld, S., & Zelenay, P. (2002). Direct methanol fuel cells: Progress in cell performance and cathode research. *Electrochimica Acta, 47,* 3741–3748.

Wang, Y., Chen, K. S., Mishler, J., Cho, S. C., & Adroher, X. C. (2011). A review of polymer electrolyte membrane fuel cells: Technology, applications, and needs on fundamental research. *Applied Energy, 88,* 981–1007. https://doi.org/10.1016/j.apenergy.2010.09.030.

Wee, J.-H. (2007). Applications of proton exchange membrane fuel cell systems. *Renewable and Sustainable Energy Reviews, 11,* 1720–1738. https://doi.org/10.1016/j.rser.2006.01.005.

Wu, Q. X., Zhao, T. S., Chen, R., & An, L. (2013). A sandwich structured membrane for direct methanol fuel cells operating with neat methanol. *Applied Energy, 106,* 301–306. https://doi.org/10.1016/j.apenergy.2013.01.016.

Wu, Q. X., Zhao, T. S., Chen, R., & Yang, W. W. (2010). Enhancement of water retention in the membrane electrode assembly for direct methanol fuel cells operating with neat methanol. *International Journal of Hydrogen Energy, 35,* 10547–10555. https://doi.org/10.1016/j.ijhydene.2010.07.178.

Wu, Q. X., Zhao, T. S., & Yang, W. W. (2011). Effect of the cathode gas diffusion layer on the water transport behavior and the performance of passive direct methanol fuel cells operating with

neat methanol. *International Journal of Heat and Mass Transfer, 54,* 1132–1143. https://doi.org/10.1016/j.ijheatmasstransfer.2010.11.009.

Wu, X., Tian, Y., Zhang, J., Zuo, W., Kong, X., Wang, J., et al. (2016). Enhanced electrochemical performance and carbon anti-coking ability of solid oxide fuel cells with silver modified nickel-yttrium stabilized zirconia anode by electroless plating. *Journal of Power Sources, 301.* https://doi.org/10.1016/j.jpowsour.2015.10.006.

Xu, C., & Faghri, A. (2010). Mass transport analysis of a passive vapor-feed direct methanol fuel cell. *Journal of Power Sources, 195,* 7011–7024. https://doi.org/10.1016/j.jpowsour.2010.05.003.

Xu, C., Faghri, A., & Li, X. (2011a). Improving the water management and cell performance for the passive vapor-feed DMFC fed with neat methanol. *International Journal of Hydrogen Energy, 36,* 8468–8477. https://doi.org/10.1016/j.ijhydene.2011.03.115.

Xu, Q., Zhao, T. S., Yang, W. W., & Chen, R. (2011b). A flow field enabling operating direct methanol fuel cells with highly concentrated methanol. *International Journal of Hydrogen Energy, 36,* 830–838. https://doi.org/10.1016/j.ijhydene.2010.09.026.

Yan, N., Pandey, J., Zeng, Y., Amirkhiz, B. S., Hua, B., Geels, N. J., et al. (2016). Developing a thermal- and coking-resistant cobalt-tungsten bimetallic anode catalyst for solid oxide fuel cells. *ACS Catalysis, 6,* 4630–4634. https://doi.org/10.1021/acscatal.6b01197.

Yan, X. H., Zhao, T. S., Zhao, G., An, L., & Zhou, X. L. (2015). A hydrophilic-hydrophobic dual-layer microporous layer enabling the improved water management of direct methanol fuel cells operating with neat methanol. *Journal of Power Sources, 294,* 232–238. https://doi.org/10.1016/j.jpowsour.2015.06.058.

Yoo, Y., & Lim, N. (2013). Performance and stability of proton conducting solid oxide fuel cells based on yttrium-doped barium cerate-zirconate thin-film electrolyte. *Journal of Power Sources, 229,* 48–57. https://doi.org/10.1016/j.jpowsour.2012.11.094.

Yu, D. M., Yoon, Y. J., Kim, T.-H., Lee, J. Y., & Hong, Y. T. (2013). Sulfonated poly(arylene ether sulfone)/sulfonated zeolite composite membrane for high temperature proton exchange membrane fuel cells. *Solid State Ionics, 233,* 55–61. https://doi.org/10.1016/j.ssi.2012.12.006.

Zainoodin, A. M., Kamarudin, S. K., & Daud, W. R. W. (2010). Electrode in direct methanol fuel cells. *International Journal of Hydrogen Energy, 35,* 4606–4621. https://doi.org/10.1016/j.ijhydene.2010.02.036.

Zhang, H., Liu, Z., Gao, S., & Wang, C. (2014). A new cathode structure for air-breathing DMFCs operated with pure methanol. *International Journal of Hydrogen Energy, 39,* 13751–13756. https://doi.org/10.1016/j.ijhydene.2014.02.145.

Zhang, Z., Yuan, W., Wang, A., Yan, Z., Tang, Y., & Tang, K. (2015). Moisturized anode and water management in a passive vapor-feed direct methanol fuel cell operated with neat methanol. *Journal of Power Sources, 297,* 33–44. https://doi.org/10.1016/j.jpowsour.2015.07.097.

Zhao, X., Yuan, W., Wu, Q., Sun, H., Luo, Z., & Fu, H. (2015). High-temperature passive direct methanol fuel cells operating with concentrated fuels. *Journal of Power Sources, 273,* 517–521. https://doi.org/10.1016/j.jpowsour.2014.09.128.

Zhiani, M., Majidi, S., Silva, V. B., & Gharibi, H. (2016). Comparison of the performance and EIS (electrochemical impedance spectroscopy) response of an activated PEMFC (proton exchange membrane fuel cell) under low and high thermal and pressure stresses. *Energy, 97,* 560–567. https://doi.org/10.1016/j.energy.2015.12.058.

Shear-Enhanced Filtration (SEF) for the Separation and Concentration of Protein

Wenxiang Zhang, Luhui Ding, and Nabil Grimi

Abstract

In this communication, we reviewed the shear-enhanced filtration (SEF) for the separation and concentration of protein. Firstly, the configuration, operation parameter, and anti-fouling capacity for different SEF modules, including rotating disk/rotor, rotating membrane, and vibratory systems, were summarized. Then, two SEF application cases (milk solution filtration and Luzerne juice filtration) were introduced. For milk solution, milk proteins were separated and concentrated. The effects of hydraulic conditions of SEF on filtration performance were investigated by response surface methodology for process parameter optimization. Pore-blocking model was utilized for better understanding the membrane fouling mechanism. Afterwards, the retentate with high milk proteins was successfully applied for cheese production. After filtration, the membrane cleaning process was also conducted by shear effect to improve the efficiency. The other aspect, during the Luzerne juice filtration process, leaf proteins were separated and concentrated. At first, the influence of filtration module structures on filtration performance was studied. Secondly, a stepwise multisite Darcy's law model (SMDM) was proposed to simulate the complex fouling process and calculate the key fouling parameters. At last, the operation conditions were optimized to enhance the filtration efficiency (permeate flux and filtration time) and separation performance (the protein rejection and protein purity in retentate). This work can give a valuable information for the application of SEF for the separation and concentration of protein solution.

Keywords

Shear-enhanced filtration (SEF) • Membrane fouling mechanism • Hydraulic condition optimization • Separation and concentration of protein • Membrane cleaning

1 Introduction

As a considerably effective, flexible (greater flexibility in design and scale-up) and economical process (energy saving and no additives and chemicals required), membrane filtration has been considered as an environmentally friendly separation technique. It has been used in many chemical processes, including purification, clarification, concentration dewatering, and so on. According to pore size, it is classified as microfiltration (MF), ultrafiltration (UF), nanofiltration (NF), and reverse osmosis (RO). Based on the configurations of filtration modules and operating conditions, membrane filtration has two conventional configurations: dead-end filtration (DF) and cross-flow filtration (CF). During membrane filtration process, flux decline caused by membrane fouling and concentration polarization would increase feeding flowrate and mean transmembrane pressure (TMP), then enhancing energy cost and bringing about non-optimal membrane utilization. Shear-enhanced filtration (SEF) can impose high shear rates and perturb the concentration polarization layer on membrane surface for alleviating the membrane fouling, afterwards decreasing filtration resistance and flux decline. Thus, compared with conventional modules, SEF module does not only improve substantially the permeate flux without a much larger inlet flow rate, but also straightening membrane selectivity (Zhang et al. 2015; Jaffrin 2008; Ding et al. 2015).

W. Zhang (✉)
College of Water Conservancy and Civil Engineering, South China Agricultural University, Guangzhou, China
e-mail: Zhangwenxiang6@hotmail.com

Department of Civil and Environmental Engineering, Faculty of Science and Technology, University of Macau, Macau, China

L. Ding · N. Grimi
ESCOM, EA 4297 TIMR, Centre de Recherch Royallieu, Sorbonne University, Université de Technologire de Compiègne, CS 60319, 60203 Compiègne Cedex, France

© Springer Nature Switzerland AG 2021
Z. Zhang et al. (eds.), *Membrane Technology Enhancement for Environmental Protection and Sustainable Industrial Growth*, Advances in Science, Technology & Innovation,
https://doi.org/10.1007/978-3-030-41295-1_9

Typical SEF systems include three types: rotating disk/rotor (Fig. 1a), rotating membrane (Fig. 1b) and vibratory systems (Fig. 1c). Rotating disk/rotor modules produce a high shear rate on the membrane by a disk rotating near a fixed circular membrane (Ding and Jaffrin 2014; Chen et al. 2019). In rotating membrane systems, rotating ceramic membranes create a high shear rate that is orders of magnitude greater than conventional filtration system and provides a high and very stable flow rate on the membrane (Ding et al. 2006). The vibratory systems involve a stack of circular organic membranes separated by gaskets and permeate collectors, installed on a vertical torsion shaft spun in azimuthal oscillations by a vibrating base (Beier et al. 2006). The shear rate on the membrane is created by the inertia of the retentate which moves at 180° out of phase and varies sinusoidally with time controlling concentration polarization and preventing membrane fouling. Compared with DF and CF, SEF has obvious advantages in the following aspects: high permeate flux, favorable membrane selectivity and concentration factor, low inlet flow rate, and great process efficiency for allowing very viscous concentrates and high-water recovery during wastewater recycle. With high shear effect, microsolute transmission and macrosolute rejection reinforce at SEF with MF or UF operation, which facilitates the separation, clarification, and concentration of protein solution. Furthermore, for wastewater treatment by NF or RO, high shear stress decreases the concentration polarization and concentration of rejected solutes at the membrane; thus, the concentration gradient and diffusive solute transfer through the membrane are lowered, as well as solutes rejection rate and permeate quality improve (Jaffrin 2008).

The separation and concentration of proteins have become a challenge for membrane filtration. Because proteins are very serious membrane foulants, meanwhile proteins also bind with other substances to form more dense fouling layers. Especially for the solution with high protein concentration, high protein foulants during its filtration process greatly reduced the filtration performance. Food processing solution contains large quantities of nutritional matters, such as proteins and polysaccharides. Membrane technology is utilized to recover these waste proteins, produce reuse water, and recycle wastewater. Conventional filtration modules are not able to sustain great permeate flux, since the shear stresses on membrane surface for DF and CF are not enough high to alleviate membrane fouling. In order to control membrane fouling and improve flux for a high filtration efficiency during membrane filtration process of protein solution, SEF has been applied for food processing solution filtration. There are some studies about both academic research and industrial applications for SEF of protein solution. This chapter on separation performance, flux behavior, and fouling mechanism for SEF can provide a valuable guideline for SEF application in protein separation and concentration.

Fig. 1 Laboratory pilot module. **a** Rotating disk module in Technological University of Compiegne (membrane area: 460 cm²), **b** multi-shaft rotating ceramic disk membrane system from Westfalia separator (total membrane area 121 cm²) (Jaffrin 2008), and **c** VSEP module from New Logic Research, Inc (membrane area: 500 cm²) (Jaffrin 2012)

2 Recent Applications for SEF of Protein Solution and Case Studies

2.1 Milk Solution Filtration

SEF has been used to separate and concentrate proteins from milk. The study, including the whole membrane industrial process, was divided into several parts: process optimization (Zhang et al. 2014), membrane fouling mechanism (Zhang and Ding 2015), and membrane cleaning (Zhang et al. 2017).

Process Optimization by Response Surface Methodology
The process optimization for SEF with UF separating protein from milk, conducted by Box–Behnken response surface methodology (BBRSM), had three parts: effluent quality, flux behavior, and energy consumption. The effects of hydraulic conditions [feed flow rate (Q), mean transmembrane pressure (TMP), and rotating speed (N)] on effluent quality, flux behavior, and energy consumption were analyzed and optimized. At first, three ultrafiltration (UF) membranes (UH005P, UH030P, and PES050) were used for experiments. PES050 with the largest pore size exhibited the lowest rejection rate, whereas UH005P had the lowest flux, because of the smallest pore size, thus UH030P, with the moderated rejection rate and permeate flux, was the most suitable membrane.

Then, A 1.5 h test was performed to analyze the kinetics of permeate flux decline for BBRSM model. The results for BBRSM design and predicted values at various hydraulic conditions were showed in Tables 1, 2 and 3. The experimental value and predicated value are very similar, due to the high fitting degree (R^2), implying BBRSM could accurately demonstrate the SEF operation. Besides, the second-degree polynomial equations (Eqs. 1–5) were fabricated to simulate the SEF process.

For effluent quality, COD and total protein are the response values. Equations (1) and (2) imply that high shear stress reduced concentration polarization, then less solutes accumulated on membrane surface or entered pores, decreasing the diffusion of solutes through the membrane. Accordingly, high shear effect inhibited the diffusion penetration phenomena. With respect to feed flow rate, its increment enhanced the diffusion penetration process through the membrane. Furthermore, the independent and interaction effects of hydraulic conditions indicated the following order: rotating speed > TMP > feed flow rate > feed flow rate × TMP. There were no clear interaction effects between feed flow rate and rotating speed/TMP and rotating speed.

With respective to flux behavior, average flux (AF) and flux decline (FD) were chosen to evaluate it. Equations (3) and (4) displayed that higher shear rates reduced

concentration polarization phenomena and foulants (milk proteins) accumulation on membrane surface, then obviously eliminating the membrane fouling. Moreover, the effect of TMP had two aspects: at low rotating speed, great TMP enhanced concentration polarization, because of low shear-enhanced back transport. However, the elevation of TMP did not conduct an increment of membrane fouling at high rotating speed, because the strong shear effect reduced the concentration polarization. In addition, with the growth of feed flow rate, more retained foulants (milk proteins) accumulated and compacted fouling layer formed.

As for energy consumption (Eq. 5), the significance of parameters on energy consumption is generally TMP > rotating speed > TMP × rotating speed > feed flow rate. TMP, the rotating speed and their interaction effect played a distinct impact on energy consumption, while energy consumption emerged a linearly elevating relationship with rotating speed and reducing clearly with TMP.

In Table 4, BBRSM was also utilized to optimize hydraulic factors for improving effluent quality and flux behavior, as well reducing energy consumption. The optimal hydraulic conditions were: $Q = 75.81$ L/h, TMP = 7 bar, and $N = 2250$ rpm. At these conditions, the highest effluent quality, best flux behavior, and moderated energy consumption were obtained.

$$\begin{aligned} COD = 9.392 &- 0.02725Q \\ &+ 0.173TMP - 0.00057N \\ &- 0.00033Q*TMP \end{aligned} \quad (1)$$

$$\begin{aligned} Total\,Protein = 3.996 &- 0.01345Q \\ &- 0.105TMP + 0.000186N \\ &+ 0.0001Q*TMP \end{aligned} \quad (2)$$

$$\begin{aligned} Flux = -6.58253 &- 0.55357Q \\ &+ 23.4652TMP + 0.024286N \\ &+ 0.022533Q*TMP \end{aligned} \quad (3)$$

$$\begin{aligned} Flux\,Decline = 23.65802 &- 0.06566Q \\ &+ 0.6484TMP - 0.00408N \\ &- 0.0054Q*TMP + 0.000117TMP*N \end{aligned}$$
$$(4)$$

$$\begin{aligned} Energy\,cost = 86.6858 &- 1.77773Q \\ &+ 18.6052TMP - 0.01157N \\ &+ 0.0492Q*TMP + 0.006381TMP*N \\ &+ 0.000123TMP*N \end{aligned}$$
$$(5)$$

W. Zhang et al.

Table 1 Experimental and predicted values of COD and total protein

Runs	Variable			Effluent quality			
	Q (L/h)	TMP (bar)	N (rpm)	COD (mg/L)		Total protein (mg/L)	
				Experimental	Predicted	Experimental	Predicted
1	60	2	1500	7.7	7.7	2.72	2.69
2	120	2	1500	8.0	8.0	2.95	2.89
3	60	7	1500	7.5	7.5	2.02	2.08
4	120	7	1500	7.7	7.7	2.28	2.31
5	60	4.5	750	8.2	8.2	3.06	2.99
6	120	4.5	750	8.4	8.4	3.31	3.27
7	60	4.5	2250	7.3	7.3	1.29	1.34
8	120	4.5	2250	7.5	7.5	1.41	1.49
9	90	2	750	8.2	8.2	3.14	3.25
10	90	7	750	7.9	7.9	2.94	2.96
11	90	2	2250	7.2	7.2	1.85	1.84
12	90	7	2250	6.9	6.9	1.04	0.94
13	90	4.5	1500	7.6	7.7	2.31	2.33
14	90	4.5	1500	7.7	7.7	2.36	2.33
15	90	4.5	1500	7.8	7.7	2.32	2.33

Table 2 Experimental and predicted values of average flux, flux decline, and permeability index

Runs	Variable			Flux behavior			
	Q (L/h)	TMP (bar)	N (rpm)	Average flux (L/m^2 h)		Flux decline	
				Experimental	Predicted	Experimental	Predicted
1	60	2	1500	41.32	44.72	12.09	12.46
2	120	2	1500	45.81	50.38	15.24	14.88
3	60	7	1500	125.11	120.54	15.26	15.62
4	120	7	1500	136.36	132.96	16.79	16.42
5	60	4.5	750	79.43	78.12	19.86	20.01
6	120	4.5	750	94.60	92.12	20.59	21.48
7	60	4.5	2250	96.70	99.19	6.41	5.53
8	120	4.5	2250	101.95	103.27	7.43	7.28
9	90	2	750	37.04	34.96	19.84	19.32
10	90	7	750	108.34	114.23	21.75	21.23
11	90	2	2250	57.02	51.14	4.02	4.54
12	90	7	2250	128.18	130.27	6.81	7.33
13	90	4.5	1500	89.64	89.99	14.14	14.19
14	90	4.5	1500	90.09	89.99	14.18	14.19
15	90	4.5	1500	90.25	89.99	14.26	14.19

Membrane Fouling Mechanisms Using Blocking Models

Membrane fouling mechanism is another important challenge for SEF of protein solution. For the sake of better reveling fouling mechanism and controlling membrane fouling, pore-blocking model (Table 5) proposed by Hermia was utilized to fit the experimental data and calculate the blocking coefficient (Zhang and Ding 2015). During the SEF of milk solution, casein micelles (200 nm) and whey proteins (10 nm) are main foulants producing complete pore blocking (Luo et al. 2012). Like a homogeneous matrix in "cherry stones," the casein micelle is fabricated by caseins and colloidal calcium phosphate nanoclusters (Kruif et al. 2012). As casein micelles have a much larger size than UF pore, they could be easily rejected, but for whey proteins, their size is similar with UF pores, thus easily causing pore blocking (Zhang and Ding 2015).

Table 3 Experimental and predicted values of energy consumption

Runs	Variable			Energy consumption (kWh/m³)	
	Q (L/h)	TMP (bar)	N (rpm)	Experimental	Predicted
1	60	2	1500	602.61	588.89
2	120	2	1500	543.55	535.73
3	60	7	1500	199.02	206.84
4	120	7	1500	182.60	196.32
5	60	4.5	750	177.81	175.35
6	120	4.5	750	149.30	140.95
7	60	4.5	2250	453.96	462.31
8	120	4.5	2250	430.58	433.04
9	90	2	750	381.31	397.49
10	90	7	750	130.36	124.99
11	90	2	2250	769.87	775.24
12	90	7	2250	342.47	326.29
13	90	4.5	1500	277.78	276.69
14	90	4.5	1500	276.39	276.69
15	90	4.5	1500	275.90	276.69

Table 4 Predicted and experimental values of the response at optimum conditions

Hydraulic condition			Effluent quality		Flux behavior		Energy consumption (kWh/m³)
Q (L/h)	TMP (bar)	N (rpm)	COD (mg/L)	Total protein (mg/L)	Average flux (L/m² h)	Flux decline	
75.81	7.00	2250.00	6.93^P	0.92^P	129.36^P	7.23^P	328.42^P
75.81	7.00	2250.00	6.86^E	0.89^E	131.21^E	7.34^E	329.89^E

E Experimental value; P Predicted value

Table 5 Four models of membrane fouling proposed by Hermia

Pore-blocking models	n	Linear equation	a	b	Physical concept	Schematic diagram
Complete pore blocking (model 1)	2	$\ln(J) = \ln(a/t) + \ln(b)$	1	$J_0 k_b A_m$	Formation of a surface deposit	
Internal pore blocking (model 2)	1.5	$J^{-1/2} = at + b$	$\dfrac{(2-n)k_s}{\sqrt{A_m}}$	$J_0^{-1/2}$	Pore blocking + surface deposit	
Intermediate pore blocking (model 3)	1	$J^{-1} = at + b$	$(2-n)k_i J_0$	J_0^{-1}	Pore constriction	
Cake formation (model 4)	0	$J^{-2} = at + b$	$(2-n)k_c J_0^2$	J_0^{-2}	Pore blocking	

As displayed in Fig. 2, pore-blocking coefficient enhances with TMP and diminishes with shear rate, showing that both higher shear rate and smaller TMP reduced membrane fouling. With high shear stress, the great shear-induced diffusion created, decreasing the deposition and agglomeration of casein micelles on membrane, as well as the cake layer fabricated by whey protein and calcium ion, thus only some pore blocking produced. Besides, higher TMP caused more obvious concentration polarization, which also reinforcing pore blocking and membrane pore

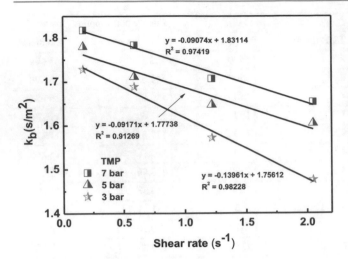

Fig. 2 Effect of shear rate on the complete pore-blocking coefficient. (Temperature = 35 °C and membrane: UH030P)

narrowing. Furthermore, foulants or blocked pores were more compact under high permeate flux and great TMP, enhancing pore-blocking degree.

On another hand, in Fig. 3, the permeability recovery after membrane cleaning has a direct relationship with TMP and shear rate. At high TMP, the permeability recovery elevates linearly with shear rate, implying that irreversible fouling eliminated with high shear effect. Besides, greater TMP enhanced the concentration polarization of casein micelles and caused the compact irreversible fouling of whey proteins on membrane surface. In general, for SEF process, a high shear rate and low TMP can sustain a low irreversible membrane fouling and stable flux operation.

SEF is considered as an important membrane technology that can contribute to pre-treat dairy wastewater and recycle valuable components such as milk proteins. However, to be efficient, it necessitates the establishment of proper methods for the assessment of membrane fouling.

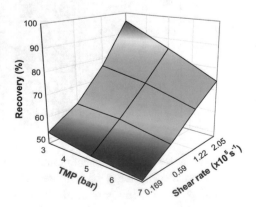

Fig. 3 Effect of experimental shear rate and TMP on membrane permeability recovery for the UH030P membrane

Four membrane blocking models proposed by Hermia were used to quantify and assess the membrane fouling of SEF observed in dairy wastewater treatment. The experiments were performed with various shear rates, mean transmembrane pressure, temperature, and membrane types. As presented in Fig. 4, good agreement between complete pore-blocking model and experimental data was found, confirming the validity of the Hermia models for assessing the membrane fouling of SEF system and that only some "sealing" of membrane pores occurs, which is due to the high shearing effect. Furthermore, the increments of shear rate, TMP, and temperature could decrease the degree of "sealing" of membrane pores and improve the filtration performance. In addition, a three-step membrane cleaning mode had achieved very satisfying results in subsequent membrane cleaning process. This work confirms that, unlike traditional filtration mode (DF and CF), SEF possesses a low degree of membrane fouling and a higher membrane permeability recovery after cleaning.

Concentration Process and Cheese Production

Three kinds of milk (skim milk, whole milk, and dairy factory whole milk) were separated and concentrated by SEF-UF (P010P) to produce high concentration protein for cheese production (Ding et al. 2016). As showed in Fig. 5, the ending VRRs were 10. Below VRR = 2, the permeate fluxes clearly reduces; above VRR = 2, permeate fluxes slightly decrease and become stable. Since at higher VRR, protein and lipid concentrations straightened, causing a higher foulant concentration and thicker fouling layer, causing a greater flux reduction. But with a high shear effect, the flux at VRR = 10 still exceeded 15 L m^{-2} h^{-1}. On the other hand, the Brix in retention elevated with VRR, as more organic matters (proteins and lipids) were separated and concentrated, when the concentration polarization improved, leading to more organic matter through the membrane. Moreover, thicker fouling layer produced by more organic matters significantly increased organic matters rejection with VRR.

After milk concentration tests, the concentrated milk proteins were used for lactic fermentation and cheese production. Figure 6 illustrates two kinds of cheeses produced by different fermentation agents [FDDVS YF-L903 (CHR HANSEN, France) and yogurt (GAZI, France)]. As shown in 6, the actual performances for these cheeses and traditional cheese have similar pH values, but clearly lower concentration for protein, total fat, and solids, implying that the concentration multiple of milk was still insufficient. In addition, the actual performance of cheeses was characterized by appearance (5 points), texture (5 points), and taste (10 points) and estimated by a French cheese company. Yogurts exhibited better performances in texture and taste, compared with FD-DVS YF-L903. However, traditional

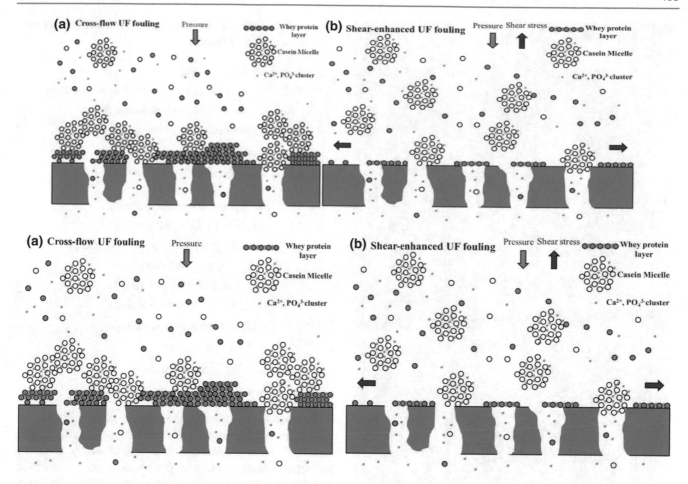

Fig. 4 Schematic diagram of fouling mechanism for wastewater treatment by cross-flow (**a**) and shear-enhanced (**b**) UF system

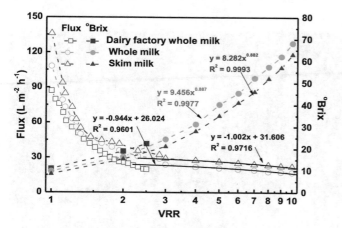

Fig. 5 Variations of permeate flux and Brix in retentate with volume reduction ratio (VRR) during concentration of dairy factory whole milk, whole milk, and skim milk by P010P at $N = 2500$ rpm and TMP = 6 bar

cheese kept a much greater total score. In general, the high concentration capacity of SEF for milk protein is highly beneficial for protein concentration and cheese protein.

Membrane Cleaning

Except for fouling alleviation, high shear stress in SEF also has high potential to assist membrane cleaning (Zhang et al. 2017). In membrane industry, water rinsing and chemical cleaning are two main cleaning technologies. For water rinsing, high shear stress effectively removed casein micelles and whey protein layer accumulating on membrane surface. With respective to chemical cleaning, cleaning agent became the foam status and the convection diffusivity improved by high shear stress, which significantly remove foulants inside membrane pores.

With high shear effect, as displayed in Fig. 7, the form status of cleaning agents (P3-ultrasil 10) containing highly effective cleaning composition, formed and kept the chemical agents on membrane surface long enough for the active compositions to enter the membrane pores and remove the foulants (Zhang et al. 2017; Gahleitner et al. 2013, 2014). Thus, cleaning agent with foam had a better cleaning efficiency. As shown in Fig. 8, the membrane permeability recovery increased with shear stress. High shear stress also strengthened the diffusivity rate of mass transfer process

Fig. 6 Pictures of cheese: **a** FD-DVS YF-L903 and **b** Yogurt

(Ding et al. 2003), promoting the interaction between cleaning agents and foulants.

$$\begin{aligned}
\mathrm{MPR}(\%) \text{ for water rinsing} \\
= 46.23 + 8.55 \times \text{shear stress} + 6.05 \times \text{temperature} \\
+ 4.32 \times \text{time} - 0.14 \times \text{shear stress} \times \text{time} \\
+ 2.45 \times \text{temperature} \times \text{time} \\
- 3.05 \times \text{shear stress}^2 \\
- 6.09 \times \text{temperature}^2 - 1.23 \times \text{time}^2
\end{aligned} \tag{6}$$

$$\begin{aligned}
\mathrm{MPR}(\%) \text{ for chemical cleaning} \\
= 92.51 + 11.55 \times \text{shear stress} \\
+ 21.36 \times \text{temperature} + 0.67 \times \text{time} \\
- 3.91 \times \text{concentration} \\
- 2.2 \times \text{shear stress} \times \text{time} \\
+ 0.33 \times \text{temperature} \times \text{time} \\
- 3.08 \times \text{time} \times \text{concentration} \\
- 4.21 \times \text{shear stress}^2 \\
- 3.7 \times \text{temperature}^2 \\
- 6.96 \times \text{time}^2 - \text{concentration}^2
\end{aligned} \tag{7}$$

In order to investigate the influence of operational conditions, including shear stress, temperature, TMP, cleaning time, and cleaning agent concentration, as well optimize the membrane cleaning process, the response surface methodology was utilized for this process optimization. The mathematical model for explaining operation condition of cleaning process and MPRs is acquired and shown in

Fig. 7 P3-ultrasil 10 solution before (left) and after (right) cleaning (shear stress = 16 Pa and time = 50 min)

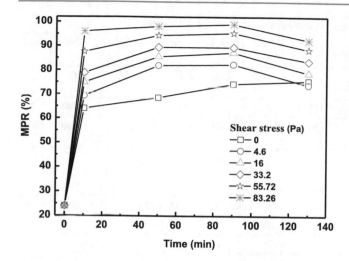

Fig. 8 MPR versus time at various shear stress for P3-ultrasil 10 (Concentration = 0.25%, TMP = 0.12 MPa, and temperature 35 °C)

Eqs. (6) and (7). Shear stress and temperature had a highly positive relationship with MPR; thus, suitable elevating operational shear stress and temperature were conducive to improve membrane cleaning efficiency. Moreover, the process parameters were optimized, while the results are showed in Table 7 and can be used for engineering application.

2.2 Leaf Protein Solution (Luzerne Juice) Filtration

As a high protein plant, Luzerne is considered as raw materials for vegetable protein products. After extraction, separation and concentration, high concentration leaf protein solution is used to produce the high-quality protein product, such as human nutrition and animal feed. Traditional leaf protein product is produced by the coagulation-thermal method. This process of high temperature may cause protein deterioration, and reducing protein quality. Thus, the separation and concentration of leaf protein product need process optimization. Due to the normal temperature operation and no phase transition, membrane technology is a sustainable and highly efficient process for protein separation and can concentrate leaf protein for vegetable protein products. However, protein is a significant membrane foulants, which can seriously reduce membrane filtration rate. Therefore, how to overcome serious membrane fouling and avoid large flux decline determines the effectiveness of membrane technology during leaf protein separation process.

Filtration Modules

Three filtration modules [dead-end filtration using laboratory Amicon cell (DA), dynamic cross filtration using rotating disk module (CRDM), and dead-end filtration using

rotating disk module (DRDM)] with MF and UF were used to investigate the filtration performance during leaf protein separation (Zhang et al. 2015). CRDM and DRDM were two different SEF types. As shown in Fig. 9, the optimized flux behavior (high flux and low flux decline) exhibits the following order: CRDM > DRDM > DA. CRDM and DRDM were able to produce the high shear rate on membrane, due to the disk equipped with vanes; thus, their concentration polarization and membrane fouling clearly reduced. Moreover, as CRDM owned an open flow channel structure, which improved the mobility of fluid on membrane, CRDM possessed a greater shear rate and lower concentration polarization than DRDM and DA. Therefore, CRDM had a greater flux and lower flux decline. The flux reduced quickly at VRR from 1 to 2, during which leaf protein deposited and adsorbed at membrane and foulant-cleaning membrane interaction formed, as well become the main fouling mechanism. Then the flux reduced slowly and fluctuated slightly, when VRR elevated from 2 to 6, because of mass transfer limited regime (Luo et al. 2012). Owing to the "self-cleaning" effect of high shear rate, membrane fouling did not deteriorate and the main fouling mechanism was foulant-deposited foulant interaction. Besides, the permeate flux of MF was higher than UF, because the larger pore size and higher permeability. However, the flux decline of MF was also clearly bigger than UF, since the size of main foulants (leaf proteins) was similar with MF pores (Marel et al. 2010), thus more serious pore blocking occurred.

As illustrated in Fig. 10, as for protein rejection, DRDM, and CRDM are obviously better than DA. Because the much higher Reynolds number regenerated by their clearly different structures, longer agitator diameter and greater shear rate (Zhu et al. 2015), more intense hydrodynamics formed, higher shear stress produced, and concentration polarization significantly reduced. This demonstrated that CRDM and DRDM could greatly improve the concentration capacity of leaf protein in Luzerne juice. The driving force of permeate flux and "secondary filtration" of fouling layer were the main factors affecting separation efficiency. For CRDM and DRDM, the fouling layer was more important. With respect to DA, the effect of permeate flux was dominant. Furthermore, owing to the open flow channel structure, the fouling layer of CRDM was thinner than DRDM; thus, its "secondary filtration" effect and protein rejection were lower. For DA, its shear rate produced by the stirring effect was least (Zhu et al. 2015); thus, DA had the most serious concentration polarization and membrane fouling. Additionally, as displayed in Fig. 10, MF demonstrates less separation performance than UF, because of larger membrane pores. Table 8 shows that retentates in CRDM and DRDM had much better protein purities than that in DA, implying that SEF could not only separate concentrate leaf protein, but also reinforce protein purity, since a membrane separation

Table 6 pH at the end of fermentation, composition and evaluation for different fermentation processes

Sample	Cheese	pH at the end of fermentation	Composition (%)			Evaluation			
			Solids	Total fat	Protein	Appearance	Texture	Taste	Total
1	FD-DVS YF-L903	4.58	29.20	7.35	7.90	2/5	1/5	2/10	5/20
2	Yaourt	4.72	29.00	7.77	7.75	2/5	2/5	3/10	7/20
3	Traditional cheese	4.60	34.93	13.10	14.00	4/5	3/5	7/10	16/20

Table 7 Optimum values of the process parameters for maximum MPR and minimum E_c

	Shear stress (Pa)	Temperature (°C)	Time (min)	Concentration (%)
Water rinsing	83.26	42.04	67.36	
Chemical cleaning	35.74	49.79	10	0.24

Fig. 9 Flux behaviors versus VRR during Luzerne juice concentration process by **a** MF at 3 bar and **b** UF at 4 bar

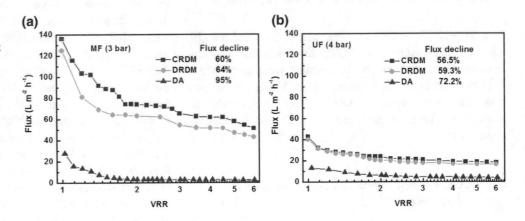

Fig. 10 Characteristics (crude protein and Brix) of permeate and retentate

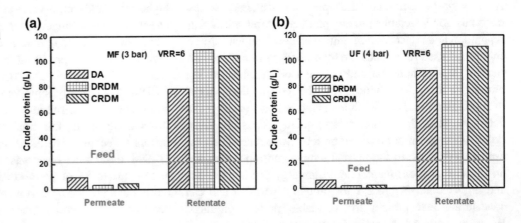

Table 8 Protein purity for various types of filtration modules

VRR = 6		Protein purity in permeate (%)	Protein purity in retentate (%)
MF	DA	14	44
	DRDM	4.6	73
	CRDM	5.5	93
UF	DA	9.3	39
	DRDM	4.8	41
	CRDM	5.3	45

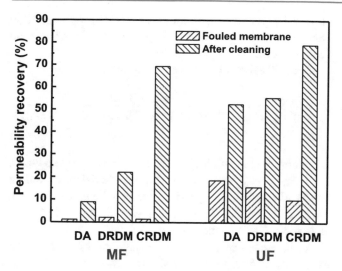

Fig. 11 Membrane cleaning for concentration tests

process could separate large solutes (leaf protein), while many small solutes passed through membrane. In comparison with DA and DRDM, CRDM had higher protein purity and best clarification effect, due to its higher shear rate and lower "secondary filtration" performance. In general, CRDM, with more advantageous performance at flux behavior and separation performance, is the best filtration selection for leaf protein separation.

As shown in Fig. 11, MF has a better membrane fouling but lower permeability recovery than UF, as the more serious pore blocking of MF caused higher irreversible fouling. It is evident that for both MF and UF, the optimal permeability recovery elevates with the order as follows: CRDM > DRDM > DA. This implied that CRDM had the

lowest irreversible fouling and recovered its membrane permeability easily, because of its excellent "self-cleaning" capacity induced by high shear effect (Zhang et al. 2014) during the concentration process. The high membrane cleaning efficiency confirms a high potential application of SEF for Luzerne juice concentration.

In this study, three types of filtration modules (DA, DRDM, and CRDM) were applied to concentrate leaf protein from Luzerne juice. Figure 12 presents that DA utilizes a traditional and simple stirring effect to create shear rate on membrane, then control concentration polarization and membrane fouling. SEF (CRDM and DRDM) with a rotating disk equipped with 6-mm-high vanes could produce a much greater shear rate than DA for reducing serious flux decline. At the same time, the high shear rate of SEF also straightened leaf protein rejection and diminishes impurities during concentration process. Moreover, it decreased membrane pore blocking produced by leaf protein and irreversible fouling, then improved permeate recovery in membrane cleaning, permitting membrane sustainable utilization. On the other hand, CRDM had the highest production efficiency. Because of its open flow channel structure, the CRDM had a better mobility for feed flow and decreases concentration polarization. In fact, SEF could operate at low feed flow rates, which were just slightly higher than permeate flow rate (about 3–5% for UF and MF) (Jaffrin 2008); thus, they did not need powerful and large pumps as in conventional DF and CF and have lower energy consumption for feed pump. In summary, in comparison with other filtration modules, the CRDM indicates many advantages and is an optimized choice for the industrial application of Luzerne juice concentration.

Fig. 12 Schematic of filtration behavior for various kinds of filtration modules

Dead-end filtration

Dynamic filtration

Flux behavior, separation performance and membrane cleaning

Conventional
stirred effect (Amicon)

Rotating
shear-effect (RDM)

Membrane Fouling Model

During the filtration process of leaf protein juice, permeate flux gradually elevates with TMP step by step. In order to better understand this stepwise multisite fouling phenomena, a stepwise multisite Darcy's law model (SMDM) is proposed to simulate the complex fouling process (Zhang et al. 2016). At the same time, the resistance coefficient and compressibility for different steps and sites could be calculated to characterize the fouling behavior parameters. The results show that the simulation of SMDM and experimental data is very similar. They could improve the understanding of fouling process for protein separation and promote membrane fouling control.

The conventional membrane fouling model [Darcy's law model (DM)] is based on the monofouling process, which assumes uniform fouling affinity scales and binding energies between foulants and membrane for fouling sites (Yang et al. 2015); thus, it is an ideal situation for the generally homogeneous membrane surface. But in the actual membrane fouling process, as displayed in Fig. 13, many kinds of active sites on membrane surface of fouling occur, and the multisite DM is can better reflect the real membrane fouling situation: not all surface sites on membrane are identical. Because foulants and the membrane surface are associated with various pits, edges, and other discontinuities (Nady et al. 2011), while a variety of site types might exist, with varying affinity scales and binding energies for fouling process. Besides, in the fouling process, all kinds of unoccupied sites are in excess, while the foulants deposit and bind preferentially to sites with greater affinity. When foulants amount rises, the fouling sites with higher binding energy sites on membrane trends to be saturated and excess foulants are compelled to deposit on other sites with lower affinity. Therefore, as illustrated in Fig. 13, the fouling process leaf proteins is a stepwise and multisite process.

A novelty membrane fouling model, stepwise multisite Darcy's law model (SMDM), is proposed to better explain fouling process of Luzerne juice.

The SMDM can be showed as follows:

$$J = \sum_{i=1}^{m} \sum_{j=1}^{n} \frac{TMP - TMP_i + |TMP - TMP_i|}{\mu[2R_m + 2^{1-\beta_j}\gamma_j(TMP - TMP_i + |TMP - TMP_i|)^{\beta_j}]}$$

(8)

where i is step number, $i = 1, 2, \ldots, m$ and j is site type j, $j = 1, 2, \ldots, n$. β_j and γ_j are resistance coefficient of fouling layer and compressibility index of fouling layer for site type. TMP_i is additional TMP during step i. R_m is hydraulic resistances (m^{-1}) of membrane.

As illustrated in Fig. 14, the simulation of SMDM and experimental data are highly similar. Therefore, SMDM is able to calculate the fouling process parameters, as well as simulate permeate flux correctly. The foulants deposited on membrane indicated the important stepwise patterns; thus, their flux behavior and membrane fouling presented various stepwise trends. With the increase of TMP, fouling elevated step by step, and fouling trends varied with different operating conditions. The SMDM was utilized to identify fouling variation and helped simulate this stepwise fouling process for the Luzerne juice filtration. Fouling process could be divided into several steps and sites, as well as their compressibility and resistance coefficients were calculated to explain the complex fouling process. Furthermore, feed composition, membrane characteristic, and hydraulic condition play an important role in fouling behavior. In this study, the effect of VRR, membrane type, and shear rate on fouling process was analyzed using SMDM. For membrane engineering, the overall effect of various factors on fouling behavior could complicate the processes. Besides, a series of long-term runs, which operated at various fouling step processes, were conducted to verify flux behavior, including flux decline and permeability loss. The results have significant implications for fouling assessment and fouling control in membrane engineering plants. SMDM can also be applied to other membrane applications as an alternative to describe the fouling mechanism.

Optimization of Operation Conditions

Luzerne juice was concentrated from 6 to 1 L for concentrating leaf protein with various operational conditions, and the results (Zhang et al. 2017) for permeate flux and protein separation are presented in Fig. 15. All permeate fluxes reduce greatly at low VRR (<2) and then decrease slowly for VRR from 2 to 5; afterwards, the fluxes reach a stable status at VRR exceeding 5. As the main foulants, leaf proteins contributed the most to the fouling resistance. According to the trend of fouling variation, the flux decline for the Luzerne juice filtration process could be divided into

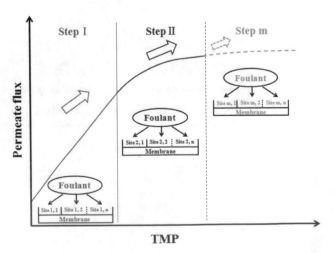

Fig. 13 A schematic illustration of SMDM for fouling process

Fig. 14 Permeate flux as a function of TMP at various rotating speeds and VRRs for MF (**a**, **c** and **e**) and UF (**b**, **d** and **f**)

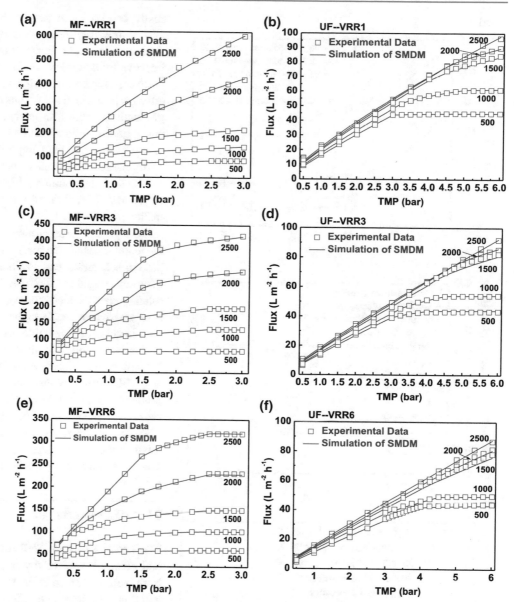

three stages (Zhang and Ding 2015): for the initial stage (VRR < 2), the leaf proteins accumulated on membrane, because of the enhancement of concentration polarization, and thus, protein foulants deposited and adsorbed on membrane rapidly; at the second stage (2 < VRR < 5), due to high shear stress effect, flux decline was not significant; in the last stage (5 < VRR), flux stabilized, and there was an equilibrium between TMP and shear rate (Luo et al. 2012). Although flux was reduced with VRR clearly, the final flux was still higher than 25 L m^{-2} h^{-1}, indicating that SEF was able to efficiently separate and concentrate proteins. For the separation performance, the crude protein concentration and Brix of permeate solution elevated by degrees with VRR. When VRR rose, concentration polarization strengthened, then higher concentration gradient for leaf proteins and other matters formed (Luo et al. 2010), enhancing transmissions of

crude proteins and soluble matters. On the other hand, the "dilution" effect decreased caused by lower flux at higher VRR, which also reduced solute concentration in permeate. Furthermore, the operational conditions exacted an important impact on flux behavior. The US100P membrane with large pore size showed a greater permeate flux, but a higher crude protein concentration and Brix in permeate. Greater shear stress generated by higher rotating speed diminished concentration polarization, and subsequent membrane fouling, while reinforcing flux. At the same time, the great shear stress also reduced the concentration gradient of proteins and soluble matters; thus, the concentrations of crude protein and Brix in permeate decreased. Besides, membrane pores expended at high temperature (Jawor and Hoek 2009), while molecule's motion straightened (Zhang et al. 2013), so flux improved. Meanwhile, an expending membrane pore size

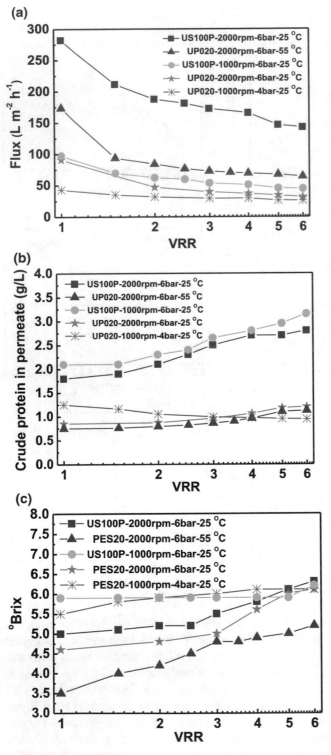

Fig. 15 Permeate flux (**a**), crude protein in permeate (**b**) and Brix in permeate (**c**) versus VRR during concentration of Luzerne juice at various operation parameters (VRR = 6)

easily caused more serious pore blocking (Hwang et al. 2007), thus improving solid rejection capacity. In general, the separation and concentration process for leaf proteins can improve by operation condition optimization.

Except for flux behavior and separation performance, the effects of operation conditions on protein purity and filtration productivity were also analyzed. As shown in Table 9, all concentration tests possess a much higher protein percentage of dry matter than feed, implying that the protein percentage of dry matter significantly improved. Besides, the great shear stress, high temperature, high TMP, and membrane with larger MWCO exhibited a better protein percentage of dry matter in retentate (Luo et al. 2010), since many impurity matters passed through membrane and promoted the purity in retentate. Furthermore, production efficiency for leaf protein separation was evaluated by a new concept of productivity (Regula et al. 2014). Table 10 illustrates that a high rotating speed effectively accelerates the concentration process and has a much higher productivity. The membrane with larger pore size could use less time to concentrate more leaf protein solution; thus, it had a higher productivity. Moreover, the higher values for shear stress, temperature, and TMP also shortened concentration time and improved productivity. In addition, high temperature and TMP also shorten concentration time and have better productivity. Therefore, operating with US100P at high TMP, high rotating speed, and high temperature can improve the UF process efficiency for Luzerne juice.

3 Conclusions

The benefits of high SEF for improving the flux behavior and membrane selectivity in protein solution filtration have been well-confirmed in the scientific literatures. For the milk solution filtration, the effect of hydraulic conditions on effluent quality, flux behavior, and energy consumption was studied by BBRSM for process optimization. The process parameter models for effluent quality, flux behavior, and energy consumption were produced to simulate the SEF of protein solution filtration. Then pore blocking models were utilized to revealing the membrane fouling mechanism. Due to the high shear stress, fouling layer was limited and pore blocking became the main fouling mechanism. Meanwhile, during the concentration process of milk solution, the concentration polarization was alleviated and most milk proteins were concentrated into the retentate solution. Cheese products were produced from the retentate solution with high

Table 9 Protein percentage of dry matter for various operation parameters

VRR = 6	Protein percentage of dry matter in permeate (%)	Protein percentage of dry matter in retentate (%)
US100P-1000 rpm-6 bar-25 °C	4.7	54
US100P-2000 rpm-6 bar-25 °C	4.5	63
UP020-1000 rpm-4 bar-25 °C	5.3	45
UP020-2000 rpm-6 bar-25 °C	4.9	50
UP020-2000 rpm-6 bar-55 °C	5.1	48

Table 10 Operation time and productivity for various operation parameters

VRR = 6 (Concentrated volume = 6 L)	Operation time (h)	Productivity (L h^{-1} m^{-2} bar^{-1})
US100P-1000 rpm-6 bar-25 °C	4.55	2.08
US100P-2000 rpm-6 bar-25 °C	1.48	6.38
UP020-1000 rpm-4 bar-25 °C	8.33	1.70
UP020-2000 rpm-6 bar-25 °C	5.92	1.59
UP020-2000 rpm-6 bar-55 °C	3.17	2.98

milk protein concentration by appropriate fermentation processing. During the membrane cleaning process, foam cleaning agent was generated by the shear effect to greatly improve cleaning efficiency, as it could penetrate into the membrane pores and react with membrane foulants in a deeper level.

As for leaf protein solution filtration, the studies about filtration module structure, membrane fouling model, and optimization of operation conditions were summarized. Among three filtration modules, CRDM exhibited a high permeate flux, great anti-fouling capacity and excellent separation performance, because of high shear effect and open flow channel structure. In order to better understand the stepwise membrane fouling process, a SMDM was proposed to simulate the complex fouling process, while calculate the key fouling parameters. Besides, with high shear stress, larger UF pore size and suitable TMP and temperature, the filtration efficiency (permeate flux and filtration time), and separation performance (protein rejection and the protein purity in retentate) could be significantly promoted. This work provides an important guidance for the application of SEF for the separation and concentration of protein solution.

Acknowledgements The authors would like to acknowledge the financial support from the National Natural Science Foundation of China (No. 51908136), Science and Technology Project of Guangzhou (201904010122), and Guangdong Natural Science Foundation of China (2017A030310540 and 2018A0303130036). The authors would like to thank Ms. Christa AOUDE for her help with the English correction.

References

Beier, S. P., Guerra, M., Garde, A., & Jonsson, G. (2006). Dynamic microfiltration with a vibrating hollow fiber membrane module: Filtration of yeast suspensions. *Desalination, 199*, 499–500.

Chen, W., Mo, J., Du, X., Zhang, Z., & Zhang, W. (2019). Biomimetic dynamic membrane for aquatic dye removal. *Water Research, 151*, 243–251.

Ding, L. H., & Jaffrin, M. Y. (2014). Benefits of high shear rate dynamic nanofiltration and reverse osmosis: A review. *Separation Science & Technology, 49*, 1953–1967.

Ding, L. H., Jaffrin, M. Y., & Luo, J. (2015). Chapter two—Dynamic filtration with rotating disks, and rotating or vibrating membranes. In *Progress in filtration & separation* (pp. 27–59).

Ding, L. H., Jaffrin, M. Y., Mellal, M., & He, G. (2006). Investigation of performances of a multishaft disk (MSD) system with overlapping ceramic membranes in microfiltration of mineral suspensions. *Journal of Membrane Science, 276*, 232–240.

Ding, L. H., Omar, A., Antoine, A., & Jaffrin, M. Y. (2003). High shear skim milk ultrafiltration using rotating disk filtration systems. *AIChE Journal, 49*(9), 2433–2441.

Ding, L. H., Zhang, W., Oulddris, A., Jaffrin, M. Y., & Bing, T. (2016). Concentration of milk proteins for producing cheese using shear-enhanced ultrafiltration technique. *Industrial and Engineering Chemistry Research, 55*, 6b–2738b.

Gahleitner, B., Loderer, C., & Fuchs, W. (2013). Chemical foam cleaning as an alternative for flux recovery in dynamic filtration processes. *Journal of Membrane Science, 431*, 19–27.

Gahleitner, B., Loderer, C., Saracino, C., Pum, D., & Fuchs, W. (2014). Chemical foam cleaning as an efficient ultrafiltration alternative for flux recovery in processes. *Journal of Membrane Science, 450*, 433–439.

Hwang, K. J., Liao, C. Y., & Tung, K. L. (2007). Analysis of particle fouling during microfiltration by use of blocking models. *Journal of Membrane Science, 287*, 287–293.

Jaffrin, M. Y. (2008). Dynamic shear-enhanced membrane filtration: A review of rotating disks, rotating membranes and vibrating systems. *Journal of Membrane Science, 324,* 7–25.

Jaffrin, M. Y. (2012). Dynamic filtration with rotating disks, and rotating and vibrating membranes: An update. *Current Opinion in Chemical Engineering, 1,* 171–177.

Jawor, A., & Hoek, E. M. V. (2009). Effects of feed water temperature on inorganic fouling of brackish water RO membranes. *Desalination, 239,* 346–359.

Kruif, C. G. D., Huppertz, T., Urban, V. S., & Petukhov, A. V. (2012). Casein micelles and their internal structure. *Advances in Colloid and Interface Science, 99,* 36–52.

Luo, J., Cao, W., Ding, L. H., & Zhu, Z. (2012). Treatment of dairy effluent by shear-enhanced membrane filtration: The role of foulants. *Separation & Purification Technology, 96,* 194–203.

Luo, J., Ding, L. H., Wan, Y., & Jaffrin, M. Y. (2012). Threshold flux for shear-enhanced nanofiltration: Experimental observation in dairy wastewater treatment. *Journal of Membrane Science, 409–410,* 276–284.

Luo, J., Ding, L. H., Wan, Y., Paullier, P., & Jaffrin, M. Y. (2010). Application of NF-RDM (nanofiltration rotating disk membrane) module under extreme hydraulic conditions for the treatment of dairy wastewater. *Chemical Engineering Journal, 163,* 307–316.

Luo, J., Ding, L. H., Wan, Y., Paullier, P., & Jaffrin, M. Y. (2012). Fouling behavior of dairy wastewater treatment by nanofiltration under shear-enhanced extreme hydraulic conditions. *Separation & Purification Technology, 88,* 79–86.

Marel, P. V. D., Zwijnenburg, A., Kemperman, A., Wessling, M., Temmink, H., & Meer, W. V. D. (2010). Influence of membrane properties on fouling in submerged membrane bioreactors. *Journal of Membrane Science, 348,* 66–74.

Nady, N., Franssen, M. C. R., Han, Z., Eldin, M. S. M., Boom, R., & Schroën, K. (2011). Modification methods for poly(arylsulfone) membranes: A mini-review focusing on surface modification. *Desalination, 275,* 1–9.

Regula, C., Carretier, E., Wyart, Y., Gésan-Guiziou, G., Vincent, A., Boudot, D., & Moulin, P. (2014). Chemical cleaning/disinfection and ageing of organic UF membranes: A review. *Water Research, 56,* 325–365.

Yang, H. C., Luo, J., Lv, Y., Shen, P., Xu, Z. K., Yang, H. C., et al. (2015). Surface engineering of polymer membranes via mussel-inspired chemistry. *Journal of Membrane Science, 483,* 42–59.

Zhang, W., & Ding, L. H. (2015). Investigation of membrane fouling mechanisms using blocking models in the case of shear-enhanced ultrafiltration. *Separation & Purification Technology, 141,* 160–169.

Zhang, W., Ding, L. H., Grimi, N., Jaffrin, M. Y., & Bing, T. (2017). Application of UF-RDM (ultafiltration rotating disk membrane) module for separation and concentration of leaf protein from alfalfa juice: Optimization of operation conditions. *Separation & Purification Technology, 175,* 365–375.

Zhang, W., Ding, L. H., Jaffrin, M. Y., Grimi, N., & Bing, T. (2016). Stepwise membrane fouling model for shear-enhanced filtration of alfalfa juice: Experimental and modeling studies. *RSC Advances, 6,* 110789–110798.

Zhang, W., Ding, L. H., Jaffrin, M. Y., & Tang, B. (2017). Membrane cleaning assisted by high shear stress for restoring ultrafiltration membranes fouled by dairy wastewater. *Chemical Engineering Journal, 325,* 457–465.

Zhang, W., Grimi, N., Jaffrin, M. Y., & Ding, L. H. (2015). Leaf protein concentration of alfalfa juice by membrane technology. *Journal of Membrane Science, 489,* 183–193.

Zhang, W., Huang, G., Jia, W., & Duan, Y. (2013). Gemini micellar enhanced ultrafiltration (GMEUF) process for the treatment of phenol wastewater. *Desalination, 311,* 31–36.

Zhang, W., Luo, J., Ding, L. H., & Jaffrin, M. Y. (2015). A review on flux decline control strategies in pressure-driven membrane processes. In: *Proceedings of the 4th International Conference on Foundations of Software Science and Computation Structures* (pp. 303–317).

Zhang, W., Zhu, Z., Jaffrin, M. Y., & Ding, L. H. (2014). Effects of hydraulic conditions on effluent quality, flux behavior, and energy consumption in a shear-enhanced membrane filtration using Box-Behnken response surface methodology. *Industrial and Engineering Chemistry Research, 53,* 7176–7185.

Zhu, Z., Mhemdi, H., Ding, L. H., Bals, O., Jaffrin, M. Y., Grimi, N., & Vorobiev, E. (2015). Dead-end dynamic ultrafiltration of juice expressed from electroporated sugar beets. *Food & Bioprocess Technology, 8,* 615–622.

Membrane-Permeation Modeling for Carbon Capture from CO_2-Rich Natural Gas

José Luiz de Medeiros, Lara de Oliveira Arinelli, and Ofélia de Queiroz F. Araújo

Abstract

This chapter contemplates two topics committed to steady-state modeling of membrane-permeation units for decarbonation of CO_2-rich natural gas at high-pressure. The first topic presents a steady-state, phenomenological, and one-dimensional distributed membrane-permeation simulation model—SPM2010—which was developed in MATLAB 2010 for rigorous simulation of CO_2 removal and natural gas purification flowsheets using membrane-permeation batteries operating with hollow-fiber membranes and parallel retentate/permeate flows. SPM2010 solves mass/momentum/energy balances of permeate and retentate non-isothermal, non-isobaric compressible flows rendering several graphical results. The second topic comprehends simulation models of membrane-permeation units appropriate for insertion in gas processing flowsheets solved by professional process simulators, such as HYSYS 10.0. These HYSYS extension models are grouped into two types, both solving mass/energy balances of permeate/retentate for hollow-fiber and spiral-wound membranes: (i) Lumped models for parallel and counter-current permeate/retentate flows using average driving forces and lumped balances; (ii) one-dimensional distributed models for parallel permeate/retentate flows using distributed driving forces and balances. The major findings of this chapter correspond to the development of two categories of membrane-permeation models (respectively, in Sects. 2 and 3), respectively, appropriate for two computing platforms—MATLAB and HYSYS—and sufficiently accurate for designing real permeation systems for CO_2 removal from CO_2-rich natural gas at high pressure. Both categories of models were calibrated with real data of CO_2 permeation batteries belonging to offshore rigs operating in the pre-salt basin in the southeast coast of Brazil.

Abbreviations

1D	One-dimensional
2D	Two-dimensional
C3+	Propane and heavier alkanes
CAM	Cellulose-acetate membrane
CC	Counter-current contact
CW	Cooling-water
DLL	Dynamic-link library
EOR	Enhanced oil recovery
EOS	Equation of state
FPSO	Floating, production, storage and offloading
GLMC	Gas-liquid membrane contactor
HCDP	Hydrocarbons dew-point
HCDPA	Hydrocarbons dew-point adjustment
HFM	Hollow-fiber membrane
ID	Internal diameter
JTE	Joule-Thomson expansion
LNG	Liquefied NG
LPG	Liquefied petroleum gas
M2F	Two-phase mixer-cooler
$MMNm^3/d$	Millions of normal m^3/d
$MMSm^3/d$	Millions of standard m^3/d
MP	Membrane-permeation
NG	Natural gas
NGL	Natural gas liquids
NRM	Newton-Raphson method
OD	Outside diameter
ODE	Ordinary differential equations
PC	Parallel contact
PFD	Process flow diagram
PHW	Pressurized-hot-water
PR-EOS	Peng-Robinson equation-of-state
PVT	Pressure-volume-temperature

J. L. de Medeiros (✉) · L. de O. Arinelli · O. de Q. F. Araújo
Escola de Química, CT, E, Federal University of Rio de Janeiro, Ilha do Fundão, Rio de Janeiro, RJ 21941-909, Brazil
e-mail: jlm@eq.ufrj.br

© Springer Nature Switzerland AG 2021
Z. Zhang et al. (eds.), *Membrane Technology Enhancement for Environmental Protection and Sustainable Industrial Growth*, Advances in Science, Technology & Innovation, https://doi.org/10.1007/978-3-030-41295-1_10

S2F	Two-phase separator
SWM	Spiral-wound membrane
TEG	Triethylene glycol
UOE	Unit operation extension
VB	Visual basic
VLE	Vapor-liquid equilibrium
WDP	Water dew-point
WDPA	Water dew-point adjustment

Nomenclature

A_{MP}	MP area (m²)
A_I, A_E	Internal and external MP heat exchange areas (m²)
$\overline{C}_P \equiv \left(\frac{\partial \overline{H}}{\partial T}\right)_{P,\underline{Z}}$	Molar heat capacity at const. P, $\underline{Z}$ (J/K mol, kJ/K mol)
D	MP shell internal diameter (m)
d_i, d_o	Internal and external HFM diameters (m)
$\overline{E}$	Total molar energy of fluid (J/mol, kJ/mol)
$\overline{E}_k^V, \overline{E}_k^L$	Retentate and permeate partial molar energy of kth species (J/mol, kJ/mol)
$\hat{f}_k^V, \hat{f}_k^L$	Retentate and permeate fugacities of species k (bar)
f_V, f_L	Retentate and permeate Darcy friction factors (dimensionless)
F	Molar flow rate of fluid (mol/s, MMNm³/d, MMSm³/d)
$\overline{H}$	Molar enthalpy of multiphase or single-phase fluid (J/mol, kJ/mol)
$\overline{H}_k$	Partial molar enthalpy of kth species (J/mol, kJ/mol)
$\overline{K}$	Molar kinetic energy of multiphase or single-phase fluid (J/mol, kJ/mol)
L	Permeate molar flow rate (mol/s, MMNm³/d, MMSm³/d)
L	Length (m)
M_M	Molar mass of multiphase or single-phase fluid (kg/mol)
nc	Number of components (species)
N_k	Species k permeation rate (mol/s, MMNm³/d, MMSm³/d)
N_M	Number of MP modules in MP battery
N_{HF}	Number of HFM's within a HFM module
P, P_V, P_L	Pressure, retentate pressure and permeate pressure (Pa, bar)
P_V^{out}, P_L^{out}	MP retentate/permeate outlet pressures (bar)
P_V^{in}, P_L^{in}	MP retentate/permeate inlet pressures (bar)
ΔP_k^{LN}	MP log mean difference of partial pressures of species k (bar)
q	Mass flow rate of multiphase or single-phase fluid (kg/s)
REC%CO$_2$	Percent recovery of CO_2
T	Temperature (K, °C)
T_L, T_V	Temperatures of permeate/retentate (K, °C)
T_V^{out}, T_L^{out}	MP retentate/permeate outlet temperatures (K, °C)
T_V^{in}, T_L^{in}	MP retentate/permeate inlet temperatures (K, °C)
ΔT_F	MP Temperature difference at the initial condition of permeate flow (°C)
$\Delta T_I^{LN}, \Delta T_E^{LN}$	MP log-mean temperature difference for internal/external heat transfers (°C)
U_I, U_E	MP internal and external heat transfer coefficients (W/m² K)
v	Axial velocity of fluid (m/s)
V	Molar flow rate of retentate (mol/s, MMNm³/d, MMSm³/d)
$\underline{X}$	Vector (nc × 1) of permeate (or liquid phase) mol fractions
$Y_k^{in}, Y_k^{out}, X_k^{out}$	Species k mol fraction in retentate/permeate inlet/outlet MP streams
$\underline{Y}$	Vector (nc × 1) of retentate (or vapor phase) mol fractions
z	MP axial position (m)
Z_M	Length (m) of HFM's in the HFM module
$\underline{Z}$	Vector (nc × 1) of total mol fractions of multiphase or single-phase fluid

Greek Symbols

Π_k	Permeance of species k (mol/s m² bar, MMNm³/d m² bar, MMSm³/d m² bar)
$\varepsilon_V, \varepsilon_L$	Retentate and permeate flow roughnesses (m)
$\eta\%$	Expander/compressor adiabatic efficiencies (%)
$\hat{\phi}_k, \phi_k$	Fugacity coefficients of species k (dimensionless)
ρ	Multiphase or single-phase fluid density (kg/m³)
$\Xi_P \equiv \left(\frac{\partial \rho}{\partial P}\right)_{T,\underline{Z}}$	Derivative of ρ with P at const. T, $\underline{Z}$ (kg/Pa m³, kg/bar m³)
$\Xi_T \equiv \left(\frac{\partial \rho}{\partial T}\right)_{P,\underline{Z}}$	Derivative of ρ with T at const. P, $\underline{Z}$ (kg/K m³)

Subscripts

i, o	Inside, outside
k	Species index

L   Liquid phase or permeate
V   Vapor phase or retentate

′   Ideal gas property
in, out Inlet, outlet
R   Residual property
V, L   Vapor, liquid or retentate, permeate

1   Introduction

Units for CO_2 removal from natural gas (NG) using membrane-permeation (MP) technology are becoming gradually more common in the context of large-scale gas processing and gas purification systems (Baker 2004). This is especially true, among several other applications, in connection to offshore rigs that have to purify high flow rates of CO_2-rich raw NG streams (from 20 to 45 mol% CO_2) producing in the retentate exportation gas (from 3 to 5 mol% CO_2) and, in the permeate, CO_2-rich product streams (from 70 to 80 mol% CO_2) for enhanced oil recovery (EOR) destinations (Ebner and Ritter 2009; Arinelli et al. 2019). Ho et al. (2006) and Bernardo et al. (2009) present complete surveys on gas processing applications of membrane-permeation technology.

Regarding decarbonation of CO_2-rich NG, it is worthwhile to notice that the chemical absorption of CO_2 with aqueous monoethanolamine (MEA) and aqueous methyl-diethanolamine (MDEA) is very mature technologies considered as benchmark options for such service (de Medeiros et al. 2013a). Nevertheless, the membrane permeation with polymeric skin-dense membranes is growing fast and it is being much more used than aqueous-amine absorption for CO_2-rich NG decarbonation at high-pressure in some niches of applications, such as, in deep waters, floating, production, storage, and offloading (FPSO) offshore platforms, where space and weight are the major concerns, and the modularity of MP units is an important advantage (Araújo and de Medeiros 2017). Other advantages of MP over aqueous amines for NG decarbonation services on offshore platforms comprise: (i) MP is a simpler process solution; (ii) MP units are smaller and lighter systems; (iii) MP is a cleaner solution with no chemical additives; (iv) MP has low fire or explosion hazards; (v) MP can execute simultaneous removal of CO_2, H_2S, and H_2O; (vi) MP has less maintenance, lower capital, and operational costs; and (vii) MP can treat NG at well-heads. On the other hand, some major comparative disadvantages of MP to aqueous-amines absorption are: (i) Decreasing CO_2–CH_4 selectivity for increasing flux; (ii) inferior economic

competitiveness at higher scales; (iii) decreasing membrane stability and resilience for increasing (T, P); (iv) degradation issues and limited lifetime of membranes; (v) MP technology is not sufficiently mature according to industrial standards (Araújo et al. 2017). Figure 1 presents the types of membrane-permeation modules and the most used ones—spiral-wound membrane (SWM) and hollow-fiber membrane (HFM)—for decarbonation of CO_2-rich NG in offshore platforms. Process configurations of MP units for NG decarbonation (Baker 2004) are shown in Fig. 2.

NG purification is one of the largest worldwide applications of gas separation. Membrane permeation has a few percent of this market but exhibits a great potential of expansion, only considering eight or nine polymeric materials that respond for 90% of applications, where cellulose-acetate membranes (CAM) are the most used for decarbonation of CO_2-rich NG under SWM as well as HFM modules. Table 1 lists some manufacturers of commercial CAM membranes for CO_2 removal from NG. Published studies have approached hundreds of new polymer materials for MP applications in the last years. Nonetheless, the harder

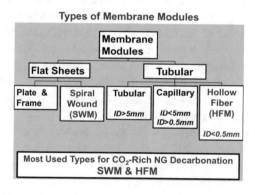

Fig. 1 Types of MP modules versus CO_2-rich NG decarbonation

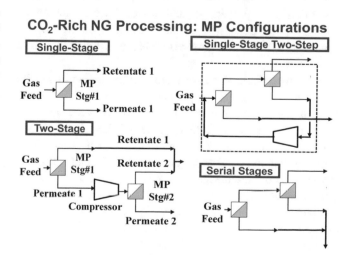

Fig. 2 Process configurations of MP units for NG decarbonation

Table 1 Manufacturers of cellulose-acetate membranes for NG processing

Manufacturer	Membrane type	Element orientation	Element $L \times D$	Element installation	Gas contact
UOP	SWM*	Horizontal	1 m × 0.2 m	Tandem elements in tubes	Cross-flow
NATCO Schlumberger	HFM*	Vertical	2 m × 0.4 m Several	Single element	Cross-flow Parallel-flow
Air liquid	HFM	Vertical Horizontal	Several	Single element	Cross-flow Parallel-flow

*SWM Spiral-wound membrane; HFM Hollow-fiber membrane

obstacle to approve new materials for commercial MP applications has to do with membrane resiliency regarding real processing conditions and membrane capability to maintain its characteristics (e.g., selectivity and capacity) through reasonable operation times.

1.1 Software for Simulation of MP Units in Gas Processing

Currently, as far as we can see, there are no commercial computational tools available for rigorous design and simulation of general membrane-permeation units. When existent, such type of software is normally developed for local and restricted ad hoc finalities of MP developers, MP manufacturers, and certain MP users. On the other hand, oil-and-gas companies, which have to operate large-capacity processing plants of CO_2-rich NG at offshore sites, are experiencing a crucial dependence on such category of simulation and design tools.

These MP modeling tools are necessary, for example, to revamp operating MP units in order to accommodate—in a new processing flowsheet—new raw NG flow rates, new (higher) CO_2 contents in raw NG, and new (stricter) CO_2 separation targets. Accurate MP models are also necessary for daily supervision of operating MP plants, particularly, regarding loss prevention and safety because membrane cartridges can burst with certain frequency during the lifetime of MP units for high-pressure NG processing (Bernardo et al. 2009).

However, since MP units have a large number of specific configurational details—e.g., geometric aspects (diameters, lengths, thicknesses), material bulk properties (density, heat capacity, operational limits), external and internal heat transfer coefficients (trans-membrane and trans-shell), permeate/retentate flow/contact configurations (parallel flows, countercurrent flows, crossed flows), permeate/retentate locations relative to the membrane (inside/outside), material surface properties (roughnesses, specific areas), permeate/retentate head-loss parameters, species permeances, etc.—the development of a truly rigorous steady-state MP simulator is a hard-core task, not counting the thermodynamic aspects of non-isothermal, non-isobaric, composition-changing permeate/retentate compressible flows, and the geometrical/mathematical issues characteristic of one-dimensional (1D) or two-dimensional (2D) frameworks.

Moreover, even if such a MP simulator could be available, several materials/structural parameters of MP units for high-pressure NG processing are not constant and expressively change with service time (Baker 2004). Some changing parameters are evidently related to the degradations that the membrane material experiences through its lifetime; namely, CO_2/CH_4 selectivity, chemical stability, structural resiliency, and mechanical stability. Such properties are known to change drastically with time in high-pressure NG processing, always toward performance deterioration, and eventually culminating with bursts due to loss of structural stability or irreversible swelling and plasticization due to excessive intake of CO_2 and H_2S into the membrane body (Bernardo et al. 2009; Ebner and Ritter 2009).

Such difficulties to model membrane permeators are also encountered with regard to other membrane-based unit operations in the context of acid-gas removal from CO_2-rich NG. This is the case, for example, of gas–liquid membrane contactors or GLMC (Marzouk et al. 2010), which are membrane operations for CO_2 removal from NG where permeation through skin-dense membranes is not the main mechanism. Instead, the high-pressure retentate gas is contacted with an alkaline solvent at similar (slight lower) pressure (e.g., aqueous-MEA or aqueous-MDEA) through a porous membrane. In this case, the selectivity is imposed by the aqueous alkaline solvent not by the membrane. The GLMC also bears analogous critical modeling aspects where flow geometry, flow/contact configurations, compressible multi-phase and single-phase flows, thermodynamic property calculations, transport phenomena, and chemical multi-reaction equilibrium play important roles.

Recently, de Medeiros et al. (2013b) presented an equilibrium-based rigorous model for simulation of gas–liquid membrane contactors (GLMC) operating with hollow-fiber porous membranes (HFM) for CO_2 removal from high-pressure NG with aqueous equimolar solutions of

MEA and MDEA. This GLMC model adopts rigorous modeling of one-dimensional (1D) single and multi-phase compressible flows of retentate and permeate, where we take the liberty to denominate the two GLMC products as retentate and permeate in an analogy with the MP case (though, literally, there is no permeation here). In a GLMC, the permeate can become a two-phase multi-reactive system due to the reactive absorption of CO_2 into the alkaline solvent phase and some trans-membrane transport of CH_4 and other light hydrocarbons that create a (initially, non-existent) vapor phase flowing jointly with the alkaline solvent. In this case, the GLMC has both retentate and permeate flows as compressible steady-state flows; where the latter is a two-phase multi-reactive equilibrium compressible flow.

Such GLMC model is evidently far beyond MP modeling in terms of complexity. But its modeling kernel can be adapted to model MP operations with similar compressible flow geometry and similar HFM configuration. By removing the GLMC formalism related to the two-phase reactive permeate of CO_2 absorption into alkaline solvents, the GLMC model of de Medeiros et al. (2013b) can be reduced to a rigorous 1D model for HFM MP units.

1.2 Outline of This Chapter

This chapter has two subjects committed to steady-state modeling of MP units for decarbonation of CO_2-rich NG. The last paragraph of Sect. 1.1 introduced one of these two subjects (Sect. 2): By suppressing the reactive two-phase nature of the permeate in the GLMC model of de Medeiros et al. (2013b), a phenomenological steady-state 1D-distributed MP simulation model—SPM2010—is developed for rigorous simulation of CO_2 removal and NG purification flowsheets using MP batteries operating with hollow-fiber membranes (HFM) and parallel (co-current) retentate and permeate flows. SPM2010 solves mass/momentum/energy balances of permeate and retentate and is numerically processed in MATLAB 2008 environment (The Mathworks) rendering several graphical results.

The second subject of this chapter (Sect. 3) comprehends simulation models of MP units appropriate for insertion in gas processing flowsheets solved by professional process simulators, such as HYSYS 8.8 and HYSYS 10.0 (ASPENTECH). These models are grouped into two types, both solving mass/energy balances of permeate and retentate for hollow-fiber membrane (HFM) and spiral-wound membranes (SWM): (i) Lumped MP models for parallel and counter-current permeate/retentate flows using average driving forces and lumped balances (Arinelli et al. 2017, 2019; Araújo et al. 2017); (ii) 1D-distributed MP models for parallel permeate/retentate flows using distributed driving forces and distributed balances.

2 Modeling of Steady-State 1D-Distributed MP Units with Hollow-Fiber Membranes and Parallel Flows: SPM2010 for MATLAB

SPM2010 models MP units assuming a fixed steady-state HFM configuration with co-current (parallel) permeate and retentate flows. Rigorous mass, energy, and momentum steady-state balances are written for compressible single-phase gas flow of both permeate (low-pressure side) and retentate (high-pressure side), where the flow geometry is 1D along the axis of permeate and retentate flows. Rigorous 1D compressible fluid flow—including thermal-compressibility effects, friction effects, and acceleration from depressurization—is supported via rigorous thermodynamic calculations with Peng–Robinson equation of state (PR-EOS) or Soave–Redlich–Kwong equation of state (SRK-EOS) (Reid et al. 1987). All thermodynamic properties—e.g., enthalpies, densities, fugacities, heat capacities, sound velocity, and differential coefficients of the density with temperature and pressure—are calculated along the flow paths with PR-EOS or SRK-EOS. Transport properties—e.g., dynamic viscosities and friction factors—are estimated with correlations for high-pressure multicomponent gas flow. The thermodynamic completeness of this MP 1D model enables it to make reasonable predictions of cooling effects which is common in MP units and characteristic of gas permeation phenomena. SPM2010 also has a built-in tool for generation of dew-point locus and bubble-point locus to evaluate condensation risks within the high-pressure retentate in consequence of the rising of retentate dew-point (thanks to the dominant permeation of lighter species) and MP cooling effects.

The MP model of SPM2010 was calibrated with operational data of offshore platforms in Brazil. The calibration parameters are basically trans-membrane permeances (e.g., for CO_2 and CH_4), but heat transfer coefficients, HFM module geometry, and HFM area per module also participated in model calibration. This work was conducted at CE-GN—Center of Excellence in Natural Gas of the Federal University of Rio de Janeiro (UFRJ), a joint venture between UFRJ and PETROBRAS, the Brazil's state oil company.

2.1 MP Unit Modeling

A MP unit is modeled as a battery of N_M parallel individual MP modules (elements or cartridges). Only hollow-fiber membrane (HFM) modules with parallel (co-current) 1D flows of retentate and permeate are considered here. All modules in the MP battery are supposed to have same spatial orientation relative to the horizontal direction. Let θ represents such angle between the module axis and the horizontal direction. Typically, θ can vary from 0° to 90°, but 90° is preferable to prevent too much contact between HFM and plausible condensates forming in the high-pressure retentate side. All MP modules are also supposed to have same geometry, same number of HFM filaments, same HFM characteristics, same permeate discharge pressure, and same gas feed (composition, flow rate, temperature, and pressure).

Figure 3 sketches a HFM parallel-flow module composed by a high-pressure retentate shell and a low-pressure permeate chamber separated by a circular plate which holds the HFM's by their open end, such that the HFM's discharge the permeate into the low-pressure chamber. The HFM module (Fig. 3) is a pressure vessel or a high-pressure shell of active length $Z_M(m)$ and internal diameter (ID) $D(m)$. Z_M represents only the axial extension available for permeation; i.e., Z_M also corresponds to the length of HFM's. The high-pressure shell contains N_{HF} parallel HFM's of length Z_M, where N_{HF} is a large dimensionless number ($\approx 10^3$–10^5). The HFM's are narrow hollow cylinders with outside diameter (OD) $d_o(m)$ and ID $d_i(m)$ whose left ends are sealed. The HFM's bundle is hold at the HFM's open ends by the rigid circular plate isolating the low-pressure permeate chamber from the high-pressure retentate shell.

The gas feed enters the high-pressure shell near the HFM's sealed ends. The retentate (V, mol/s) and permeate (L, mol/s) outlets are located on the opposed side of the shell (Fig. 3). The axial spatial coordinate $z(m) \in [0, Z_M]$ sweeps the high-pressure shell longitudinally and also sweeps the inner 1D space of HFM's (Fig. 3). The permeate (L, mol/s) is a low-pressure gas which has permeated through the HFM walls and flows in the membrane inner space toward the low-pressure permeate chamber at pressure P_L^{out} (bar).

Retentate and permeate molar flow rates (mol/s) of species k are represented by V_k and L_k, respectively. Retentate and permeate pressures (P_V and P_L) are typically of 40–

55 bar and 1–4 bar (absolute), and albeit MP plants can also operate with higher permeate pressure entailing lower driving force. Figure 4 sketches a typical SPM2010 MP flowsheet for CO_2 removal from a NG with 12 mol% CO_2 using three MP stages (no recycles) and auxiliary compressors and coolers. In SPM2010 retentate and permeate streams deriving from a feed xx are automatically named V@xx and L@xx, respectively, (see Fig. 4 for stream naming in SPM2010).

2.1.1 Model Equations for Steady-State HFM Parallel-Flow MP Module

Consider Fig. 3 of a steady-state HFM parallel-flow MP module in SPM2010. Both permeate and retentate flow axially and parallelly along the z-axis. The high-pressure retentate gas flows in the space between the HFM's and the shell, leaving the module through the retentate nozzle. Permeate and retentate streams are labeled, respectively, L and V, which are also the algebraic symbols of the respective molar flow rates (mol/s). Permeate and retentate properties and parameters are also labeled with subscripts/superscripts L and V, whereas the properties of the kth species have a subscript k ($k = 1...nc$). Strict SI units are used, where T, P, flow rate and molar energy appear in the equations using strict SI units (K), (Pa), (kg/s), and (J/mol), but can appear as graphical/tabulated results using more convenient secondary SI units (°C), (bar), (MMNm^3/d), and (kJ/mol). Hence, the retentate and permeate temperatures (K, °C), pressures (Pa, bar), densities (kg/m^3), mass flow rates (kg/s), and flow sections (m^2) are, respectively, represented as T_V, P_V, ρ_V, q_V, S_V, T_L, P_L, ρ_L, q_L, S_L, while the respective flow velocities (m/s) are written from the mass flow rates as $q_V/(\rho_V \cdot S_V)$, $q_L/(\rho_L \cdot S_L)$. Other symbols follow: M_k, M_V, M_L represent molar masses (kg/mol) of species k, and of retentate and permeate streams; T_E represents the external temperature (K, °C); g is gravity acceleration (9.81 m/s^2); $S = \pi D^2/4$ is shell section (m^2); $a = N_{HF} \cdot d_o^2/D^2$ is the ratio of HFM transfer area per unit of shell volume (m^2/m^3); Ω and Ω_E stand for retentate/permeate and retentate-outside heat transfer coefficients (W m^{-2} K^{-1}); $h(z)$ is the elevation (m) as function of z (m); Ψ_V, Ψ_L represent shear stresses (Pa) at contact surfaces for retentate and permeate.

For kth species, its trans-membrane flux (mol s^{-1} m^{-2}), its permeance (mol s^{-1} m^{-2} bar^{-1}) and its retentate/permeate fugacities (bar) are written, respectively, as N_k, Π_k, $\hat{f}_k^V$, $\hat{f}_k^L$. Trans-membrane fluxes (positive direction V → L) are represented in Eq. (2.1), while Eqs. (2.2) and (2.3) express the mass balances of kth species for retentate and permeate streams. It worth noting that trans-membrane fluxes are driven by retentate/permeate differences of species fugacities (i.e., are appropriate for high-pressure applications). One-dimensional compressible flow momentum

Hollow-Fiber Parallel-Flow Module

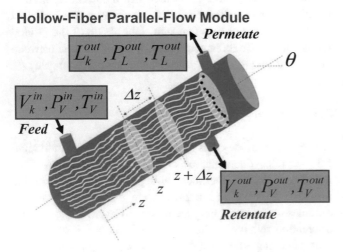

Fig. 3 Parallel-flow HFM module in SPM2010

Fig. 4 SPM2010: three-staged MP flowsheet with rules of stream naming

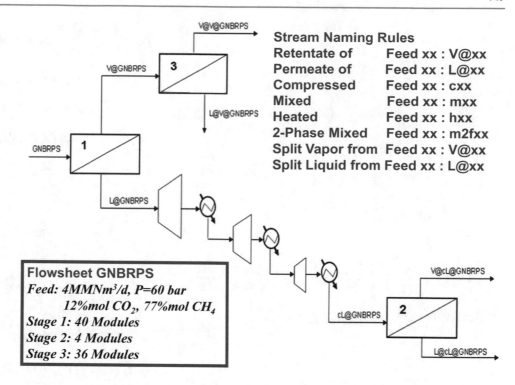

Stream Naming Rules

Retentate of	Feed xx : V@xx
Permeate of	Feed xx : L@xx
Compressed	Feed xx : cxx
Mixed	Feed xx : mxx
Heated	Feed xx : hxx
2-Phase Mixed	Feed xx : m2fxx
Split Vapor from	Feed xx : V@xx
Split Liquid from	Feed xx : L@xx

Flowsheet GNBRPS
Feed: 4MMNm³/d, P=60 bar
        12%mol CO₂, 77%mol CH₄
Stage 1: 40 Modules
Stage 2: 4 Modules
Stage 3: 36 Modules

balances are written for retentate and permeate streams in Eqs. (2.4) and (2.5) in order to express momentum changes as temperature and pressure effects. In the same way, temperature and pressure effects are written as consequences of energy changes of retentate and permeate in the 1D energy balance relationships Eqs. (2.6) and (2.7). Retentate/permeate momentum changes derive from velocity changes, gravity force, and shear stress, while retentate/permeate energy changes result from velocity changes, gravity work, shear stress work, trans-membrane mass/energy transfers, and heat transfers (permeate/retentate and retentate-outside).

Equations (2.2) to (2.7) represent a system of $2nc + 4$ nonlinear ordinary differential equations (ODE) with one independent variable (z) for $2nc + 4$ dependent variables of the HFM model; namely, temperature, pressure, and species molar flow rates of retentate and permeate; i.e., T_V, T_L, P_V, P_L, V_k ($k = 1...nc$), and L_k ($k = 1...nc$).

Equations (2.1) and (2.8) to (2.21) represent subsidiary relationships for calculating terms within the set of $2nc + 4$ ODE's. For example, the retentate/permeate shear stresses at contact surfaces are shown in Eqs. (2.10) and (2.11), respectively, where the respective dimensionless Darcy friction factors (f_V, f_L) are predicted by the universal formula of Churchill (1977) using the respective Reynolds numbers (Re$_V$, Re$_L$), the respective surface roughnesses ($\varepsilon_V, \varepsilon_L$) and the respective hydraulic diameters (D_V, D_L) defined in Eq. (2.12). Hydraulic diameters and Reynolds numbers of retentate/permeate are written in terms of the respective flow

perimeters ($\wp_V, \wp_L$) defined in Eq. (2.13). Retentate/permeate dynamic viscosities (μ_V, μ_L) for the Reynolds numbers (Re$_V$, Re$_L$) are estimated via the correlation of Chung et al. (1988) for multicomponent gases at high-pressure as shown in Eq. (2.14).

Mass flow rates (kg/s) of retentate and permeate (q_V, q_L) are shown in Eq. (2.15), while Eq. (2.16) expresses the total trans-membrane mass flux q_{TM} (kg s^{-1} m^{-2}).

Partial molar *energy* of kth species in retentate and permeate ($\overline{E}_k^V, \overline{E}_k^L$) include kth species partial molar enthalpy and partial molar kinetic and potential energies as given in Eqs. (2.8) and (2.9). Density differential coefficients with (T, P) are written in Eq. (2.17) for permeate and retentate. Retentate (V) and permeate (L) partial molar enthalpies of kth species, molar heat capacities at constant pressure, densities, and species fugacities are, respectively, given in Eqs. (2.18), (2.19), (2.20), and (2.21). Thermodynamic properties in Eqs. (2.1)–(2.16) were isolated in Eqs. (2.17)–(2.21) with the respective dependencies on the model dependent variables. They are calculated directly (e.g., ρ_V, ρ_L) from an appropriate equation of state (EOS) for high-pressure CO₂-rich NG such as the Peng–Robinson EOS (PR-EOS), or from a two-step procedure using ideal gas properties [e.g., kth species ideal gas enthalpy $\overline{H}_k'(T)$ and stream ideal gas enthalpy $\overline{H}'(T)$] and the corresponding correcting residual properties [e.g., $\overline{H}_V^R(T_V, P_V, \underline{V})$, $\overline{H}_L^R(T_L, P_L, \underline{L})$] calculated with PR-EOS.

$$N_k = \Pi_k \left(\hat{f}_k^{\mathrm{V}} - \hat{f}_k^{\mathrm{L}} \right) \qquad (2.1)$$

$$\frac{\partial V_k}{\partial z} = -SaN_k \quad (k = 1 \ldots nc) \qquad (2.2)$$

$$\frac{\partial L_k}{\partial z} = SaN_k \quad (k = 1 \ldots nc) \qquad (2.3)$$

$$\left\{ 1 - \Gamma_{P_{\mathrm{V}}} \left(\frac{q_{\mathrm{V}}}{\rho_{\mathrm{V}} S_{\mathrm{V}}} \right)^2 \right\} \frac{\partial P_{\mathrm{V}}}{\partial z} + \left\{ -\Gamma_{T_{\mathrm{V}}} \left(\frac{q_{\mathrm{V}}}{\rho_{\mathrm{V}} S_{\mathrm{V}}} \right)^2 \right\} \frac{\partial T_{\mathrm{V}}}{\partial z}$$
$$= \frac{Saq}{S_{\mathrm{V}}} \left(\frac{q_{\mathrm{V}}}{\rho_{\mathrm{V}} S_{\mathrm{V}}} \right) - \rho_{\mathrm{V}} g \sin(\theta) - \frac{\Psi_{\mathrm{V}}(\pi D + Sa)}{S_{\mathrm{V}}} \qquad (2.4)$$

$$\left\{ 1 - \Gamma_{P_{\mathrm{L}}} \left(\frac{q_{\mathrm{L}}}{\rho_{\mathrm{L}} S_{\mathrm{L}}} \right)^2 \right\} \frac{\partial P_{\mathrm{L}}}{\partial z} + \left\{ -\Gamma_{T_{\mathrm{L}}} \left(\frac{q_{\mathrm{L}}}{\rho_{\mathrm{L}} S_{\mathrm{L}}} \right)^2 \right\} \frac{\partial T_{\mathrm{L}}}{\partial z}$$
$$= -\frac{Saq}{S_{\mathrm{L}}} \left(\frac{q_{\mathrm{L}}}{\rho_{\mathrm{L}} S_{\mathrm{L}}} \right) - \rho_{\mathrm{L}} g \sin(\theta) - \frac{\Psi_{\mathrm{L}} Sa}{S_{\mathrm{L}}} \qquad (2.5)$$

$$\left\{ 1 - \Gamma_{P_{\mathrm{V}}} \left(\frac{q_{\mathrm{V}}}{\rho_{\mathrm{V}} S_{\mathrm{V}}} \right)^2 + \frac{T_{\mathrm{V}} \Gamma_{T_{\mathrm{V}}}}{\rho_{\mathrm{V}}} \right\} \frac{\partial P_{\mathrm{V}}}{\partial z} + \left\{ \frac{\rho \bar{C}_P^{\mathrm{V}}}{M_{\mathrm{V}}} - \Gamma_{T_{\mathrm{V}}} \left(\frac{q_{\mathrm{V}}}{\rho_{\mathrm{V}} S_{\mathrm{V}}} \right)^2 \right\} \frac{\partial T_{\mathrm{V}}}{\partial z}$$
$$= -\rho_{\mathrm{V}} g \sin(\theta) + \rho_{\mathrm{V}} \left(\frac{Saq}{q_{\mathrm{V}}} \right) \left(\frac{q_{\mathrm{V}}}{\rho_{\mathrm{V}} S_{\mathrm{V}}} \right)^2$$
$$+ \frac{\Omega_{\mathrm{E}} \pi D \rho_{\mathrm{V}}}{q_{\mathrm{V}}} (T_{\mathrm{E}} - T_{\mathrm{V}}) - \frac{\Omega Sa \rho_{\mathrm{V}}}{q_{\mathrm{V}}} (T_{\mathrm{V}} - T_{\mathrm{L}}) \qquad (2.6)$$

$$\left\{ 1 - \Gamma_{P_{\mathrm{L}}} \left(\frac{q_{\mathrm{L}}}{\rho_{\mathrm{L}} S_{\mathrm{L}}} \right)^2 + \frac{T_{\mathrm{L}} \Gamma_{T_{\mathrm{L}}}}{\rho_{\mathrm{L}}} \right\} \frac{\partial P_{\mathrm{L}}}{\partial z}$$
$$+ \left\{ \frac{\rho \bar{C}_P^{\mathrm{L}}}{M_{\mathrm{L}}} - \Gamma_{T_{\mathrm{L}}} \left(\frac{q_{\mathrm{L}}}{\rho_{\mathrm{L}} S_{\mathrm{L}}} \right)^2 \right\} \frac{\partial T_{\mathrm{L}}}{\partial z}$$
$$= -\rho_{\mathrm{L}} g \sin(\theta) - \rho_{\mathrm{L}} \left(\frac{Saq}{q_{\mathrm{L}}} \right) \left(\frac{q_{\mathrm{L}}}{\rho_{\mathrm{L}} S_{\mathrm{L}}} \right) \qquad (2.7)$$
$$+ \frac{\Omega Sa \rho_{\mathrm{L}}}{q_{\mathrm{L}}} (T_{\mathrm{V}} - T_{\mathrm{L}})$$
$$+ \frac{Sa \rho_{\mathrm{L}}}{q_{\mathrm{L}}} \sum_k N_k \left(E_k^{\mathrm{V}} - E_k^{\mathrm{L}} \right)$$

$$\bar{E}_k^{\mathrm{V}} = \bar{H}_k^{\mathrm{V}} + \frac{M_k}{2} \left(\frac{q_{\mathrm{V}}}{\rho_{\mathrm{V}} S_{\mathrm{V}}} \right)^2 + M_k . g . h(z) \qquad (2.8)$$

$$\bar{E}_k^{\mathrm{L}} = \bar{H}_k^{\mathrm{L}} + \frac{M_k}{2} \left(\frac{q_{\mathrm{L}}}{\rho_{\mathrm{L}} S_{\mathrm{L}}} \right)^2 + M_k . g . h(z) \qquad (2.9)$$

$$\Psi_{\mathrm{V}} = \frac{1}{8} f_{\mathrm{V}} \left(\frac{q_{\mathrm{V}}^2}{\rho_{\mathrm{V}} S_{\mathrm{V}}^2} \right), f_{\mathrm{V}} = f_{\mathrm{V}}(\mathrm{Re}_{\mathrm{V}}, \frac{\varepsilon_{\mathrm{V}}}{D_{\mathrm{V}}}), \mathrm{Re}_{\mathrm{V}} = \frac{4q_{\mathrm{V}}}{\wp_{\mathrm{V}} \mu_{\mathrm{V}}} \qquad (2.10)$$

$$\Psi_{\mathrm{L}} = \frac{1}{8} f_{\mathrm{L}} \left(\frac{q_{\mathrm{L}}^2}{\rho_{\mathrm{L}} S_{\mathrm{L}}^2} \right), f_{\mathrm{L}} = f_{\mathrm{L}}(\mathrm{Re}_{\mathrm{L}}, \frac{\varepsilon_{\mathrm{L}}}{D_{\mathrm{L}}}), \mathrm{Re}_{\mathrm{L}} = \frac{4q_{\mathrm{L}}}{\wp_{\mathrm{L}} \mu_{\mathrm{L}}} \qquad (2.11)$$

$$D_{\mathrm{V}} = 4S_{\mathrm{V}}/\wp_{\mathrm{V}}, D_{\mathrm{L}} = 4S_{\mathrm{L}}/\wp_{\mathrm{L}} \qquad (2.12)$$

$$\wp_{\mathrm{V}} = \pi D + N_{\mathrm{HF}} \pi d_{\mathrm{o}}, \wp_{\mathrm{L}} = N_{\mathrm{HF}} \pi d_{\mathrm{i}} \qquad (2.13)$$

$$\mu_{\mathrm{V}}(T_{\mathrm{V}}, P_{\mathrm{V}}, \underline{V}), \mu_{\mathrm{L}}(T_{\mathrm{L}}, P_{\mathrm{L}}, \underline{L}) \qquad (2.14)$$

$$q_{\mathrm{V}} = \sum_{k=1}^{nc} M_k . V_k, q_{\mathrm{L}} = \sum_{k=1}^{nc} M_k . L_k \qquad (2.15)$$

$$q_{\mathrm{TM}} = \sum_{k=1}^{nc} M_k . N_k \qquad (2.16)$$

$$\Gamma_{T_{\mathrm{V}}} = \left(\frac{\partial \rho_{\mathrm{V}}}{\partial T_{\mathrm{V}}} \right)_{P_{\mathrm{V}}, \underline{V}}, \Gamma_{T_{\mathrm{L}}} = \left(\frac{\partial \rho_{\mathrm{L}}}{\partial T_{\mathrm{L}}} \right)_{P_{\mathrm{L}}, \underline{L}},$$
$$\Gamma_{P_{\mathrm{V}}} = \left(\frac{\partial \rho_{\mathrm{V}}}{\partial P_{\mathrm{V}}} \right)_{T_{\mathrm{V}}, \underline{V}}, \Gamma_{P_{\mathrm{L}}} = \left(\frac{\partial \rho_{\mathrm{L}}}{\partial P_{\mathrm{L}}} \right)_{T_{\mathrm{L}}, \underline{L}} \qquad (2.17)$$

$$\bar{H}_k^{\mathrm{V}}(T_{\mathrm{V}}, P_{\mathrm{V}}, \underline{V}), \bar{H}_k^{\mathrm{L}}(T_{\mathrm{L}}, P_{\mathrm{L}}, \underline{L}) \quad (k = 1 \ldots nc) \qquad (2.18)$$

$$\bar{C}_P^{\mathrm{V}}(T_{\mathrm{V}}, P_{\mathrm{V}}, \underline{V}), \bar{C}_P^{\mathrm{L}}(T_{\mathrm{L}}, P_{\mathrm{L}}, \underline{L}) \qquad (2.19)$$

$$\rho_{\mathrm{V}}(T_{\mathrm{V}}, P_{\mathrm{V}}, \underline{V}), \rho_{\mathrm{L}}(T_{\mathrm{L}}, P_{\mathrm{L}}, \underline{L}) \qquad (2.20)$$

$$\hat{f}_k^{\mathrm{V}}(T_{\mathrm{V}}, P_{\mathrm{V}}, \underline{V}), \hat{f}_k^{\mathrm{L}}(T_{\mathrm{L}}, P_{\mathrm{L}}, \underline{L}) \quad (k = 1 \ldots nc) \qquad (2.21)$$

2.1.2 Parameters of Parallel-Flow HFM Module

Three classes of geometrical and/or physical parameters have to be specified for substitution in the HFM model in Eqs. (2.1) to (2.21); namely: (i) HFM OD (d_{o}), HFM ID (d_{i}), and retentate/permeate flow roughnesses ($\varepsilon_{\mathrm{V}}, \varepsilon_{\mathrm{L}}$); (ii) module ID ($D$) and other module parameters θ, Z_{M}, N_{HF}, T_{E}, Ω, Ω_{E}; and (iii) kth species equivalent HFM permeance $\Pi_k(k = 1 \ldots nc)$.

Species equivalent HFM permeances are critical parameters defining both the selectivity of the MP module and its gas processing capacity in terms of the trans-membrane flux (mol s^{-1} m^{-2}). Equivalent HFM permeances allow the HFM parallel-flow model to approximately reproduce the responses of a real MP battery processing CO_2-rich NG using, for example, cellulose-acetate SWM elements (Table 1). Equivalent HFM premeances are supposed to be constant parameters. In the present case, Table 2 shows values of equivalent HFM permeances and heat transfer coefficients for CO_2 removal from CO_2-rich NG in the MP applications of Sect. 2. They were estimated with data of SWM MP units processing NG with $\approx$20 mol% CO_2, so that a parallel-flow horizontal ($\theta = 0°$) HFM module (10 m 0.2 m) with $Z_{\mathrm{M}} = 10$ m, $N_{\mathrm{HF}} = 110{,}000$ and approximate

Table 2 Equivalent HFM permeances, heat transfer coefficients, and HFM and module parameters for MP applications in Sect. 2 (HFM parallel-flow MP model)

Species	Permeance (mol s^{-1} m^{-2} Pa)	Internal and external heat transfer coefficients	HFM parameters and HFM module parameters
CO_2	1.33×10^{-8}	$\Omega = 13$ W m^{-2} K^{-1} $\Omega_E = 13$ W m^{-2} K^{-1}	Membrane: CAM
CH_4	4.61×10^{-10}		$d_o = 0.502$ mm $d_i = 0.5$ mm $\varepsilon_V = \varepsilon_L = 4.57 \times 10^{-2}$ mm
N_2	$\approx 4 \times 10^{-10}$		
C_2H_6	$\approx 1 \times 10^{-11}$		$D = 0.2$ m $Z_M = 10$ m $T_E = 25$ °C $\theta = 0°$ $N_{HF} = 110{,}000$ Module area: 1734.79 m^2
C_3H_8	$\approx 1 \times 10^{-12}$		
iC_4H_{10}, C_4H_{10}, C_5H_{12}	$\approx 1 \times 10^{-13}$		
C_6H_{14} to C_8H_{18}	$\approx 1 \times 10^{-14}$		
C_9H_{20} and heavier	$\approx 1 \times 10^{-15}$		

transfer area of 1735 m^2 would behave similarly as a horizontal SWM tube (10 m × 0.2 m) with 10 SWM cylindrical cartridges (1 m × 0.2 m). Table 2 presents values of the estimated species HFM equivalent permeances and internal/external heat transfer coefficients, besides the other HFM and module parameters considered in this Sect. 2.

2.1.3 Species Parameters for Calculations of Thermodynamic and Transport Properties

PR-EOS and SRK-EOS are used with ideal gas properties (enthalpy of formation, isobaric heat capacity) and species constants and species critical data (T_{Ck}, P_{Ck}, ω_k, M_k) from Reid et al. (1987) in order to obtain retentate and permeate thermodynamic properties in Eqs. (2.17) to (2.21). Dynamic viscosities (μ_V, μ_L) are predicted from Chung et al. (1988) model and also use the same species constants.

2.1.4 Process Specifications for Simulation of HFM Battery

Besides the HFM parameters (Sect. 2.1.2) and species parameters (Sect. 2.1.3), MP process specifications have to be supplied for simulation of a HFM unit. These specifications comprehend four classes of data: (i) Battery size: number of MP modules (N_M); (ii) permeate outlet pressure (P_L^{out}); (iii) MP battery gas feed: molar flow rate ($F^{Feed} = V^{in}$), species mol fractions ($Y^{Feed} = V^{in} \div V^{in}$), pressure ($P^{Feed} = P_V^{in}$) and temperature ($T^{Feed} = T_V^{in}$); (iv) head-loss ($\Delta P^{HEX}$) and final temperature of auxiliary heaters/coolers (T^{Heater}, T^{Cooler}); (v) outlet pressure and adiabatic efficiencies ($\eta\%$) of compressors; (vi) temperature of two-phase mixer-coolers M2F (T^{M2F}).

2.1.5 Numerical Procedure: Simulation of Parallel-Flow HFM Battery

Simulation of a single parallel-flow HFM module with a gas feed having divided flow rate F^{FEED}/N_M is perfectly

equivalent to the simulation of an entire battery with N_M modules and feed flow rate F^{FEED}. To solve a single parallel-flow HFM module, the set of $2nc + 4$ ODE's represented by Eqs. (2.2) to (2.7) have to be numerically integrated along the z-axis from $z = 0$ to $z = Z_M$. This gives the MP outlet values of T_V, P_V, T_L, P_L, L_k, and V_k. Several subsidiary algebraic relationships—Eqs. (2.1), (2.8) to (2.16)—and several thermodynamic/transport property predictors—Eqs. (2.14) and (2.17) to (2.21)—are embedded into Eqs. (2.2) to (2.7) and are also calculated for every integration step.

Thus, a MP HFM simulation with a NG composed by 15 species will entail the spatial integration of 34 ODE's. Besides its large size, another particular concern associated to the HFM ODE system has to do with its high stiffness degree due to the coexistence of ODE's of very different natures and scaling like mass, momentum, and energy balances for both retentate and permeate. In this regard, the MATLAB ODE solver ODE15s has been particularly efficient in solving Eqs. (2.2) to (2.7) with several specifications. ODE15s is used in all applications of Sect. 2.

2.1.6 Boundary Conditions for Simulation of HFM Module

For the retentate state variables (V_k, T_V, P_V), Eqs. (2.2) to (2.7) configure an initial value problem which needs only the initial condition related to the known gas feed stream at $z = 0$ as shown in Eq. (2.22). On the other hand, the situation is different for the permeate pressure in the vector of permeate state variables (L_k, T_L, P_L), which must satisfy a boundary-value problem expressed by the boundary conditions in Eqs. (2.23) and (2.24).

$$z = 0 : V_k = F^{FEED} \cdot Y_k^{FEED}/N_M \quad (k = 1 \ldots nc);$$
$$P_V = P^{FEED}; T_V = T^{FEED} \tag{2.22}$$

$$z = 0 : L_k = 0 \quad (k = 1 \ldots nc)$$
$$P_L = P_L^{(0)}, T_L = T_L^{(0)} \tag{2.23}$$

$$z = Z_M : \quad P_L = P_L^{out} \tag{2.24}$$

The initial permeate temperature at $z = 0$, $T_L^{(0)}$, is irrelevant and can be assumed with any value since the initial condition of all permeate species flow rate is zero ($L_k^{(z = 0)} = 0$), which turns the initial permeate into a fluid with zero isobaric heat capacity and, consequently, with a promptly adjustable temperature even if $T_L^{(0)}$ is badly chosen. On the other hand, the initial permeate pressure $P_L^{(z = 0)}$ has a relevant physical role because the permeate must flow through the inner HFM space toward the permeate outlet at P_L^{OUT} which is an important MP process specification. In other words, $P_L^{(z = 0)}$ must be greater than P_L^{OUT} and must be adjusted so that Eq. (2.24) is satisfied. In SPM2010, a shooting strategy is used to adjust $P_L^{(z = 0)}$ in order to achieve the specified P_L^{OUT} at the permeate outlet. Normally, an initial $P_L^{(z = 0)}$ guess is generated by assuming isothermal laminar gas flow through the inner HFM space with an estimated flow rate taking into account the expected average behavior of MP modules for CO_2-rich NG processing and the mol% of CO_2 in the feed gas.

2.2 CO2-Rich Natural Gas Feed for Simulation of Membrane-Permeation Units

A dehydrated CO_2-rich NG is chosen as gas feed for MP simulation with the HFM model. It is a NG with 60 mol% CO_2 labeled as G60 with data in Table 3. Figures 5, 6, 7, 8, and 9 depict vapor–liquid equilibrium (VLE) loci of G60 rendered by SPM2010 using PR-EOS. Figure 5 shows the VLE locus of G60 on plane $T \times P$, while Fig. 6 shows the $T \times P$ G60 VLE locus superposing the single-phase molar isobaric heat capacity $\overline{C}_P$ map under constant composition. Figures 7, 8, and 9 are similar plots onto the $T \times P$ G60 VLE locus, respectively, superposing the single-phase density ρ map, the single-phase sound speed map, and the single-phase isobaric expansivity $\Gamma_T = \left(\frac{\partial \rho}{\partial T} \right)_{P, \underline{Z}}$ map. All maps were created under constant G60 composition. The property maps let clear that G60 is modeled in SPM2010

with appropriate PR-EOS thermodynamic framework, including second-order properties (e.g., Γ_T) for the HFM model Eqs. (2.2) to (2.7). All forthcoming process and MP simulations adopted PR-EOS for thermodynamic modeling.

2.3 Process FGc40p20: Production of Fuel-Gas with 40 mol% CO2 from G60

This example #1 aims at producing the maximum possible flow rate of fuel-gas (FG) with 40 mol% CO_2 from 1.5 MMNm³/d of G60 at $T = 55\ °C$ and $P = 53$ bar. This FG is appropriated for burning in some gas-fired turbines and this is the primary target. A secondary target is to produce the CO_2-rich permeate at highest possible mol% CO_2 and at highest possible pressure in order to reduce hydrocarbon losses and compression power necessary to dispatch this fluid for enhanced oil recovery (EOR). Thus, a first attempt is to generate the CO_2-rich permeate at $P = 20$ bar. Process specifications (Sect. 2.1.4) comprise: $N_M = 38$ (per battery); $P_L^{out} = 20$ bar; (iii) G60 feed (F^{Feed}, $\underline{Y}^{Feed}$, P^{Feed}, T^{Feed}) in Table 3; (iv) $\Delta P^{HEX} = 0.5$ bar, $T^{Heater} = 45\ °C$, $T^{Cooler} = 30\ °C$; (v) compressor adiabatic efficiency $\eta\% = 75\%$; (vi) the temperature of the two-phase mixer-cooler M2F is $T^{M2F} = 5\ °C$.

The flowsheet FGc40p20 proposes to produce 40 mol% CO_2 FG using two serial MP batteries (Fig. 2) with $N_M = 38$ HFM 10 m × 0.2 m elements each (Table 2) both with permeate at $P_L^{out} = 20$ bar, but the second MP battery (MP#2) is used only if the first MP battery (MP#1) does not accomplish the targets. Figure 10 shows the flowsheet FGc40p20 using the rules of stream names as shown in Fig. 4. Since the load of CO_2 removal is not too high (from 60 to 40 mol%), it seems reasonable to produce permeate at high-pressure of $P_L^{out} = 20$ bar. Moreover, since MP operations generate temperature falls, FGc40p20 firstly apply a hydrocarbon dew-point adjustment (HCDPA) of G60 by cooling it to $T = 5\ °C$ in a two-phase mixer-cooler (M2F) followed by a two-phase separator (S2F) for liquid separation. M2F + S2F produce a small condensation of natural gas liquids (NGL) reducing the split vapor flow rate to 1.4581 MMNm³/d. As a result, the split vapor V@m2fG60 has a hydrocarbon dew-point (HCDP) of 5 °C at $P = 53$ bar and its dew-point locus retrocedes toward lower

Table 3 Data of G60 gas feed for simulation of HFM MP unit	Flow rate F^{Feed}	T^{Feed}, P^{Feed}	Species molar fractions $\underline{Y}^{Feed}$
	774.64 mol/s 96,067 kg/h 1.5 MMNm³/d	$T^{Feed} = 55\ °C$ $P^{Feed} = 53$ bar	$CO_2 = 60.006\%$ mol; $CH_4 = 33.983\%$ mol; $C_2H_6 = 2.39\%$ mol $C_3H_8 = 1.41\%$ mol; $N_2 = 0.85\%$ mol; $iC_4H_{10} = 0.25\%$ mol $C_4H_{10} = 0.45\%$ mol; $iC_5H_{12} = 0.13\%$ mol; $C_5H_{12} = 0.16\%$ mol $C_6H_{14} = 0.14\%$ mol; $C_7H_{16} = 0.07\%$ mol; $C_8H_{18} = 0.08\%$ mol $C_9H_{20} = 0.04\%$ mol; $C_{10}H_{22} = 0.02\%$ mol; $C_{11}H_{24} = 0.02\%$ mol

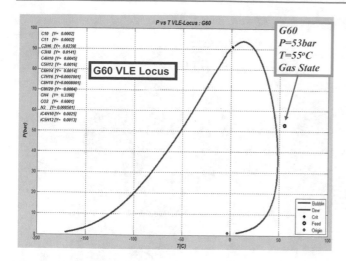

Fig. 5 $T \times P$ G60 VLE locus

temperatures as shown in Fig. 11 (compare with the "fatter" G60 dew-point locus as shown in Fig. 5).

The split vapor V@m2fG60 at $T = 5$ °C is heated to $T = 45$ °C becoming hV@m2fG60 which feeds the first MP stage (MP#1). Since $\Delta P^{HEX} = 0.5$ bar, the inlet pressure of hV@m2fG60 in MP#1 is $P_V^{in} = 52.5$ bar. The MP#1 retentate and permeate are, respectively, V@hV@m2fG60 and L@hV@m2fG60. The head-loss on the retentate path—computed by the retentate momentum balance Eq. (2.4)—reduces the outlet pressure of V@hV@m2fG60 to $P_V^{out} =$

51.84 bar (Fig. 12). Stream results of the dew-point adjustment of G60 via M2F + S2F (5 °C Flash) and MP#1 permeation results are shown in Fig. 12, where mol% CH_4, mol% CO_2, and flow rate (MMNm^3/d) are underlined in red. It is seen that only the first MP stage (MP#1) of FGc40p20 is sufficient to accomplish both targets: (i) The retentate V@hV@m2fG60 fuel-gas ($P = 51.84$ bar, $T = 31.8$ °C) has 39.7 mol% CO_2, 50.3 mol% CH_4, and 0.777 MMNm^3/d ($\approx 50\%$ of G60); and (ii) the permeate L@hV@m2fG60 ($P = 20$ bar, $T = 31.7$ °C) has 82.8 mol% CO_2, 16.8 mol% CH_4, and 0.681 MMNm^3/d.

Figure 13 depicts MP#1 axial profiles: (A) mol% in retentate (V); (B) mol% in permeate (L); (C) T_V, T_L (°C). In Fig. 13a, the retentate mol% CO_2 decreases monotonously until 39.7 mol% at the outlet, while the permeate mol% CO_2 (Fig. 13b) decreases smoothly from 93 mol% (the very first generated permeate) toward 82.8 mol% at the permeate outlet. The underlying reason for such %CO_2 fall in the permeate is the increasing CH_4 permeation in MP#1 due to the simultaneous decrease of CO_2 driving force and increase of CH_4 driving force. Figure 13c depicts the retentate/permeate temperature profiles. Temperature firstly falls in the permeate due to Joule–Thomson effects associated with CO_2 permeation; then retentate/permeate heat transfer takes place "breaking" the permeate cooling at the same time slowly cooling the retentate. The retentate has a much greater heat capacity than the permeate, especially in the first fourth of the HFM's where the permeate has a low flow rate

Fig. 6 $T \times P$ G60 VLE locus with single-phase $\overline{C}_P$ (kJ/mol K) map superposed

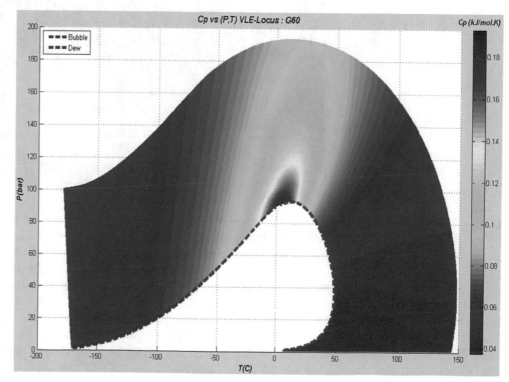

Fig. 7 $T \times P$ G60 VLE locus
with single-phase density ρ
(kg/m^3) map superposed

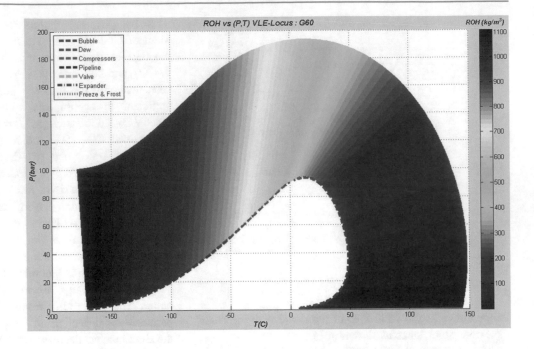

Fig. 8 $T \times P$ G60 VLE locus
with single-phase sound speed
(m/s) map superposed

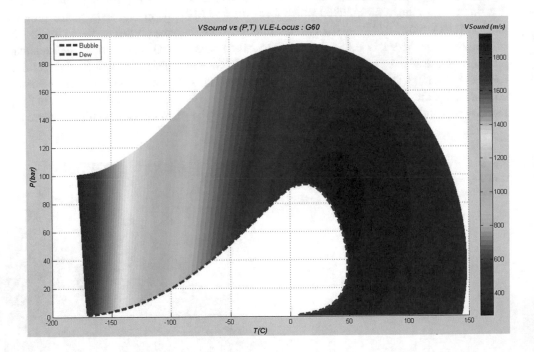

($z \leq 2.5$ m). Since the temperature of the MP#1 feed
hV@m2fG60 is 45 °C, the MP#1 permeation generated
13.3 °C of cooling of retentate and permeate
($T_V^{out} \approx T_L^{out} = 31.7$ °C).

Figure 14 depicts the retentate path on plane $T \times P$. In
order to investigate the occurrence of condensation in the
retentate path, the dew-point (HCDP) locus of the MP#1
feed (blue line) and the dew-point (HCDP) locus of the
retentate product (dashed-magenta line) are also plotted. It
can be seen that the retentate dew-point locus becomes
gradually hotter due to the loss of light species CO_2 and CH_4
that are transferred to the permeate. In other words, the
retentate becomes gradually heavier on its path through
MP#1, consequently, its dew-point locus gradually moves
from the blue line to the dashed-magenta line. At the same
time, the retentate cools down due to Joule–Thomson effects
associated to the permeation of CH_4 and mostly of CO_2. This
is seen clearly in Fig. 14 (see the magnified view at the
right); i.e., the retentate path and the retentate dew-point
locus move in opposed directions toward each other. If they

Fig. 9 $T \times P$ G60 VLE locus with single-phase $\Gamma_T = \left(\frac{\partial \rho}{\partial T}\right)_{P,\underline{Z}}$ (kg/m³ K) map

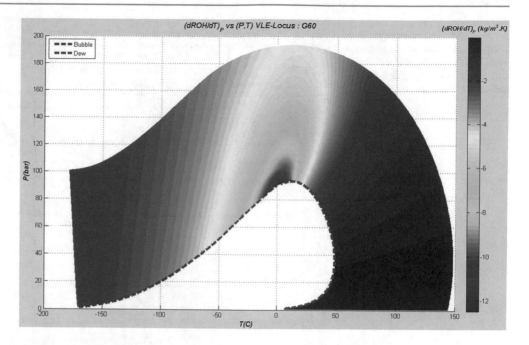

Fig. 10 Process FGc40p20 for G60: 40 mol% CO_2 FG with permeate at $P = 20$ bar

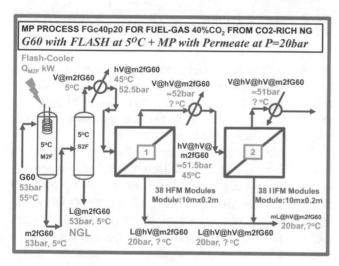

encounter each other and the retentate path crosses its dew-point locus, condensation occurs. In the present case, the initial hot feed temperature of 45 °C was devised to prevent such encounter. The temperature difference between the retentate path and its dew-point locus is ≈40 °C at the beginning of the permeation (Fig. 11) and was reduced at the end of permeation to 18 °C (Fig. 14), a safe operational margin to avoid the inconvenience of such condensation which can damage HFM's.

Figure 15 depicts a summary of results and a diagnosis of the performance of process FGc40p20. In brief, process FGc40p20 was reasonably successful. Starting with 1.5 MMNm³/d of G60, both goals—production of 0.777 MMNm³/d of 40 mol% CO_2 FG and high-pressure ($P = 20$ bar) permeate with high 82.8 mol% of CO_2—were attained. But there are other alternatives for processing G60. The next section explores a different one.

2.4 Process FGc20p7: Production of Fuel-Gas with 20 mol% CO_2 from G60

Example #2 aims at producing the maximum possible flow rate of a fuel-gas (FG) with less CO_2 than the previous case in Sect. 2.3. That is, new target#1 is to produce a better FG with 20 mol% CO_2 from 1.5 MMNm³/d of G60 at $T = 55$ °C and $P = 53$ bar. New target#2 is to produce CO_2-rich permeate at highest possible mol% CO_2 and at highest possible pressure in order to reduce CH_4 losses and compression power demanded for EOR. But it is necessary to recognize that a FG with 20 mol% CO_2 is a harder target than the previous one (40 mol% CO_2) in Sect. 2.3. Consequently, the permeate pressure has to be lower to increase the MP driving force. This is the reason to try $P = 7$ bar as permeate outlet pressure. Moreover, the attained FG (20 mol% CO_2) flow rate in this case should be lower than the previous one in Sect. 2.3 and the permeate mol% CO_2 is likely to be lower than the counterpart in Sect. 2.3; i.e., other things constant, greater CH_4 losses to the permeate are expected in this case since the driving force is much bigger. Thus, in order to compensate the selectivity loss caused by a higher driving force, a lesser MP area is recommendable now (i.e., less HFM 10 m × 0.2 m elements are used) for same feed flow rate of Sect. 2.3. This helps to re-erect the CO_2/CH_4 selectivity in order to produce permeate with

Fig. 11 FGc40p20:
hV@m2fG60 VLE locus (MP#1
feed) after dew-point adjustment
in M2F/S2F (flash) at $T = 5$ °C;
dew-point locus retrocedes to
colder temperatures

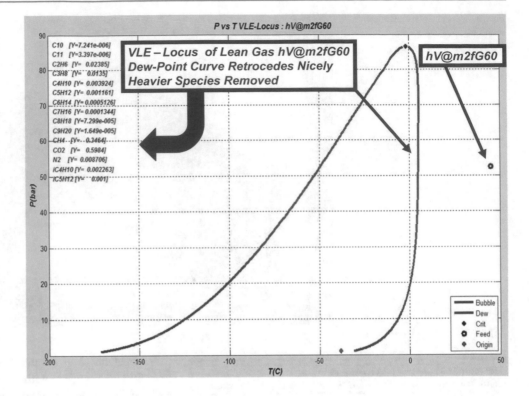

better mol% CO_2 than the previous case in Sect. 2.3. On the other hand, if less HFM elements are to be used for same gas feed—i.e., each HFM 10 m × 0.2 m element has a greater gas flow rate to re-erect selectivity—it also should be expected a greater head-loss through the HFM elements in the first MP stage. In other words, the retentate from the first MP stage are expected to attain a lower outlet pressure (P_V^{out}) relatively to the same result in Sect. 2.3. Besides, thanks to the greater expansion to $P_L^{out} = 7$ bar, the temperature fall through the first MP stage will also be greater due to a deeper Joule–Thomson effect. Process specifications (Sect. 2.1.4) are the same of Sect. 2.3, except that lower P_L^{out} and less HFM modules are used: $N_M = 20$ (per battery); $P_L^{out} = 7$ bar; (iii) G60 feed (F^{Feed}, $\underline{Y}^{Feed}$, P^{Feed}, T^{Feed}) in Table 3; (iv) $\Delta P^{HEX} = 0.5$ bar, $T^{Heater} = 45$ °C, $T^{Cooler} = 30$ °C; (v) compressor adiabatic efficiency $\eta\% = 75\%$; (vi) two-phase mixer-cooler M2F at $T^{M2F} = 5$ °C.

Thus, the flowsheet FGc20p7 proposes producing 20 mol % CO_2 FG via two smaller serial MP batteries (Fig. 2) with $N_M = 20$ HFM 10 m × 0.2 m elements each (Table 2) both with permeate outlet at $P_L^{out} = 7$ bar. Again, the second MP battery (MP#2) is used only if the first MP battery (MP#1) does not accomplish targets. Flowsheet FGc20p7 is shown in Fig. 16 using the rules in Fig. 4 for stream names. Since this deeper MP operation generates a greater cooling than the previous case in Sect. 2.3, FGc20p7 firstly apply the same hydrocarbon dew-point adjustment (HCDPA) of G60 by cooling it to $T = 5$ °C in the same two-phase mixer-cooler (M2F) followed by a two-phase separator (S2F) for liquid

separation. The same small condensation of NGL occurs reducing the split vapor flow rate to 1.4581 MMNm³/d. As a result, the split vapor V@m2fG60 has a hydrocarbon dew-point of 5 °C at $P = 53$ bar and its dew-point locus retrocedes toward lower temperatures (Fig. 11).

The split vapor V@m2fG60 at $T = 5$ °C is heated to $T = 45$ °C becoming hV@m2fG60 which feeds the first MP stage (MP#1). Since $\Delta P^{HEX} = 0.5$ bar, the inlet pressure of hV@m2fG60 in MP#1 is $P_V^{in} = 52.5$ bar. The MP#1 retentate and permeate are, respectively, V@hV@m2fG60 and L@hV@m2fG60. The head-loss on the retentate path— computed by the retentate momentum balance Eq. (2.4)— reduces the outlet pressure of V@hV@m2fG60 to a lower value (see first paragraph of Sect. 2.4); i.e., $P_V^{out} = 51.42$ bar (Fig. 17). Figure 17 depicts stream results of the HCDPA of G60 via M2F + S2F (5 °C Flash) and permeation results of MP#1, where mol% CH_4, mol% CO_2, and flow rate (MMNm³/d) are underlined in red. Again, only MP#1 of FGc20p7 is sufficient to attain both goals: (i) The retentate V@hV@m2fG60 fuel-gas ($P = 51.42$ bar, $T = 22.19$ °C) has 19.365 mol% CO_2, 68.081 mol% CH_4, and 0.6273 MMNm³/d ($\approx$42% of G60); and (ii) the permeate L@hV@m2fG60 ($P = 7$ bar, $T = 21.97$ °C) has 90.404 mol % CO_2, 9.395 mol% CH_4, and 0.8308 MMNm³/d.

Figure 18 depicts MP#1 axial profiles of FGc20p7: (A) mol% in retentate (V); (B) mol% in permeate (L); and (C) T_V, T_L (°C). In Fig. 18a, the retentate mol% CO_2 decreases monotonously until 19.365 mol% at the outlet, while the permeate mol% CO_2 (Fig. 18b) decreases

G60 with FLASH at 5°C + MP with Permeate at P=20bar

Stream	G60 VAPOR	m2fG60 LIQ+VAP	V@m2fG60 VAPOR	L@m2fG60 LIQUID	hV@m2fG60 VAPOR	V@hV@m2fG60 VAPOR	L@hV@m2fG60 VAPOR
C10 [J1] (%mol)	0.02	0.02	0.001	0.691	0.001	0.001	0
C11 [J1] (%mol)	0.02	0.02	0	0.704	0	0.001	0
C2H6 [V] (%mol)	2.39	2.39	2.385	2.588	2.385	4.474	0.001
C3H8 [V] (%mol)	1.41	1.41	1.35	3.489	1.35	2.534	0
C4H10 [V] (%mol)	0.45	0.45	0.392	2.455	0.392	0.736	0
C5H12 [V] (%mol)	0.16	0.16	0.116	1.686	0.116	0.218	0
C6H14 [V] (%mol)	0.14	0.14	0.051	3.227	0.051	0.096	0
C7H16 [V] (%mol)	0.07	0.07	0.013	2.038	0.013	0.025	0
C8H18 [V] (%mol)	0.08	0.08	0.007	2.609	0.007	0.014	0
C9H20 [V] (%mol)	0.04	0.04	0.002	1.374	0.002	0.003	0
CH4 [V] (%mol)	33.983	33.983	34.641	11.098	34.641	50.273	16.807
CO2 [V] (%mol)	60.006	60.006	59.844	65.652	59.844	39.7	82.827
N2 [V] (%mol)	0.85	0.85	0.871	0.138	0.871	1.314	0.365
iC4H10[V] (%mol)	0.25	0.25	0.226	1.077	0.226	0.425	0
iC5H12[V] (%mol)	0.13	0.13	0.1	1.174	0.1	0.188	0
T(C)	55	5	5	5	45	31.843	31.743
P(bar)	53	53	53	53	52.5	51.84	19.99
kg/mol	0.034449	0.034449	0.034036	0.048795	0.034036	0.029466	0.039251
mol/s	774.64	774.64	752.99	21.646	752.99	401.28	351.71
kgph	96067	96067	92264	3802.3	92264	42566	49698
MMm3/d	1.5	1.5	1.4581	0.041914	1.4581	0.77703	0.68105
V (m3/mol)	0.00042891	0.00029281	0.00029929	6.7507e-005	0.00041346	0.00039932	0.0011535
Density(kg/m3)	80.317	117.65	113.72	722.82	82.32	73.79	34.029
%molVapor	100	97.206	100	0	100	100	100
Z	0.83322	0.67108	0.68593	0.15472	0.82064	0.81636	0.9096
Viscosity(Pa.s)	1.793e-005	4.8993e-005	3.6634e-005	0.00018827	1.7352e-005	1.6857e-005	1.5454e-005
HFLUX(kW)	-2.0742e+005	-2.098e+005	-2.029e+005	-6896.6	-2.0119e+005	-81911	-1.1932e+005
SFLUX(kW/K)	-50.426	-58.337	-54.217	-4.1197	-48.422	-35.159	-11.606
HFLUX-T0*SFLUX(kW)	-1.9238e+005	-1.924e+005	-1.8673e+005	-5668.3	-1.8676e+005	-71428	-1.1586e+005
H(kJ/mol)	-267.76	-270.83	-269.46	-318.62	-267.19	-204.12	-339.25

Fig. 12 Process FGc40p20 with feed G60: results of HCDPA of G60 via flash (M2F + S2F) at $T = 5$ °C and permeation results of MP#1 battery

smoothly from 97 mol% (the very first generated permeate) toward 90.404 mol% at the permeate outlet. As before, the underlying reason is the increasing CH_4 permeation in MP#1 due to simultaneous decrease of CO_2 driving force. Figure 18c depicts the retentate/permeate temperature profiles. Temperature firstly falls in permeate due to CO_2 permeation Joule–Thomson effects; then retentate/permeate heat transfer takes place "breaking" the permeate cooling, and slowly cooling down the retentate which has a much greater heat capacity. Since the initial temperature of the MP#1 feed hV@m2fG60 is 45 °C, the permeation in the first MP stage of FGc20p7 produced a more intense retentate/permeate cooling of 23 °C ($T_V^{out} \approx T_L^{out} = 22\,°C$) as discussed in the first paragraph of Sect. 2.4.

Figure 19 depicts the retentate path on plane $T \times P$. To investigate the occurrence of condensation on the retentate path, the dew-point (HCDP) locus of the MP#1 feed (blue line) and the dew-point (HCDP) locus of the retentate product (dashed-magenta line) are also plotted. It can be seen that the retentate dew-point locus becomes gradually hotter due to the loss of light species CO_2 and CH_4 via permeation. In other words, the retentate becomes gradually heavier on its path through MP#1 and its dew-point locus moves from the blue line to the dashed-magenta line. Simultaneously, the retentate cools down due to Joule–Thomson effects associated with the permeation of CH_4 and mostly of CO_2. This is clearly seen in Fig. 19; i.e., the retentate path and the retentate dew-point locus move toward each other. If they encounter and the retentate path crosses its dew-point locus, condensation occurs. In FGc20p7, the initial hot feed temperature of 45 °C was again sufficient to prevent such encounter, though a lesser margin of only 6 °C remained for the occurrence of condensation at the end of the retentate flow path (compare with the analogous wider margin of 18 °

Fig. 13 FGc40p20 MP#1 axial profiles: **a** mol% in V; **b** mol% in L; **c** T (°C)

(a) *G60 with FLASH at 5ºC + MP with Permeate at P=20bar*

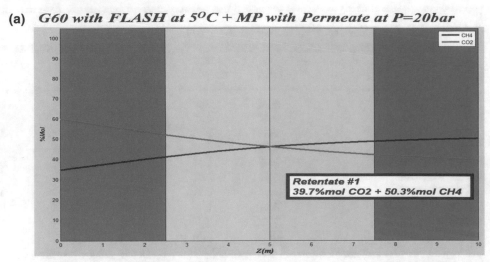

Retentate #1
39.7%mol CO2 + 50.3%mol CH4

(b) *G60 with FLASH at 5ºC + MP with Permeate at P=20bar*

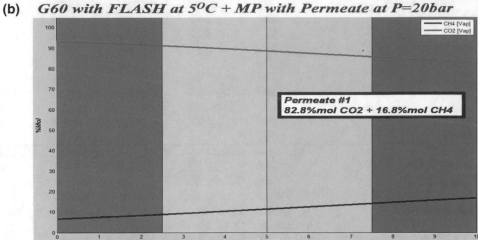

Permeate #1
82.8%mol CO2 + 16.8%mol CH4

(c) *G60 with FLASH at 5ºC + MP with Permeate at P=20bar*

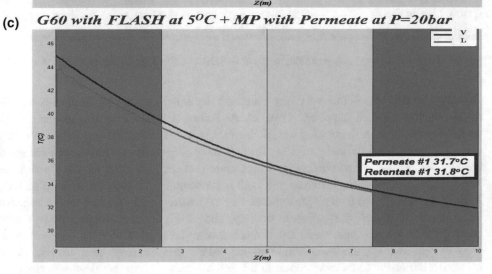

Permeate #1 31.7°C
Retentate #1 31.8°C

C as shown in Fig. 14 for process FGc40p20). The temperature difference between the retentate and its HCDP locus was again ≈40 °C at the beginning of permeation (Fig. 11) and was reduced to 6 °C at the end of permeation (Fig. 19),

a narrow but acceptable margin to avoid retentate condensation.

Figure 20 presents a summary of results and a diagnosis of the performance of process FGc20p7. In brief, the process

Fig. 14 FGc40p20 retentate path on plane $T \times P$. Blue line: feed dew-point locus; dashed-magenta line: final retentate dew-point locus; black-dot line: retentate path. Path does not reach dew-point locus: no condensation predicted for MP#1

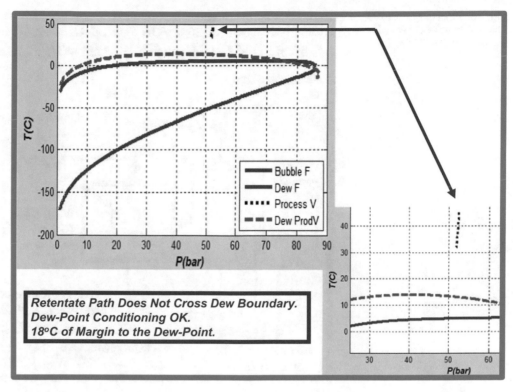

Retentate Path Does Not Cross Dew Boundary.
Dew-Point Conditioning OK.
18°C of Margin to the Dew-Point.

Fig. 15 Process FGc40p20 diagnosis: targets accomplished; MP#2 not necessary

G60 with FLASH at 5°C + MP with Permeate at P=20bar

Diagnosis: [Successful]

Battery #1 with 38 HFM Modules Tight (only Stage 1 Used)   [OK]

Retentate Product (Fuel-Gas) 39.7%CO2+50.3%CH4+4.74%C2+2.5%C3   [OK]

Retentate Flow Rate (Fuel-Gas) 777030 Nm³/d =833909 Sm³/d   [OK]

No Retentate Condensation in MP#1   [OK]

NGL Produced: 3.8ton/h with 65.7%CO2+11.1%CH4+2.6%C2+3.5%C3+3.5%C4

Permeate P=20bar: 0.6811MMNm³/d with 82.8%CO2 + 16.8%CH4

Flash Pre-Conditioning T=5°C, 53bar: Necessary to avoid MP condensation.

HFM Modules: 38
N_{HF}:   110,000 Hollow-Fibers
MP#1 Area: 38*1734.79 = 65,992 m²
MP#1 Feed: 1.4581MMNm³/d,   P=52.5bar,   T=45°C
MP#1 Retentate: 0.777030 MMNm³/d,   P=51.84bar,   T=31.8°C
MP#1 Permeate: 0.6811MMNm³/d,   P=20bar,   T=31.7°C
External T: 25°C
Internal and External MP Heat Transfer Coeficients: 13W/m².K
Permeances: CO2: 1.332.10⁻⁸ mol/s.m².Pa, CH4: 4.614.10⁻¹⁰ mol/s.m².Pa

Fig. 16 Process FGc20p7 for G60: 20 mol% CO_2 FG with permeate at $P = 7$ bar

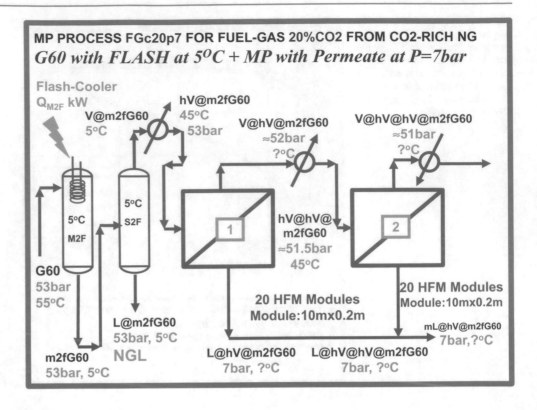

FGc20p7 was also successful. Starting with 1.5 MMNm³/d of G60, both goals—production of 0.62725 MMNm³/d of 20 mol% CO_2 FG and medium-pressure ($P = 7$ bar) permeate with excellent 90.404 mol% of CO_2—were attained. Naturally, the production of 20 mol% CO_2 FG of FGc20p7 is inferior to the counterpart of FGc40p20 with 40 mol% CO_2, but the former FG is of better quality comparatively to the latter.

2.5 Remarks

Section 2 demonstrated the capabilities of software SPM2010 for simulating real MP flowsheets for the purification of CO_2-rich NG. SPM2010 models MP units as a set of HFM co-current parallel modules using full thermodynamic modeling via appropriate equations of state (e.g., PR-EOS) for permeate and retentate non-isothermal flows. Component mass, energy, and momentum balances for multicomponent axial compressible flow are solved simultaneously for both permeate and retentate in order to predict MP separation, temperature, and pressure axial distributions and respective heat effects. Besides the MP units, several other peripheral units are also available in SPM2010 with full thermodynamic equilibrium modeling (e.g., flashes, M2F, S2F, compressors, expanders, exchangers, membrane contactors, etc.). SPM2010 can also be applied to different MP configurations

via calibration procedures. In the present case, the HFM-based model of SPM2010 was previously calibrated using real NG processing data of SWM cross-flow MP units operating CO_2 removal from CO_2-rich NG in offshore oil-and-gas fields in Brazil.

MP2010 is also under continuous expansion in order to install new developments to improve its capabilities for simulation of MP units for CO_2 removal of CO_2-rich natural gas. Some examples of future expansions are the following: (i) To adopt permeances that are functions of the local retentate temperature; (ii) to adopt permeances that are functions of the local retentate temperature and also of the retentate fugacity of CO_2 in order to mimic membrane plasticization effects caused by high CO_2 fugacity; (iii) to install a distributed MP model for counter-current flows of permeate and retentate using a boundary-value problem framework.

3 HYSYS Implementation of Membrane-Permeation Units

Concerning CO_2-rich NG processing in offshore platforms, membrane permeation is one of many steps of gas purification. Therefore, for simulation and assessments of such complex processes comprising MP, specific membrane models must be developed to be inserted in the simulator

G60 with FLASH at 5°C + MP with Permeate at P=7bar

Stream	G60 VAPOR	m2fG60 LIQ+VAP	V@m2fG60 VAPOR	L@m2fG60 LIQUID	hV@m2fG60 VAPOR	V@hV@m2fG60 VAPOR	L@hV@m2fG60 VAPOR
C10 [J1](%mol)	0.02	0.02	0.001	0.691	0.001	0.002	0
C11 [J1](%mol)	0.02	0.02	0	0.704	0	0.001	0
C2H6 [V](%mol)	2.39	2.39	2.385	2.588	2.385	5.542	0
C3H8 [V](%mol)	1.41	1.41	1.35	3.489	1.35	3.139	0
C4H10[V](%mol)	0.45	0.45	0.392	2.455	0.392	0.912	0
C5H12[V](%mol)	0.16	0.16	0.116	1.686	0.116	0.27	0
C6H14[V](%mol)	0.14	0.14	0.051	3.227	0.051	0.119	0
C7H16[V](%mol)	0.07	0.07	0.013	2.038	0.013	0.031	0
C8H18[V](%mol)	0.08	0.08	0.007	2.609	0.007	0.017	0
C9H20[V](%mol)	0.04	0.04	0.002	1.374	0.002	0.004	0
CH4 [V](%mol)	33.983	33.983	34.641	11.098	34.641	68.081	9.395
CO2 [V](%mol)	60.006	60.006	59.844	65.652	59.844	19.365	90.404
N2 [V](%mol)	0.85	0.85	0.871	0.138	0.871	1.758	0.2
iC4H10[V](%mol)	0.25	0.25	0.226	1.077	0.226	0.526	0
iC5H12[V](%mol)	0.13	0.13	0.1	1.174	0.1	0.232	0
T(C)	55	5	5	5	45	22.187	21.968
P(bar)	53	53	53	53	52.5	51.419	6.9907
kg/mol	0.034449	0.034449	0.034036	0.048795	0.034036	0.024349	0.04135
mol/s	774.64	774.64	752.99	21.646	752.99	323.93	429.06
kgph	96067	96067	92264	3802.3	92264	28394	63870
MMm3/d	1.5	1.5	1.4581	0.041914	1.4581	0.62725	0.83083
V(m3/mol)	0.00042891	0.00029281	0.00029929	6.7507e-005	0.00041346	0.00039352	0.0033805
Density(kg/m3)	80.317	117.65	113.72	722.82	82.32	61.873	12.232
%molVapor	100	97.206	100	0	100	100	100
Z	0.83322	0.67108	0.68593	0.15472	0.82064	0.82407	0.96316
Viscosity(Pa.s)	1.793e-005	4.8993e-005	3.6634e-005	0.00018827	1.7352e-005	1.612e-005	1.4418e-005
HFLUX(kW)	-2.0742e+005	-2.098e+005	-2.029e+005	-6896.6	-2.0119e+005	-45265	-1.5594e+005
SFLUX(kW/K)	-50.426	-58.337	-54.217	-4.1197	-48.422	-35.378	-8.3726
HFLUX-TO*SFLUX(kW)	-1.9238e+005	-1.924e+005	-1.8673e+005	-5668.3	-1.8676e+005	-34717	-1.5344e+005
H(kJ/mol)	-267.76	-270.83	-269.46	-318.62	-267.19	-139.74	-363.44

Fig. 17 Process FGc20p7 for feed G60: results of HCDPA of G60 via flash (M2F + S2F) at $T = 5$ °C and permeation results of MP#1 battery

process flow diagrams (PFDs). In the case of HYSYS professional software simulation, these customizable modules are called unit operation extensions (UOEs).

Arinelli et al. (2017) developed a steady-state MP unit for simulation in HYSYS: MP-UOE. The model consists of a lumped short-cut method that makes an analogy with shell and tube heat exchangers, where the driving force of MP is the log mean of species partial pressure differences in membrane extremities. Overall mass balance and component mass balance equations complete the permeation model. MP-UOE component permeances were calibrated with real data from offshore platforms with CAM decarbonating CO_2-rich NG, therefore, despite being a short-cut method, it reproduces with good agreement real NG separation results. On the other hand, permeances were adjusted as constant average values, thus the independent of temperature and

CO_2 fugacity (the main permeating component), which is a simplification of the real operation. After permeation calculations, the energy balance around MP module finishes MP-UOE algorithm, considering a default value for the difference between product temperatures. The extension admits both HFM and SWM types, for countercurrent or parallel flows. MP-UOE was used in a variety of CO_2-rich NG decarbonating studies of the authors, with different processing scenarios and configurations (Araújo et al. 2017; Arinelli et al. 2017, 2019).

In this section, MP-UOE is further developed with improvements regarding the energy balance, which is now defined locally for each stream inside MP, generating two new model categories: (i) Lumped MP models for parallel and counter-current permeate/retentate flows using average driving forces and lumped balances (MPx-UOE); and

Fig. 18 FGc20p7 MP#1 axial
profiles: **a** mol% in V; **b** mol% in
L; **c** T (°C)

(a)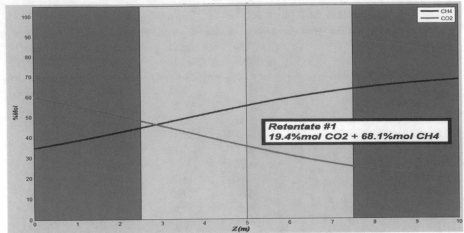

(b)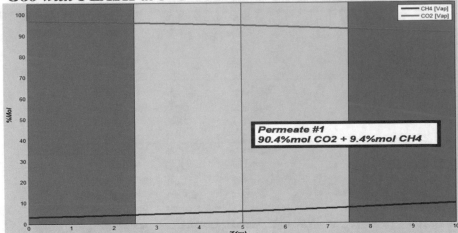

(c)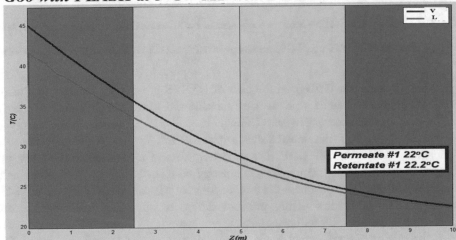

Fig. 19 FGc20p7 retentate path on plane $T \times P$. Blue line: feed dew-point locus; dashed-magenta line: final retentate dew-point locus; black-dotted line: retentate path. The path does not reach dew-point: no condensation predicted for MP#1

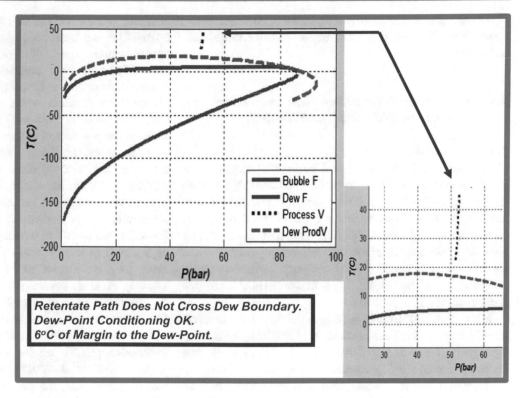

Retentate Path Does Not Cross Dew Boundary.
Dew-Point Conditioning OK.
6°C of Margin to the Dew-Point.

Fig. 20 Process FGc20p7 diagnosis: targets accomplished; MP#2 not necessary

G60 with FLASH at 5°C + MP with Permeate at P=7bar

Diagnosis: [Successful]

Battery #1 with 20 Modules Tight (only Stage #1 Used)   [OK]

Retentate Product (Fuel-Gas) 19.4%CO2+68.1%CH4+5.54%C2+3.14%C3   [OK]

Retentate Flow Rate (Fuel-Gas) 627250 Nm³/d = 673165 Sm³/d   [OK]

No Retentate Condensation   [OK]

NGL Produced: 3.8ton/h with 65.7%CO2+11.1%CH4+2.6%C2+3.5%C3+3.5%C4

Permeate P=7bar: 0.83083MMNm³/d with 90.4%CO2+9.4%CH4

Flash Pre-Conditioning T=5°C, 53bar: Necessary to avoid MP condensation

HFM Modules:	20
N_{HF} :	110,000 Hollow-Fibers
MP#1 Area:	20*1734.79 = 34,695.8 m²

Feed: 1.4581MMNm³/d,   P=52.5bar,   T=45°C
Retentate: 0.627250 MMNm³/d,   P=51.4bar,   T=22.2°C
Permeate: 0.83083MMNm³/d,   P=7bar,   T=22.0°C
External T: 25°C
Internal and External Heat Transfer Coefficients: 13W/m².K
Permeances: CO2: 1.332.10⁻⁸ mol/s.m².Pa, CH4: 4.614.10⁻¹⁰ mol/s.m².Pa

(ii) 1D-distributed MP models for parallel permeate/retentate flows using distributed driving forces and distributed balances (MPd-UOE).

3.1 Membrane-Permeation Unit Operation Extensions: MPx-UOE and MPd-UOE

3.1.1 Premises

UOEs were developed with Visual Basic (VB) programming language, generating DLLs to be installed in HYSYS. They are loaded in HYSYS as customized operations, and after installation, their icons appear on the HYSYS operations palette. MPx-UOE and MPd-UOE have their own property window to set specifications such as design parameters and operational conditions. The property windows were designed in View Editor, a software available in the Aspentech HYSYS package.

MPx-UOE and MPd-UOE both simulate steady-state MP units using a short-cut method to calculate the species k transmembrane molar fluxes, N_k (MMSm^3/m^2 d), that needs calibration of permeances. The model draws an analogy between membrane units and shell and tube heat exchangers, where the retentate would flow in the shell, and permeate, in the tube. The permeation driving force is the log mean difference of partial pressures of species k, ΔP_k^{LN} (bar). The difference between the two units is that MPx-UOE is a lumped model that considers the membrane unit as one block with one feed and two products, and thus the short-cut method is applied for this block and the fluids paths through the membrane unit are not assessed.

Differently from MPx-UOE, MPd-UOE is a distributed model; i.e., profiles of dependent variables are obtained throughout the MP unit. MPd-UOE divides the membrane unit into smaller membrane cells of same permeation area, consecutively, applying the MP algorithm for each element. The outlet condition of first MP cell is calculated based on the specified inlet condition and will be the inlet condition of the subsequent cell, and from then on until it completes the entire membrane unit. Therefore, if the number of MP elements is high enough, the fluids paths can be attained. The short-cut method error decreases for more distributed simulations, as the number of elements increases and the size of permeation cells decreases, thus calculations are more accurate.

Permeation area A_{MP} (m^2), retentate and permeate pressures P_V^{out}, P_L^{out} (bar), and membrane type—HFM or SWM, as depicted in Fig. 1—must be selected by user in both MPx-UOE and MPd-UOE property windows. The extensions automatically retrieve feed data—molar composition $\underline{Y}^{in}$, molar flow rate V^{in} (MMSm^3/d), temperature T_V^{in} (K), pressure (bar) P_V^{in}, and molar enthalpy $\overline{H}_V^{in}$—from the

material stream connected to the unit operation. For MPx-UOE, the user must also select the contact type—counter-current contact (CC) or parallel contact (PC)—while for MPd-UOE, only parallel contact type is admitted. Transmembrane molar fluxes (N_k) are considered positive in the direction retentate → permeate. In MPd-UOE, the permeate head-loss is fixed as 0.1 bar and equally distributed through the permeation elements.

Permeances Π_k of main species involved in CO_2 separation from NG are defined in the MP models for both HFM and SWM, as shown in Table 4, yet can be set otherwise in the UOE property window. Table 4 values were calibrated in element with real MP separation data of pre-salt offshore NG processing with CAM. Permeances of H_2S and H_2O were estimated as equal to the CO_2 value, since they are known to be high for skin-dense CAM, showing good adherence when compared to the real data. N_2 permeance was estimated as similar to the CH_4 value. C3+ permeances are small, so they were estimated from the C_2H_6 value with reduction of 90% per additional C atom. Permeation of C5+ species is negligible with CAM. Such as in Arinelli et al. (2017), despite being calibrated with real operation data, the permeances were adjusted as constant average values, independent of temperature, and CO_2 fugacity.

Retentate and permeate temperatures, T_V^{out}, T_L^{out} (K), are calculated via energy balance equations for both streams, considering the partial molar enthalpies of species permeating from retentate to permeate ($N_k A_{MP} \langle \overline{H}_k \rangle$), and external and internal heat exchanges. Considering the shell and tube analogy of the short-cut method, the external heat transfer in MP is between retentate and the vicinity, while the internal heat transfer is between retentate and permeate streams. The external temperature T_E is defined as 25 °C in the MP models but can be set otherwise in the UOE property window. Overall, the heat transfer coefficients for internal and external heat exchanges are defined as 5 and 2 W/m^2 K, respectively.

For determination of log mean of temperature differences in membrane extremities for internal heat transfer calculation, temperature of permeate at the beginning of permeation is needed. Since it is unknown, the parameter ΔT_F was created, where $\Delta T_F = T_V^{out} - T_L^{in}$ for counter-current contact type, and $\Delta T_F = T_V^{in} - T_L^{in}$ for parallel contact type. ΔT_F has a default value of 3 °C in the extensions yet can be changed by the user in the UOE property window. In MPd-UOE, ΔT_F specification is only valid for the first membrane element; it can be calculated for the next elements as $\Delta T_F = T_V^{in} - T_L^{in}$, where the inlet streams are the outlet streams of the previous element. In Sect. 3.2.4, a sensitivity analysis is conducted to assess the impact of ΔT_F specification for both MPx-UOE and MPd-UOE.

Table 4 Permeances in MPx-UOE and MPd-UOE

Component	Permeance HF (Π_k) (MMSm³/d m² bar)	Permeance SW (Π_k) (MMSm³/d m² bar)
CO_2	2.77E−6	1.95E−5
CH_4	2.77E−7	2.16E−6
C_2H_6	9.57E−9	6.75E−8
H_2S	2.77E−6	1.95E−5
H_2O	2.77E−6	1.95E−5
N_2	3.07E−7	2.16E−6
C_3H_8	9.57E−10	6.75E−9
iC_4H_{10}	9.57E−11	6.75E−10
C_4H_{10}	9.57E−11	6.75E−10
C5+	9.57E−12	6.75E−11

For the distributed model, as the number of elements increases, the area of each membrane element decreases, and the log means used in the short-cut method approximate to the respective arithmetic means. Therefore, in MPd-UOE, it was considered that for a number of elements equal or higher than 10, and the log means are replaced by arithmetic means.

3.1.2 Lumped Model Algorithm: MPx-UOE

The algorithm for MPx-UOE model comprises five steps, which are described below: [S1] Input data; [S2] Parameters for energy balance; [S3] Initial values for NRM; [S4] Lumped permeation and energy balance calculations; [S5] Returning product data to simulation.

[S1] Input data: Feed temperature, pressure, molar flow, composition, and enthalpy are rescued from feed stream in HYSYS PFD in Eq. (3.1a). Permeation area, product pressures, and contact type are defined by the user in the UOE property window—Eq. (3.1b). Default values for species k permeances are depicted in Table 4, yet the user can specify otherwise in the UOE property window, as depicted in Eq. (3.1c).

$$T_V^{in}, P_V^{in}, V^{in}, \underline{Y}^{in}, \overline{H}_V^{in} \text{ from simulation environment} \quad (3.1a)$$

$$A_{MP}, P_V^{out}, P_L^{out}, \text{Contact defined by user} \quad (3.1b)$$

$$\Pi_k \text{ default from Table 4 or defined by user} \quad (3.1c)$$

[S2] Parameters for energy balance: ΔT_F and external temperature (T_E) both have default value specified in MPx-UOE—Eqs. (3.2a) and (3.2b), respectively—however, the user can set other values in UOE property window. Internal and external overall heat transfer coefficients (U_I and U_E) are defined in Eqs. (3.2c) and (3.2d), respectively. The internal area for heat transfer is equal to the defined permeation area via Eq. (3.2e). Equation (3.2f) shows the relation between the external and internal areas for heat transfer.

$$\Delta T_F = 3\,°C \text{ (default) or defined by user} \quad (3.2a)$$

$$T_E = 25\,°C \text{ (default) or defined by user} \quad (3.2b)$$

$$U_I = 5\,W/m^2\,K \quad (3.2c)$$

$$U_E = 2\,W/m^2\,K \quad (3.2d)$$

$$A_I = A_{MP} \quad (3.2e)$$

$$A_E = A_I/276 \quad (3.2f)$$

[S3] Initial values for NRM: Eqs. (3.3a) to (3.3j) set the initial values of species trans-membrane molar fluxes. Initial values of retentate and permeate temperatures are defined respectively by Eqs. (3.3k) and (3.3m) if contact is CC, or by Eqs. (3.3n) and (3.3o) if contact is PC.

$$N_{CO_2} = 0.5 * V_{CO_2}^{in}/A_{MP} \quad (3.3a)$$

$$N_{CH_4} = 0.075 * V_{CH_4}^{in}/A_{MP} \quad (3.3b)$$

$$N_{C_2H_6} = 0.01 * V_{C_2H_6}^{in}/A_{MP} \quad (3.3c)$$

$$N_{C_3H_8} = 0.005 * V_{C_3H_8}^{in}/A_{MP} \quad (3.3d)$$

$$N_{i-C_4H_{10}} = 0.0015 * V_{i-C_4H_{10}}^{in}/A_{MP} \quad (3.3e)$$

$$N_{n-C_4H_{10}} = 0.0015 * V_{n-C_4H_{10}}^{in}/A_{MP} \quad (3.3f)$$

$$N_{H_2O} = 0.5 * V_{H_2O}^{in}/A_{MP} \quad (3.3g)$$

$$N_{H_2S} = 0.5 * V_{H_2S}^{in}/A_{MP} \quad (3.3h)$$

$$N_{N_2} = 0.075 * V_{N_2}^{in}/A_{MP} \quad (3.3i)$$

$$N_{k \neq CO_2,CH_4,C_2H_6,C_3H_8,iC_4H_{10},nC_4H_{10},H_2O,H_2S,N_2} = 0.0001 * V_k^{in}/A_{MP} \quad (k = 1...nc) \quad (3.3j)$$

If Contact=Counter-Current Then

$$T_V^{out} = T_V^{in} - 10 \tag{3.3k}$$

$$T_L^{out} = T_V^{in} - 5 \tag{3.3m}$$

Elseif Contact=Parallel Then

$$T_V^{out(0)} = T_V^{in(0)} - 5 \tag{3.3n}$$

$$T_L^{out(0)} = T_V^{in(0)} - 15 \tag{3.3o}$$

End if

[S4] Lumped permeation and energy balance calculations: The Newton–Raphson method (NRM) is applied for the target equations described in Eqs. (3.4a) to (3.4c), which represent the transmembrane molar fluxes of species k, and the energy balance equations for retentate and permeate streams, respectively. The driving force in Eq. (3.4a) is the log mean of partial pressure differences of species k (ΔP_k^{LN}), which varies accordingly to the contact type: if CC, it is defined by Eq. (3.4d); if PC, by Eq. (3.4f); where $P_V^{in}Y_k^{in}$, $P_V^{out}Y_k^{out}$, $P_L^{out}X_k^{out}$ represent the partial pressures of species k in feed, retentate, and permeate, respectively. The same methodology applies to the log mean of temperature differences for internal heat transfer (ΔT_I^{LN}): if contact is counter-current, Eq. (3.4e) is applied; if it is parallel, Eq. (3.4g) is selected. Equations (3.4h) and (3.4i) represent the log mean of temperature differences between the retentate and the external area, and the log mean of partial molar enthalpies of species k in retentate stream, respectively. Equations (3.4j) to (3.4p) are applied to calculate the molar flow rates and molar compositions of retentate and permeate streams ($V^{out}, L^{out}, \underline{Y}^{out}, \underline{X}^{out}$, respectively). The molar enthalpies of retentate and permeate streams ($\overline{H}_V^{out}$ and $\overline{H}_L^{out}$) are obtained via flash($T_V^{out}, P_V^{out}, \underline{Y}^{out}$) and flash($T_L^{out}, P_L^{out}, \underline{X}^{out}$), respectively. In summary, the MPx-UOE model comprises a system of $7nc + 6$ nonlinear equations Eqs. (3.4a) to (3.4p), to be numerically solved by NRM for $7nc + 6$ variables $N_k, \Delta P_k^{LN}, T_V^{out}, T_L^{out}, \Delta T_I^{LN}, \Delta T_E^{LN}, \langle \overline{H}_k \rangle$, $L_k^{out}, L^{out}, X_k^{out}, V_k^{out}, V^{out}, Y_k^{out}$.

Target Equations:

$$N_k - \Pi_k \Delta P_k^{LN} = 0 \quad (k = 1...nc) \tag{3.4a}$$

$$\begin{aligned} V^{out}\overline{H}_V^{out} - V^{in}\overline{H}_V^{in} \\ - U_E A_E \Delta T_E^{LN} + U_I A_I \Delta T_I^{LN} \\ + \sum_{k=1}^{nc} N_k A_I \langle \overline{H}_k \rangle = 0 \end{aligned} \tag{3.4b}$$

$$L^{out}\overline{H}_L^{out} - U_I A_I \Delta T_I^{LN} - \sum_{k=1}^{nc} N_k A_I \langle \overline{H}_k \rangle = 0 \tag{3.4c}$$

Auxiliary Equations:
If Contact=Counter-Current Then

$$\Delta P_k^{LN} = \left(\frac{\left(P_V^{in}Y_k^{in} - P_L^{out}X_k^{out}\right) - \left(P_V^{out}Y_k^{out}\right)}{\ln\left(\frac{P_V^{in}Y_k^{in} - P_L^{out}X_k^{out}}{P_V^{out}Y_k^{out}}\right)} \right) \quad (k = 1...nc)$$

$$\tag{3.4d}$$

$$\Delta T_I^{LN} = \left[\frac{\Delta T_F - T_V^{in} + T_L^{out}}{\ln\left(\frac{\Delta T_F}{T_V^{in} - T_L^{out}}\right)} \right] \tag{3.4e}$$

Elseif Contact=Parallel Then

$$\Delta P_k^{LN} = \left(\frac{\left(P_V^{in}Y_k^{in}\right) - \left(P_V^{out}Y_k^{out} - P_L^{out}X_k^{out}\right)}{\ln\left(\frac{P_V^{in}Y_k^{in}}{P_V^{out}Y_k^{out} - P_L^{out}X_k^{out}}\right)} \right) \quad (k = 1...nc)$$

$$\tag{3.4f}$$

$$\Delta T_I^{LN} = \left[\frac{\Delta T_F - T_V^{out} + T_L^{out}}{\ln\left(\frac{\Delta T_F}{T_V^{out} - T_L^{out}}\right)} \right] \tag{3.4g}$$

End if

$$\Delta T_E^{LN} = \left[\frac{T_V^{in} - T_V^{out}}{\ln\left(\frac{T_E - T_V^{out}}{T_E - T_V^{in}}\right)} \right] \tag{3.4h}$$

$$\langle \overline{H}_k \rangle = \frac{\overline{H}_{V_k}^{out} - \overline{H}_{V_k}^{in}}{\ln\left(\frac{\overline{H}_{V_k}^{out}}{\overline{H}_{V_k}^{in}}\right)} \quad (k = 1...nc) \tag{3.4i}$$

$$L_k^{out} = N_k A_{MP} \quad (k = 1...nc) \tag{3.4j}$$

$$L^{out} = \sum_{k}^{nc} L_k^{out} \tag{3.4k}$$

$$X_k^{out} = \frac{L_k^{out}}{L^{out}} \quad (k = 1...nc) \tag{3.4m}$$

$$V_k^{out} = V^{in}Y_k^{in} - N_k A_{MP} \quad (k = 1...nc) \tag{3.4n}$$

$$V^{out} = \sum_{k}^{nc} V_k^{out} \tag{3.4o}$$

$$Y_k^{out} = \frac{V_k^{out}}{V^{out}} \quad (k = 1...nc) \tag{3.4p}$$

[S5] Returning product data to simulation: Data of retentate and permeate streams are pasted onto the product streams of MPx-UOE in HYSYS PFD via Eqs. (3.5a) and (3.5b).

$$\text{Retentate Stream}: T_V^{\text{out}}, P_V^{\text{out}}, V^{\text{out}}, \underline{Y}^{\text{out}} \qquad (3.5a)$$

$$\text{Permeate Stream}: T_L^{\text{out}}, P_L^{\text{out}}, L^{\text{out}}, \underline{X}^{\text{out}} \qquad (3.5b)$$

3.1.3 Distributed Model Algorithm: MPd-UOE

The algorithm for MPd-UOE model comprises six steps, which are described below: [S1] Input data; [S2] Adjusting input parameters for first permeation element; [S3] Parameters for energy balance; [S4] Initial values for first permeation element NRM; [S5] Distributed permeation and energy balance calculations loop; [S6] Returning product data to simulation.

[S1] Input data: Feed temperature, pressure, molar flow, composition, and enthalpy are rescued from feed stream in HYSYS PFD in Eq. (3.6a). Total permeation area, product pressures, and contact type are defined by the user in the UOE property window—Eq. (3.6b). Note that differently from MPx-UOE, the contact type is not a specification, since the MPd-UOE model is valid only for parallel MP. Default values for species k permeances are depicted in Table 4, yet the user can specify otherwise in the UOE property window, as depicted in Eq. (3.6c).

$$T_V^{\text{in}}, P_V^{\text{in}}, V^{\text{in}}, \underline{Y}^{\text{in}}, \overline{H}_V^{\text{in}} \text{ from simulation environment} \quad (3.6a)$$

$$A_{\text{MP}}, P_V^{\text{out}}, P_L^{\text{out}}, \text{n\_elements defined by user} \qquad (3.6b)$$

$$\Pi_k \text{ default from Table 4 or defined by user} \qquad (3.6c)$$

[S2] Adjusting input parameters for first permeation element: Feed conditions are set as main inlet parameters for the first permeation element in Eq. (3.7a). Since there is no second inlet stream in the MP module, the molar flow and composition for the first element are set to zero in Eq. (3.7b). Head-loss in retentate stream is linearly distributed through the permeation elements, so for the first element, the outlet retentate pressure is set by Eq. (3.7c). For the permeate stream, the final outlet pressure was specified in Eq. (3.6b), yet there is no inlet permeate stream, so the head-loss is selected and fixed as 0.1 bar, and linearly distributed through the permeation elements; thus, the outlet permeate pressure for the first element is given by Eq. (3.7d). Permeation elements are equally distributed in the total permeation area also defined in Eq. (3.6b), so for each element, the permeation area is a fraction of the total specification, as shown in Eq. (3.7e).

$$T_V^{\text{in}^{(0)}} = T_V^{\text{in}}, P_V^{\text{in}^{(0)}} = P_V^{\text{in}},$$
$$V^{\text{in}^{(0)}} = V^{\text{in}}, \underline{Y}^{\text{in}^{(0)}} = \underline{Y}^{\text{in}}, \overline{H}_V^{\text{in}^{(0)}} = \overline{H}_V^{\text{in}} \qquad (3.7a)$$

$$L^{\text{in}^{(0)}} = 0, \underline{X}^{\text{in}^{(0)}} = \underline{0} \qquad (3.7b)$$

$$P_V^{\text{out}^{(0)}} = P_V^{\text{in}} - (P_V^{\text{in}} - P_V^{\text{out}})/\text{n\_elements} \qquad (3.7c)$$

$$P_L^{\text{out}^{(0)}} = P_L^{\text{out}} + (0.1/\text{n\_elements}) * (\text{n\_elements} - 1) \qquad (3.7d)$$

$$A_{\text{MP}} = A_{\text{MP}}/\text{n\_elements} \qquad (3.7e)$$

[S3] Parameters for energy balance: $\Delta T_F^{(0)}$ and external temperature (T_E) both have default values specified in MPd-UOE—Eqs. (3.8a) and (3.8b), respectively—however, the user can set other values in UOE property window. Note that differently from MPx-UOE algorithm, in MPd-UOE, the ΔT_F specification is set as the value for the first permeation element only ($\Delta T_F^{(0)}$), since it is calculated for the next elements in step [S5]. Internal and external overall heat transfer coefficients (U_I and U_E) are defined in Eqs. (3.8c) and (3.8d), respectively. The internal area for heat transfer is equal to the permeation area of each element, via Eq. (3.8e). Equation (3.8f) shows the relation between the external and internal areas for heat transfer.

$$\Delta T_F^{(0)} = 3\,°\text{C (default) or defined by user} \qquad (3.8a)$$

$$T_E = 25\,°\text{C (default) or defined by user} \qquad (3.8b)$$

$$U_I = 5\,\text{W/m}^2\,\text{K} \qquad (3.8c)$$

$$U_E = 2\,\text{W/m}^2\,\text{K} \qquad (3.8d)$$

$$A_I = A_{\text{MP}} \qquad (3.8e)$$

$$A_E = A_I/276 \qquad (3.8f)$$

[S4] Initial values for first permeation element NRM: Eqs. (3.9a) to (3.9j) set initial values of species transmembrane molar fluxes for the first permeation area, which depend on the number of membrane elements chosen by the user; the more distributed, the lower the permeation area for each element, yet the higher the number of elements, which is quadratic, so the lower is the permeation flux. Equations (3.9k) and (3.9m), respectively, define initial values of retentate and permeate temperatures for the first permeation area.

$$N_{\text{CO}_2}^{(0)} = 0.5 * V_{\text{CO}_2}^{\text{in}^{(0)}}/(A_{\text{MP}} * \text{n\_elements}^2) \qquad (3.9a)$$

$$N_{\text{CH}_4}^{(0)} = 0.075 * V_{\text{CH}_4}^{\text{in}^{(0)}}/(A_{\text{MP}} * \text{n\_elements}^2) \qquad (3.9b)$$

$$N_{\text{C}_2\text{H}_6}^{(0)} = 0.01 * V_{\text{C}_2\text{H}_6}^{\text{in}^{(0)}}/(A_{\text{MP}} * \text{n\_elements}^2) \qquad (3.9c)$$

$$N_{\text{C}_3\text{H}_8}^{(0)} = 0.005 * V_{\text{C}_3\text{H}_8}^{\text{in}^{(0)}}/(A_{\text{MP}} * \text{n\_elements}^2) \qquad (3.9d)$$

$$N_{i-C_4H_{10}}^{(0)} = 0.0015 * V_{i-C_4H_{10}}^{in^{(0)}}/(A_{MP} * \text{n\_elements}^2) \quad (3.9e)$$

$$N_{n-C_4H_{10}}^{(0)} = 0.0015 * V_{n-C_4H_{10}}^{in^{(0)}}/(A_{MP} * \text{n\_elements}^2) \quad (3.9f)$$

$$N_{H_2O}^{(0)} = 0.5 * V_{H_2O}^{in^{(0)}}/(A_{MP} * \text{n\_elements}^2) \quad (3.9g)$$

$$N_{H_2S}^{(0)} = 0.5 * V_{H_2S}^{in^{(0)}}/(A_{MP} * \text{n\_elements}^2) \quad (3.9h)$$

$$N_{N_2}^{(0)} = 0.075 * V_{N_2}^{in^{(0)}}/(A_{MP} * \text{n\_elements}^2) \quad (3.9i)$$

$$N_{k \neq CO_2,CH_4,C_2H_6,C_3H_8,iC_4H_{10},nC_4H_{10},H_2O,H_2S,N_2}^{(0)}$$
$$= 10^{-4} * V_k^{in^{(0)}} \div (A_{MP} * \text{n\_elements}^2) \quad (k = 1 \ldots nc)$$
$$(3.9j)$$

$$T_V^{out^{(0)}} = T_V^{in^{(0)}} - 5 \quad (3.9k)$$

$$T_L^{out^{(0)}} = T_V^{in^{(0)}} - 15 \quad (3.9m)$$

[S5] Distributed permeation and energy balance calculations loop: The distributed model comprises n_elements loops, starting from index $m = 0$ until (n_elements − 1). The NRM is applied for the target equations of the current MP element, described in Eqs. (3.10a) to (3.10c), which represent the transmembrane molar fluxes of species k, and the energy balance equations for retentate and permeate streams, respectively. As the number of elements selected by the user increases, the area of each membrane element decreases, and the log mean approximates to the arithmetic mean. Therefore, the log means of MPx-UOE algorithm ($\Delta P_k^{LN}, \Delta T_I^{LN}, \Delta T_E^{LN}, \langle \bar{H}_k \rangle$) are valid in MPd-UOE for n_elements < 10; for higher values, they are replaced by the respective arithmetic means. This procedure is described by Eqs. (3.10d) to (3.10k). Equations (3.10m) to (3.10r) are applied to calculate the molar flow rates and molar compositions of outlet streams of the current element ($V^{out^{(m)}}, L^{out^{(m)}}, \underline{Y}^{out^{(m)}}, \underline{X}^{out^{(m)}}$). The respective molar enthalpies ($\bar{H}_V^{out^{(m)}}$ and $\bar{H}_L^{out^{(m)}}$) are obtained via flash($T_V^{out^{(m)}}$, $P_V^{out^{(m)}}, \underline{Y}^{out^{(m)}}$) and flash($T_L^{out^{(m)}}, P_L^{out^{(m)}}, \underline{X}^{out^{(m)}}$). The model comprises a system of $7nc + 6$ nonlinear equations Eqs. (3.10a) to (3.10r), to be numerically solved by NRM for $7nc + 6$ variables $N_k^{(m)}, \Delta P_k^{LN}, T_V^{out^{(m)}}, T_L^{out^{(m)}}, \Delta T_I^{LN}, \Delta T_E^{LN}$, $\langle \bar{H}_k \rangle, L_k^{out^{(m)}}, L^{out^{(m)}}, X_k^{out^{(m)}}, V_k^{out^{(m)}}, V^{out^{(m)}}, Y_k^{out^{(m)}}$ for each MP element. After finding the NRM solution for the current element, Eqs. (3.10s) to (3.10y) are applied to set parameters and initial values for the next element as follows: (i) retentate and permeate streams from current element respectively become main and second feed of the next element—Eqs. (3.10s) to (3.10u); (ii) ΔT_F of the next element is calculated as the difference between the temperatures of the two feed

streams—Eq. (3.10v); and (iii) initial values for NRM of the next element are set for $N_k, T_V^{out}, T_L^{out}$ variables—Eqs. (3.10x) to (3.10y). Then, the algorithm loops again for the next m MP element.

For $m = 0$ to $(n\_ elements - 1)$
　　　　　　NRM Block Begins

Target Equations:

$$N_k^{(m)} - \Pi_k \Delta P_k^{LN} = 0 \quad (k = 1 \ldots nc) \quad (3.10a)$$

$$V^{out^{(m)}} \bar{H}_V^{out^{(m)}} - V^{in^{(m)}} \bar{H}_V^{in^{(m)}}$$
$$- U_E A_E \Delta T_E^{LN} + U_I A_I \Delta T_I^{LN}$$
$$+ \sum_{k=1}^{nc} N_k^{(m)} A_I \langle \bar{H}_k \rangle = 0 \quad (3.10b)$$

$$L^{out^{(m)}} \bar{H}_L^{out^{(m)}} - L^{in^{(m)}} \bar{H}_L^{in^{(m)}}$$
$$- U_I A_I \Delta T_I^{LN} - \sum_{k=1}^{nc} N_k^{(m)} A_I \langle \bar{H}_k \rangle = 0 \quad (3.10c)$$

Auxiliary Equations:
If $n\_elements \geq 10$ then

$$\Delta P_k^{LN} \approx \Delta P_k^{Arith.}$$
$$= \left(\frac{\left(P_V^{in^{(m)}} Y_k^{in^{(m)}} \right) + \left(P_V^{out^{(m)}} Y_k^{out^{(m)}} - P_L^{out^{(m)}} X_k^{out^{(m)}} \right)}{2} \right)$$
$$(k = 1 \ldots nc)$$
$$(3.10d)$$

$$\Delta T_I^{LN} \approx \Delta T_I^{Arith.}$$
$$= \left[\frac{\Delta T_F^{(m)} + (T_V^{out^{(m)}} - T_L^{out^{(m)}})}{2} \right] \quad (3.10e)$$

$$\Delta T_E^{LN} \approx \Delta T_E^{Arith.}$$
$$= \left[\frac{(T_E - T_V^{in^{(m)}}) + (T_E - T_V^{out^{(m)}})}{2} \right] \quad (3.10f)$$

$$\langle \bar{H}_k \rangle = \frac{\bar{H}_{V_k}^{out^{(m)}} + \bar{H}_{V_k}^{in^{(m)}}}{2} \quad (3.10g)$$
$$(k = 1 \ldots nc)$$

Else

$$\Delta P_k^{LN} = \left(\frac{\left(P_V^{in^{(m)}} Y_k^{in^{(m)}} \right) - \left(P_V^{out^{(m)}} Y_k^{out^{(m)}} - P_L^{out^{(m)}} X_k^{out^{(m)}} \right)}{\ln \left(\frac{P_V^{in^{(m)}} Y_k^{in^{(m)}}}{P_V^{out^{(m)}} Y_k^{out^{(m)}} - P_L^{out^{(m)}} X_k^{out^{(m)}}} \right)} \right)$$
$$(k = 1 \ldots nc)$$
$$(3.10h)$$

$$\Delta T_{\mathrm{I}}^{\mathrm{LN}} = \left[\frac{\Delta T_{\mathrm{F}}^{(m)} - T_{\mathrm{V}}^{\mathrm{out}(m)} + T_{\mathrm{L}}^{\mathrm{out}(m)}}{\ln\left(\frac{\Delta T_{\mathrm{F}}^{(m)}}{T_{\mathrm{V}}^{\mathrm{out}(m)} - T_{\mathrm{L}}^{\mathrm{out}(m)}}\right)} \right] \tag{3.10i}$$

$$\Delta T_{\mathrm{E}}^{\mathrm{LN}} = \left[\frac{T_{\mathrm{V}}^{\mathrm{in}(m)} - T_{\mathrm{V}}^{\mathrm{out}(m)}}{\ln\left(\frac{T_{\mathrm{E}} - T_{\mathrm{V}}^{\mathrm{out}(m)}}{T_{\mathrm{E}} - T_{\mathrm{V}}^{\mathrm{in}(m)}}\right)} \right] \tag{3.10j}$$

$$\langle \bar{H}_k \rangle = \frac{\bar{H}_{V_k}^{\mathrm{out}(m)} - \bar{H}_{V_k}^{\mathrm{in}(m)}}{\ln\left(\frac{\bar{H}_{V_k}^{\mathrm{out}(m)}}{\bar{H}_{V_k}^{\mathrm{in}(m)}}\right)} \tag{3.10k}$$
$$(k = 1 \ldots nc)$$

End if

$$L_k^{\mathrm{out}(m)} = L^{\mathrm{in}(m)} X_k^{\mathrm{in}(m)} + N_k^{(m)} A_{\mathrm{MP}} \tag{3.10m}$$
$$(k = 1 \ldots nc)$$

$$L^{\mathrm{out}(m)} = \sum_k^{nc} L_k^{\mathrm{out}(m)} \tag{3.10n}$$

$$X_k^{\mathrm{out}(m)} = \frac{L_k^{\mathrm{out}(m)}}{L^{\mathrm{out}(m)}} \tag{3.10o}$$
$$(k = 1 \ldots nc)$$

$$V_k^{\mathrm{out}(m)} = V^{\mathrm{in}(m)} Y_k^{\mathrm{in}(m)} - N_k^{(m)} A_{\mathrm{MP}} \tag{3.10p}$$
$$(k = 1 \ldots nc)$$

$$V^{\mathrm{out}(m)} = \sum_k^{nc} V_k^{\mathrm{out}(m)} \tag{3.10q}$$

$$Y_k^{\mathrm{out}(m)} = \frac{V_k^{\mathrm{out}(m)}}{V^{\mathrm{out}(m)}} \quad (k = 1 \ldots nc) \tag{3.10r}$$

NRM Block Ends

Adjusting parameters and initial values for next element:
If $m < (n\_elements - 1)$ Then

$$V^{\mathrm{in}(m+1)} = V^{\mathrm{out}(m)}, L^{\mathrm{in}(m+1)} = L^{\mathrm{out}(m)},$$
$$P_{\mathrm{V}}^{\mathrm{in}(m+1)} = P_{\mathrm{V}}^{\mathrm{out}(m)}, P_{\mathrm{L}}^{\mathrm{in}(m+1)} = P_{\mathrm{L}}^{\mathrm{out}(m)} \tag{3.10s}$$

$$T_{\mathrm{V}}^{\mathrm{in}(m+1)} = T_{\mathrm{V}}^{\mathrm{out}(m)}, T_{\mathrm{L}}^{\mathrm{in}(m+1)} = T_{\mathrm{L}}^{\mathrm{out}(m)},$$
$$\bar{H}_{\mathrm{V}}^{\mathrm{in}(m+1)} = \bar{H}_{\mathrm{V}}^{\mathrm{out}(m)}, \bar{H}_{\mathrm{L}}^{\mathrm{in}(m+1)} = \bar{H}_{\mathrm{L}}^{\mathrm{out}(m)} \tag{3.10t}$$

$$Y_k^{\mathrm{in}(m+1)} = Y_k^{\mathrm{out}(m)}, X_k^{\mathrm{in}(m+1)} = X_k^{\mathrm{out}(m)} \tag{3.10u}$$
$$(k = 1 \ldots nc)$$

$$\Delta T_{\mathrm{F}}^{(m+1)} = T_{\mathrm{V}}^{\mathrm{in}(m+1)} - T_{\mathrm{L}}^{\mathrm{in}(m+1)} \tag{3.10v}$$

$$N_k^{(m+1)} = 0.3 * L_k^{\mathrm{in}(m+1)} / (m+1) \tag{3.10w}$$
$$(k = 1 \ldots nc)$$

$$T_{\mathrm{V}}^{\mathrm{out}(m+1)} = T_{\mathrm{V}}^{\mathrm{in}(m+1)}$$
$$- (5/n\_elements) * (m+2) \tag{3.10x}$$

$$T_{\mathrm{L}}^{\mathrm{out}(m+1)} = T_{\mathrm{V}}^{\mathrm{out}(m+1)} - 10 \tag{3.10y}$$

End if
Next m

[S6] Returning product data to simulation: Data of retentate and permeate streams of the final permeation element are pasted onto the product streams of MPd-UOE in the HYSYS PFD via Eqs. (3.11a) and (3.11b).

$$\text{Retentate Stream}: T_{\mathrm{V}}^{\mathrm{out}}, P_{\mathrm{V}}^{\mathrm{out}}, V^{\mathrm{out}}, \underline{Y}^{\mathrm{out}} \tag{3.11a}$$

$$\text{Permeate Stream}: T_{\mathrm{L}}^{\mathrm{out}}, P_{\mathrm{L}}^{\mathrm{out}}, L^{\mathrm{out}}, \underline{X}^{\mathrm{out}} \tag{3.11b}$$

3.2 Models Performance for CO$_2$-Rich Natural Gas Processing

3.2.1 Premises

MPx-UOE and MPd-UOE were applied to simulate CO$_2$ removal from a hypothetical CO$_2$-rich NG after dehydration for water dew-point adjustment (WDPA) and hydrocarbon dew-point adjustment (HCDPA) on offshore platforms. Table 5 shows the NG feed conditions used in all simulations. MP cases were simulated in HYSYS v8.8 with PR-EOS, which is indicated as thermodynamic modeling of NG processing operations. All optional parameters of MPx-UOE and MPd-UOE were used as default values, except for ΔT_{F} regarding the sensitivity analysis in Sect. 3.2.4. Retentate pressure was set according to a fixed head-loss of 1 bar per MP stage. Permeate pressure was chosen as 4 bar in all simulations. Both counter-current and parallel contact types were evaluated for MPx-UOE. Permeation areas defined for each stage configuration in Sect. 3.2.2 are maintained for the next sections simulations. Head-loss of heat exchangers was fixed at 0.5 bar. Cooling-water (CW) was used in compressors intercoolers, reducing gas temperature to 45 °C. Pressurized hot water (PHW) produced in gas turbines waste heat recovery units in the platform was used as heating utility.

3.2.2 Stage Configuration: MPx-UOE

Different process configurations can be used in MP modules to capture CO$_2$ from NG as shown in Fig. 2. Three configurations were selected for evaluation with MPx-UOE: (i) one single MP stage (Fig. 21a); (ii) two serial MP stages (Fig. 21b); and (iii) one MP stage followed by a second

Table 5 Feed conditions of CO_2-rich natural gas

Parameter	Value
Vapor fraction	1.00
Temperature (°C)	62.00
Pressure (bar)	45.00
Molar flow (MMSm^3/d)	12.00
%CO_2	45.23
%CH_4	42.22
%C_2H_6	6.03
%C_3H_8	4.02
%i-C_4H_{10}	0.50
%n-C_4H_{10}	1.01
%i-C_5H_{12}	0.26
%n-C_5H_{12}	0.19
%n-C_6H_{14}	0.24
%n-C_7H_{16}	0.03
%n-C_8H_{18}	0.01
%N_2	0.26
ppmH_2O	1.00

stage for the first permeate stream with recycle of second retentate to the first stage (Fig. 21c). To compare the three possibilities, MPx-UOE was used with CC contact, HFM, and default values of parameters described in Sect. 3.1.2. For two serial stage configuration, the permeation area of the second stage was set as half of the area set for the first stage. Moreover, a heater was added between the two stages to

avoid condensation in the second stage, which would happen otherwise for this contact type—temperature drop in retentate stream is higher for CC than for PC. For the two-stage configuration with recycle, the permeation area of the second stage was set as 1/3 of the area set for the first stage.

Product streams results for each configuration are depicted in Table 6. Comparing both two-stage configurations, the recycled scheme produces a final retentate stream richer in methane (73 mol% vs. 67 mol%) and with higher molar flow rate ($\approx$ +28%) than the serial configuration. Consequently, the final permeate of the recycled configuration is richer in CO_2 (91 mol% vs. 74 mol%) and has lower molar flow rate ($\approx$ −19%) than the serial counterpart. The one stage configuration has the worst results: there is condensation inside the membrane unit (retentate stream is 1% condensed) and it produces retentate with lowest methane content and permeate with lowest CO_2 content (66 mol% and 73 mol%, respectively), though close to the two serial stages results.

Total permeation area, methane loss, CO_2 capture, and power demand of all configurations are displayed in Fig. 22, including a fourth case comprising two parallel contact serial stages (with no inter-stage heater, since condensation is absent) for comparison with contact types. Methane loss is the lowest for the two recycled stages configuration (10.5% against 38.3%, 36.0%, and 37.5% for one CC stage, two serial CC stages, and two serial PC stages, respectively), as a result of methane recovery from the first permeate stream in the second MP stage, producing final retentate and permeate with better quality, as discussed in Table 6. However, this

Fig. 21 Membrane process configurations: **a** one single stage; **b** two serial stages; **c** two stages with recycle

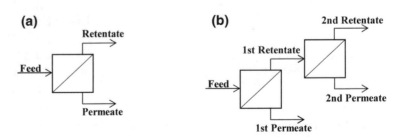

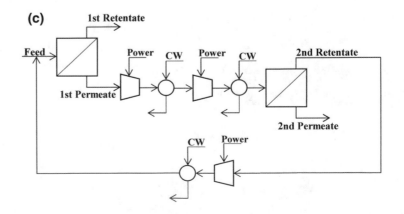

Table 6 Product streams results for counter-current MPx-UOE stage configuration cases

Parameter	Two serial stages		Two recycled stages		One stage	
	Retentate	Permeate	Retentate	Permeate	Retentate	Permeate
Vapor fraction	1.00	1.00	1.00	1.00	0.99	1.00
Temperature (°C)	58.14	54.60	38.60	48.43	39.77	54.57
Pressure (bar)	42.50	4.00	44.00	4.00	43.00	4.00
Molar flow (MMSm^3/d)	4.87	7.13	6.22	5.78	4.75	7.25
%CO_2	3.00	74.08	3.00	90.72	3.00	72.89
%CH_4	66.59	25.57	72.87	9.21	65.85	26.74
%C_2H_6	14.63	0.16	11.63	0.00	14.97	0.18
%C_3H_8	9.89	0.01	7.76	0.00	10.14	0.01
%i-C_4H_{10}	1.24	0.00	0.97	0.00	1.27	0.00
%n-C_4H_{10}	2.48	0.00	1.94	0.00	2.54	0.00
%i-C_5H_{12}	0.63	0.00	0.50	0.00	0.65	0.00
%n-C_5H_{12}	0.47	0.00	0.37	0.00	0.48	0.00
%n-C_6H_{14}	0.59	0.00	0.47	0.00	0.61	0.00
%n-C_7H_{16}	0.06	0.00	0.05	0.00	0.07	0.00
%n-C_8H_{18}	0.02	0.00	0.01	0.00	0.02	0.00
%N_2	0.39	0.17	0.44	0.07	0.39	0.18
ppmH_2O	0.07	1.64	0.07	2.01	0.07	1.61

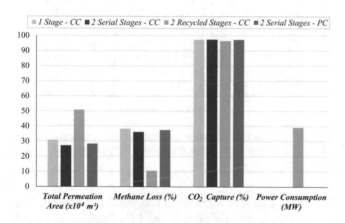

Fig. 22 Total permeation area, methane loss, CO_2 capture, and power consumption of process configurations for counter-current and parallel MPx-UOE

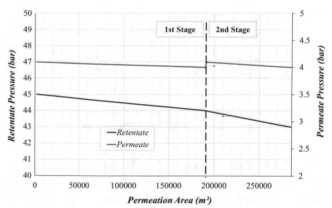

Fig. 23 Retentate and permeate pressure profiles through MPd-UOE for two serial stages with parallel contact

result comes with a price: there is a power consumption of almost 40 MW for compression of first permeate to the second MP stage and for recycle of second retentate to the first MP stage, absent in all other cases. Moreover, the total permeation area is the highest of all (508,800 m^2 against 308,000 m^2, 270,450 m^2, and 284,550 m^2 for one CC stage, two serial CC stages, and two serial PC stages, respectively). The other three configurations show similar results in Fig. 22, with two counter-current serial stages being slightly best, followed by the parallel counterpart. CO_2 capture is approximately 97% on molar basis for all cases. It is important to remember that the permeances are independent

of CO_2 fugacity, and since the composition in each stage differs considerably, the permeation area set for each stage is super or subdimensioned.

3.2.3 Profiles: MPd-UOE

To evaluate the performance of the distributed model MPd-UOE, the two serial parallel stages configuration was simulated with 100 permeation elements in the first stage (189,700 m^2) and 50 permeation elements in the second stage (94,850 m^2). Figures 23, 24, 25 and 26 display pressure, temperature, molar flow rate, and molar composition profiles for retentate and permeate streams along both MP stages. Pressure profiles in Fig. 23 show linear head-loss

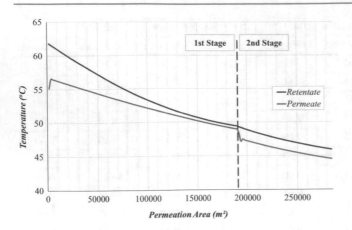

Fig. 24 Retentate and permeate temperature profiles through MPd-UOE for two serial stages with parallel contact

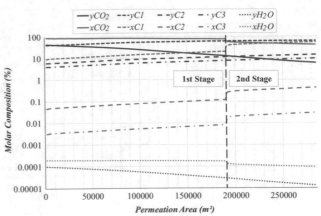

Fig. 26 Retentate and permeate main component molar compositions through MPd-UOE for two serial stages with parallel contact

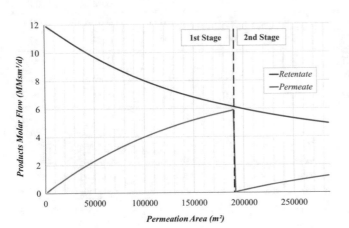

Fig. 25 Retentate and permeate molar flow rate profiles through MPd-UOE for two serial stages with parallel contact

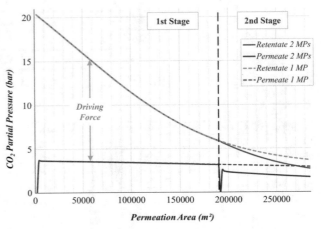

Fig. 27 CO_2 partial pressure in retentate and permeate through MPd-UOE for two serial stages and for one single stage, with parallel contact type

through membrane stages—1 bar for retentate and 0.1 bar for permeate, per stage—as described in Sect. 3.1.3. Temperature profiles in Fig. 24 are both smoothly decreasing, with the exception of a small deviation of about 1.5 °C in the beginning of each stage. This oscillation is a result of the $\Delta T_F = T_V^{in} - T_L^{in}$ specification with default value of 3 °C in the first permeation element of each parallel stage, since the temperature of the permeate stream in MP inlet is unknown. The effect of this parameter in MPd-UOE results is discussed in Sect. 3.2.4.

Molar flow rates and compositions in Figs. 25 and 26 also display smooth profiles, with a deviation in permeate in the stage change due to the withdrawal of the first permeate stream. Figure 26 shows that CO_2 and H_2O contents decrease in the retentate stream while hydrocarbons contents increase, due to the higher permeation fluxes of the first two through the membrane. In the permeate stream, CO_2 and H_2O contents are higher at the beginning of each stage, as a result of high inlet driving force, with both contents decreasing slightly through the membrane, as the driving

force is reduced, and the other components permeate. Figure 27 shows the driving force of CO_2 decreasing through the membrane due to permeation. The CO_2 driving force for a case with one single parallel stage (simulated with 150 elements in MPd-UOE) is also shown in Fig. 27. For PC type, one could think that separating the permeation in two stages would make no difference. However, since a permeate stream is withdrawn in the first stage, there is a sudden increase of CO_2 driving force at the beginning of the second stage, enhancing the overall MP operation.

In MPd-UOE, the simplification of constant component permeances impacts the profiles of temperature, molar flow, and compositions through the membrane unit (Figs. 24, 25 and 26). The permeance of CO_2, for example, would be higher in the beginning, where the partial pressure in retentate is higher, decreasing with the permeation of CO_2 along the unit. Therefore, the profiles would be more incisive in the beginning of permeation and smoothing toward the end. This simplification could be easily overcome by implementing permeances dependent of temperature and CO_2 fugacity in MPd-UOE model, updating the values for each permeation element. However, the real operation data used for calibration of permeances in MPx-UOE and MPd-UOE is not enough for this purpose. Thus, other literature or experimental data could be implemented for the adjustment of equations to correct the component permeances according to the temperature and CO_2 fugacity.

3.2.4 Sensitivity Analysis: MPx-UOE and MPd-UOE

In this section, three sensitivity analyses are conducted: (i) on the ΔT_F specification for MPx-UOE; (ii) on the number of permeation elements for MPd-UOE; and (iii) on the ΔT_F specification for MPd-UOE. For the sake of simplicity, MP configuration for all sensitivity analyses were conducted in one single stage MP: for CC MPx-UOE, the configuration chosen is one single stage from Sect. 3.2.2, with 308,000 m^2 of permeation area; in the case of PC, both MPx-UOE and MPd-UOE adopted one single stage from Sect. 3.2.3, with 284,550 m^2 of permeation area.

Figure 28 displays the results obtained for product temperatures with both CC and PC MPx-UOE, varying ΔT_F specification from 0.02 to 20 °C. The first notable characteristic is the opposite behavior between contact types: for parallel MP, the product temperatures converge for higher ΔT_F values, with retentate temperature above permeate temperature; while for counter-current MP, product temperatures diverge for higher ΔT_F values, always with higher permeate temperature than the retentate counterpart. Moreover, the amplitude of product temperatures difference is notably higher for CC MP, achieving $\approx$25 °C of temperature difference for the highest ΔT_F analyzed, while for PC, the highest temperature difference is only $\approx$7 °C, for the lowest

ΔT_F analyzed. Therefore, the ΔT_F specification clearly impacts more the counter-current MP operation. The default value for this variable in the MP models is 3 °C, which gives good average results for both contact types.

Figures 29 and 30 show, respectively, product molar flow rates, product CO_2, and methane molar compositions versus the number of permeation elements selected for MPd-UOE (with default $\Delta T_F = 3$ °C). The outcome is that the distributed MP model rapidly converges to constant values for products results as the number of permeation elements increases, with less than 1% of variation in all output variables for n_elements $\geq$ 5. The first conclusion is that the short-cut method adopted in the MP models presents great performance even for few permeation elements, or just one —as in MPx-UOE—with low variations against the more rigorous distributed simulations. For one permeation element only, the average deviation was 0.7% for all analyzed parameters (products molar flow rates, molar compositions, and temperatures), with the highest oscillation value of 6.3% for CO_2 content in the retentate (which in absolute values represent only 0.01 on CO_2 molar fraction). On the other hand, the consideration of constant component permeances also contributes to this reduced variation between MPx-UOE and MPd-UOE results. If the permeance values were corrected according to the temperature and CO_2 fugacity along the membrane, the deviation of the lumped model results to the distributed more rigorous model results would be more expressive, since in the latter, the correction would be applied to each membrane element. Another important conclusion obtained with Figs. 29 and 30 is that the approximation of log means by arithmetic means in MPd-UOE algorithm (Sect. 3.1.3) for n_elements $\geq$ 10 was smooth, since results converge to practically constant values for n_elements $\geq$ 5, before the change of means

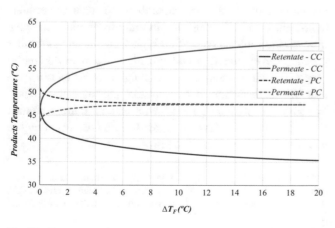

Fig. 28 Retentate and permeate temperatures in MPx-UOE versus ΔT_F specification for one single counter-current stage and one single parallel stage

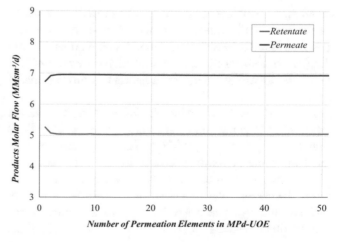

Fig. 29 Retentate and permeate molar flow rates in MPd-UOE versus the number of permeation elements selected by user for one single parallel stage

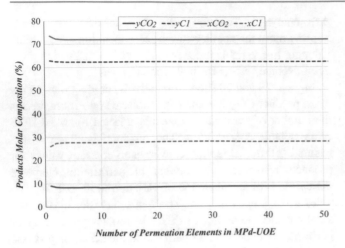

Fig. 30 Retentate and permeate main molar compositions in MPd-UOE versus the number of permeation elements selected by user for one single parallel stage

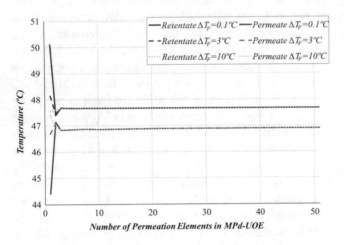

Fig. 31 Retentate and permeate final temperatures in MPd-UOE versus the number of permeation elements selected by user for one single parallel stage with $\Delta T_F = \{0.1\,°C, 3\,°C, 10\,°C\}$

calculation. Therefore, a great suggestion to achieve accurate results with easy convergence with MPd-UOE is to use n_elements = 10, since calculations and derivatives are simpler with arithmetic means, and higher number of elements is excessive, with no important gains in model accuracy.

Figure 31 displays product temperature profiles versus the number of permeation elements selected for MPd-UOE with $\Delta T_F = \{0.1\,°C, 3\,°C, 10\,°C\}$. The temperature profiles show the same behavior discussed for Figs. 29 and 30 of rapid convergence to constant values as the number of elements increases. As discussed in MPx-UOE sensitivity analysis, the value selected for ΔT_F clearly impacts the non-distributed model (n_elements = 1). Since MPd-UOE

only admits parallel contact type, products temperature variations are mainly for lower ΔT_F values, as already observed in MPx-UOE model for PC MP. The higher oscillation observed in MPd-UOE was 5% for both product temperatures, for $\Delta T_F = 0.1\,°C$ and n_elements = 1. For the other ΔT_F values, this oscillation was reduced to only $\approx$1%. However, as the number of permeation elements increases, the influence of this specification becomes meaningless: for n_elements $\geq$ 3, the final product temperature oscillations reduce to less than 0.1% for all ΔT_F.

3.3 Remarks

Section 3 describes the algorithm of a lumped model of membrane permeation for HYSYS: MPx-UOE. A distributed model was also developed, MPd-UOE, dividing the MP unit into smaller cells and applying MPx-UOE methodology for each cell consecutively, thus reproducing stream profiles inside the membrane. The MP models were calibrated using real NG processing operation data for CO_2 removal from CO_2-rich NG in offshore oil-and-gas fields in Brazil.

Both extensions were evaluated for CO_2-rich NG decarbonation simulations in HYSYS with PR-EOS. Different MP process configurations were investigated with MPx-UOE, concluding that the two-stage scheme with recycle of retentate led to minimum methane loss, yet at the cost of power consumption for compression and higher permeation area, as already stated, otherwise, in MP literature. MPd-UOE successfully represented smooth profiles of temperature, pressure, molar flow rates, and compositions through the membrane unit. Comparisons between both extensions indicate that the lumped model obtained results in good agreement with the distributed more accurate model results, with small deviations.

MPx-UOE and MPd-UOE models can be improved with the admission of permeance equations dependent of retentate temperature and CO_2 fugacity, in order to mimic membrane plasticization effects caused by high CO_2 fugacity. Another possible development would include a retentate dew-point check in each element of MPd-UOE, in order to warn the user if condensation occurs inside the membrane unit. Moreover, a distributed model with counter-current contact type can also be developed as MPd-UOE, involving a boundary-value problem framework.

Acknowledgements Authors acknowledge financial support from Petrobras S.A. (0050.0096933.15.9). J.L. de Medeiros and O.Q.F. Araújo also acknowledge financial support from CNPq-Brazil (311076/2017-3).

References

Araújo, O. Q. F., & de Medeiros, J. L. (2017). Carbon capture and storage technologies: Present scenario and drivers of innovation. *Current Opinion in Chemical Engineering, 17*, 22–34. https://doi.org/10.1016/j.coche.2017.05.004.

Araújo, O. Q. F., Reis, A. C., de Medeiros, J. L., Nascimento, J. F., Grava, W. M., & Musse, A. P. S. (2017). Comparative analysis of separation technologies for processing carbon dioxide rich natural gas in ultra-deepwater oil fields. *Journal of Cleaner Production, 155*, 12–22. https://doi.org/10.1016/j.jclepro.2016.06.073.

Arinelli, L. O., de Medeiros, J. L., de Melo, D. C., Teixeira, A. M., Brigagão, G. V., Passarelli, F. M., et al. (2019). Carbon capture and high-capacity supercritical fluid processing with supersonic separator: Natural gas with ultra-high CO_2 content. *Journal of Natural Gas Science and Engineering, 66*, 265–283. https://doi.org/10.1016/j.jngse.2019.04.004.

Arinelli, L. O., Trotta, T. A. F., Teixeira, A. M., de Medeiros, J. L., & Araújo, O. Q. F. (2017). Offshore processing of CO_2 rich natural gas with supersonic separator versus conventional routes. *Journal of Natural Gas Science and Engineering, 46*, 199–221. https://doi.org/10.1016/j.jngse.2017.07.010.

Baker, R. (2004). *Membrane technology and applications* (2nd ed.). England: Wiley.

Bernardo, P., Drioli, E., & Golemme, G. (2009). Membrane gas separation: A review/state of the art. *Industrial & Engineering Chemistry Research, 48*, 4638–4663. https://doi.org/10.1021/ie8019032.

Chung, T. H., Ajlan, M., Lee, L. L., & Starling, K. E. (1988). Generalized multi-parameter correlation for nonpolar and polar fluid transport properties. *Industrial & Engineering Chemistry Research, 27*, 671–679. https://doi.org/10.1021/ie00076a024.

Churchill, S. (1977). Friction-factor equation spans all fluid-flow regimes. *Chemical Engineering, 84*, 24, 91–92.

de Medeiros, J. L., Barbosa, L. C., & Araújo, O. Q. F. (2013a). Equilibrium approach for CO_2 and H_2S absorption with aqueous solutions of alkanolamines: Theory and parameter estimation. *Industrial & Engineering Chemistry Research, 52*, 9203–9226. https://doi.org/10.1021/ie302558b.

de Medeiros, J. L., Nakao, A., Grava, W. M., Nascimento, J. F., & Araújo, O. Q. F. (2013b). Simulation of an offshore natural gas purification process for CO_2 removal with gas—Liquid contactors employing aqueous solutions of ethanolamines. *Industrial & Engineering Chemistry Research, 52*, 7074–7089. https://doi.org/10.1021/ie302507n.

Ebner, A. D., & Ritter, J. A. (2009). State-of-the-art adsorption and membrane separation processes for carbon dioxide production from carbon dioxide emitting industries. *Separation Science & Technology, 44*, 1273. https://doi.org/10.1080/01496390902733314.

Ho, M. T., Allinson, G., & Wiley, D. E. (2006). Comparison of CO_2 separation options for geo-sequestration: Are membranes competitive? *Desalination, 192*, 288–295. https://doi.org/10.1016/j.desal.2005.04.135.

Marzouk, S. A. M., Al-Marzouqi, M. H., El-Naas, M. H., Abdullatif, N., & Ismail, Z. M. (2010). Removal of carbon dioxide from pressurized CO_2–CH_4 gas mixture using hollow fiber membrane contactors. *Journal of Membrane Science, 351*, 21. https://doi.org/10.1016/j.memsci.2010.01.023.

Reid, R. C., Prausnitz, J. M., & Poling, B. E. (1987). *The properties of gases and liquids* (4th ed.). New York: McGraw-Hill.

Polyhydroxyalkanoates (PHAs) for the Fabrication of Filtration Membranes

Pacôme Tomietto, Patrick Loulergue, Lydie Paugam, and Jean-Luc Audic

Abstract

Undoubtedly, in our current society, the development of more sustainable materials has to be considered in many applications. In this chapter, the interest of new potential biomaterials intended for the fabrication of filtration membranes is discussed. A focus is made on the polyhydroxyalkanoates (PHAs) polymers family. These biobased and biodegradable polyesters have gained attention in the past few years thanks to their versatile properties. Up to date, they have shown promising results for the fabrication of pervaporation and liquid filtration membranes. The membrane performances could be tuned by the use of PHAs having different comonomer contents and by the addition of proper additives. By discussing what has been developed in other application areas, such as biomedical and packaging, some insights are suggested in order to improve the overall PHAs-based membranes properties. Hence, the first part will deal with the membrane technologies and the current polymeric materials used. A second part will highlight the interest of biopolymers. Then, the properties and potential applications of PHAs in the membrane manufacture will be discussed.

1 Introduction

1.1 Membrane Technologies

The term membrane refers here to synthetic membranes used in separation processes. It can be defined as a selective barrier between two mediums which allows to separate and/or to concentrate components from a solution or a suspension when applying a driving force. Membrane technologies are mainly recognized as energy-efficient processes, compared to thermal processes which involve a phase change (Macedonio and Drioli 2017). They can be applied either for separation in liquid or gaseous phase. This book chapter will mostly focus on membrane-based processes treating liquid feeds. These techniques can be classified as a function of their driving force (Table 1).

Pressure-driven processes, with the flux related to the transmembrane pressure, refer to the most industrially implemented membranes processes. For instance, about 80% of desalination plants currently use the reverse osmosis (RO) technology (Macedonio and Drioli 2017). The concentration-driven processes work with a concentration difference between the two separated media and have a flux related to the diffusion coefficient of the different species involved in the separation. Then, thermally-driven processes work thanks to a temperature difference inducing a vapor pressure difference between both sides of the membrane (Cath 2010). In the case of electrically-driven processes, an electrical potential difference is applied and the flux is related to the mobilities, in the membrane, of the charged particles involved. Further in this chapter, pressure-driven processes and pervaporation will be mainly discussed as potential applications for PHAs-based membranes.

It may be pointed out that membrane technologies are quite recent and have required several development steps before being of industrial interest. Some of the key historical developments of separation membrane technologies are mentioned in Fig. 1.

To get an industrial interest, membranes filtration must present a high permeation flux coupled with a high rejection. Historically, it has been possible thanks to the works of Loeb and Sourirajan (1963; Loeb 1981), in the 1960s, who developed a new membrane fabrication process called phase inversion. This process has led to integral asymmetric structures showing high permeation flux and high rejection thus leading to the first uses of membranes at industrial

P. Tomietto · P. Loulergue (✉) · L. Paugam · J.-L. Audic
Univ Rennes, Ecole Nationale Supérieure de Chimie de Rennes, CNRS, ISCR – UMR 6226, F-35000 Rennes, France
e-mail: patrick.loulergue.1@univ-rennes1.fr

© Springer Nature Switzerland AG 2021
Z. Zhang et al. (eds.), *Membrane Technology Enhancement for Environmental Protection and Sustainable Industrial Growth*, Advances in Science, Technology & Innovation,
https://doi.org/10.1007/978-3-030-41295-1_11

Table 1 Current membrane processes, for liquid separation, and their associated driving forces

Pressure-driven processes	Microfiltration Ultrafiltration Nanofiltration Reverse osmosis
Concentration-driven processes	Pervaporation Forward osmosis Dialysis
Thermally-driven processes	Membrane distillation Thermopervaporation
Electrically-driven process	Electrodialysis

Based on Strathmann (1981)

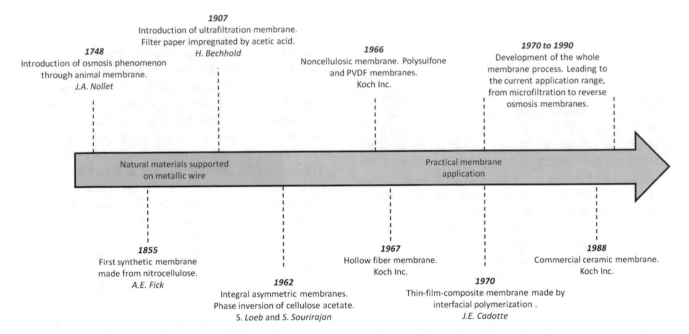

Fig. 1 Key historical development of separation membranes technologies. Based on Ahmed et al. (2017)

scale. Since then and until the 1990s several improvements have been made, from the development of new membrane materials to the evolution of the whole membrane process. From that moment, the development of new membrane materials has been intensively studied, but no major breakthrough has been made at the industrial scale. Hence, there is a constant interest to find new materials offering stability, selectivity, permeance and with a good scale-up potential (Hennessy et al. 2017). For the following of this chapter, the focus is made on the most widespread membrane materials, the polymeric materials.

1.2 Current Polymeric Materials for Membrane Applications

Commercial membranes can be made of various materials, from inorganic to organic, but 95% of the total industrial

market consists of polymeric membranes (Hennessy et al. 2017). The polymers present several advantages, whether for their use in membrane technologies or for other applications: they are cheap, easily shaped, lightweight, and resistant. Polymers are critical in our modern life and the membrane technologies are no exception. Today, a wide variety of polymers with different properties is used for membrane production. Table 2 gives some examples of common polymers used in membrane production and their applications in separation processes.

However, these conventional polymeric materials suffer from some major drawbacks concerning their environmental impact (Zhu et al. 2016). Indeed, it can be noted that none of the mentioned materials in Table 2 is biobased or biodegradable, except for the cellulose-based polymers that are partially biobased. It means that most of these polymers are produced from petroleum, which is a non-renewable resource, and are facing problem regarding their end of life

Table 2 Commercially available hydrophilic and hydrophobic polymers for membrane production and their applications

Hydrophilic polymers			
Cellulose acetate	RO, UF, MF, GS	Polyamide, aliphatic	MF
Cellulose nitrate	MF	Polyamide, aromatic	RO, NF, UF
Cellulose regenerated	UF	Polyacrylonitrile	UF
Polyvinyl alcohol	PV		
Hydrophobic polymers			
Polysulfone	UF, GS	Polysiloxane	NF, GS, PV
Polyethersulfone	UF, MF	Polyvinylidene fluoride	UF, MF, MD
Polytetrafluoroethylene	MF, MD, GS	Polyimide	NF, GS
Polyethylene	MF	Polycarbonate	RO, NF, MF
Polypropylene	MF, MD	Polyvinyl chloride	UF, MF

Adapted from Ahmed et al. (2017), Ladewig and Al-Shaeli (2017)

RO reverse osmosis, *NF* nanofiltration, *UF* ultrafiltration, *MF* microfiltration, *MD* membrane distillation, *GS* gas separation, *PV* pervaporation

management. Hence, polymers have created 6.3 billion tonnes of plastic waste since the plastic revolution. About 79% of plastics end in landfills which results in up to 2.41 million tonnes of plastic waste entering the oceans via rivers every year (The future of plastic 2018). And, when the degraded plastic becomes invisible to the eye, the micro-plastics and nano-plastics become a new issue for the marine life (Gigault et al. 2016). Thus, in our current society evolving in a new circular economy, the environmental impact is a new criterion that must be taken into account for the choice of materials.

That is why there is a recent gaining interest for the development of more sustainable materials intended for membrane technologies (Galiano et al. 2018). In the life cycle assessment studies about membrane separation systems, the environmental cost and end of life management of the membrane material were taken into account (Hancock et al. 2012; Martins et al. 2017; Ioannou-Ttofa et al. 2016). They agree to state that the main cause of the environmental impact is the energy consumption during the operation phase. However, Ioannou-Tofta et al. concluded that the materials used for the overall process, including the polymeric membrane, are also relevant to the overall environmental impact (Ioannou-Ttofa et al. 2016).

A part of the solution would be the replacement of these conventional polymers by alternative materials with similar physicochemical properties but without being petro-based and time persistent. To do so, biopolymers are of interest.

1.3 Biopolymers in Daily Life

Biopolymers are in opposition to most of the conventional polymers regarding their production and end of life management. According to European Bioplastics, a biopolymer is either biobased, biodegradable, or features both properties (Bioplastics 2019). Biobased means that the material is fully or partly derived from biomass. Biodegradable means that the material must be converted into water and carbon dioxide by the action of microorganisms within a given period.

Biopolymers present the advantages of using renewable resources so that they provide a reduced carbon footprint and are not time persistent which avoids the environmental pollutions. The biopolymers have been of growing interest for the past decades and are now implemented in a multitude of sectors, from packaging to automotive.

Biobased polymers may be classified as a function of their production way. Figure 2 shows how these biopolymers can be produced. Three production paths exist: the extraction from plants or animals, the polymerization of biobased monomers or the fermentation of microorganisms.

Not mentioned in Fig. 2, there are also biobased but non-biodegradable polymers such as bio-polyethylene or bio-polyethylene terephthalate which are partially or completely synthesized from renewable resources. Finally, the last family is composed of biodegradable polymers obtained from non-renewable compounds such as polybutylene adipate terephthalate (PBAT, known as ecoflex®) or polycaprolactone.

1.4 Biopolymers for Membranes

Biopolymers are a field of interest in membrane technologies for different reasons: firstly, because they are an alternative to the conventional petro-based and non-biodegradable polymeric materials, and secondly because they could bring new application opportunities due to their wealth of functionalities and properties. A few biopolymers have already been investigated for membrane fabrication. Table 3

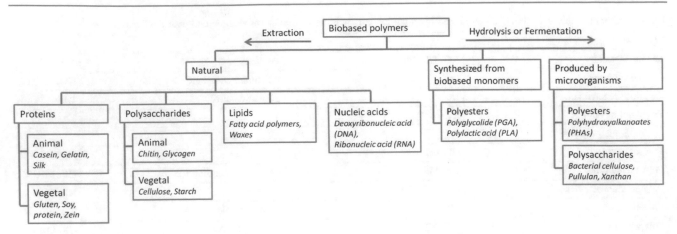

Fig. 2 Schematic representation of biobased polymers and their production [adapted from Chaunier et al. (2018)]

Table 3 Examples of biopolymers studied in the literature for the membrane fabrication

Biopolymer	Characteristics	Membranes specificities	Polymer disadvantages
Cellulose and its derivatives	Biobased	Already used for commercial membranes (cf Table 2)	Need chemical modification of cellulose
Polyvinyl alcohol (PVA)	Biodegradable		Water soluble. Must be crosslinked
Polylactic acid (PLA)	Biobased, biodegradable, good processability	For MF or PV applications	Produced by chemical reactions from lactic acid
Polybutylene succinate (PBS)	Biobased, biodegradable	Blended with cellulose acetate and polyethersulfone. For water treatment applications	Produced by chemical reactions from succinic acid
Alginate	Biobased, biodegradable, hydrophilicity, chelation behavior	Crosslinked with salt or multilayered membranes with chitosan. For GS or PV applications	Water soluble. Must be crosslinked
Chitosan	Biobased, biodegradable, abundant, antibacterial property	Dense membranes for PV or GS. High flux composite membranes for RO or NF	Produced by chemical modification of chitin
Starch	Biobased, biodegradable, cheap	Blended with other biopolymer such as chitosan or cellulose acetate. Antibacterial membranes	Swellable in water. Brittleness
Collagen and sericin	Biobased, biodegradable, hydrophilic	Crosslinked membranes for PV	Water soluble. Must be crosslinked

Based on Galiano et al. (2018)

lists some of these biopolymers with their associated characteristics.

In addition to the environmental point of view, it can be seen that some biopolymers offer the advantage to be cheap and naturally abundant. Another approach consists of choosing a targeted biopolymer in order to take advantage of its special characteristic. For instance, thanks to its ability of chelating to metal ions, alginate has been used for to fabricate membranes showing enhanced copper adsorption capacity (Paiva et al. 2012).

Another attractive area for the development of new membrane materials is the use of copolymers (Ladewig and Al-Shaeli 2017; Zhang et al. 2018). Indeed, owing to their myriad chemical functionalities, copolymers are very interesting for the fabrication of membranes with tailored properties.

There is one family of polymers, not cited in Table 3, belonging to the biopolymers and copolymers, the polyhydroxyalkanoates (PHAs). Hence, by combining the environmental advantage of biopolymers and versatile properties of copolymers, PHAs are of prime interest for membrane filtration applications.

2 Polyhydroxyalkanoates (PHAs)

2.1 PHAs Properties

In 1925, the French scientist Lemoigne discovered the presence of polyhydroxybutyrate polymer granules inside bacteria acting as energy storage (Chee et al. 2010; Dawes and Senior 1973). Since then, the polymers produced by bacteria and belonging to the PHAs family have been of growing interest. They are biobased and biodegradable linear polyesters and have versatile properties due to their structural diversity (Fig. 3).

Polyhydroxyalkanoates are made of hydroxyalkanoate (HA) units with various lengths and functional contents. Depending on the number of carbon atoms in the HA unit, PHAs may be classified into three groups: short-chain length (scl), medium-chain length (mcl), and long-chain length (lcl). The scl-PHAs refer to PHAs with less than 6 carbons in the HA (like 3HB, 3HV, and 4HB), mcl-PHAs have 6 to 14 carbons (like 3HHx) and lcl-PHAs, the less common group, have more than 14 carbons. There are homopolymers, such as polyhydroxybutyrate (P(3HB)), and copolymers, such as polyhydroxybutyrate-co-hydroxyvalerate (P(3HB3HV)). At the end, the macromolecule chain composition and molecular weight strongly depend on their way of production and they can be produced by a wide variety of bacteria under various growth conditions (Steinbüchel and Hein 2001). Until today, more than 150 different PHAs have been identified (Li et al. 2016). The molecular composition diversity of PHAs leads to a wide range of properties (Table 4).

P(3HB) is the most widespread PHA and presents similar relaxation temperatures and crystallinity ratio to polypropylene but has a much higher brittleness. Its lower mechanical properties can be explained by the presence of rigid amorphous fractions between the crystalline parts which renders it more rigid (Bugnicourt et al. 2014). Nonetheless, the presence of comonomers decreases the crystallinity and consequently improves the elasticity of the material. It has been explained that some comonomer units do not crystallize in the 3HB units and act as defects in the P (3HB) crystal network (Doi 1995). It is interesting to note that the intracellular PHAs granules, before extraction from bacteria, are completely amorphous and protected by a layer of proteins and phospholipids (Jendrossek et al. 1996). The

protecting surface is lost during the granules extraction, and the resulting impurities initiate the nucleation of crystallites. Another thermal property of PHAs is their poor melt stability. Indeed, their thermal degradation is near to their melting temperature (Mohanty et al. 2002). Hence, the chain scission mechanism involved during the thermal degradation can lead to a decrease of the molecular mass and may change the polymer properties (Grassie et al. 1984a, b, c).

PHAs have been investigated for their barrier properties (Modi et al. 2011; Miguel et al. 1997; Corre et al. 2012). Compared to common polymers like PP or polystyrene (PS), some PHAs grades exhibit 10 times lower oxygen permeabilities. In terms of water vapor permeability, PHAs have a slightly higher permeability (Corre et al. 2012). Its good barrier properties are explained by their high crystallinity degree. Among the PHAs commercial grades, it was shown that higher PHAs purities, with lower additives, lead to higher barrier behaviors. Compared to polylactic acid (PLA), which is a biobased polyester too, PHAs can be 6 times less permeable to water vapor and 11 times less permeable to oxygen (Corre et al. 2012).

PHAs are insoluble in water and in most common organic solvents except for chloroform ($CHCl_3$) and few other halogenated solvents (Jacquel et al. 2007). The good solubility in chloroform could be explained by a polar interaction between the chlorine and the carbonyl carbon, associated with a hydrogen bonding between the hydrogen on chloroform and the carbonyl oxygen (Fig. 4). The poor solubility in other solvents can be partially explained by the Hansen solubility parameters (Jacquel et al. 2007; Terada and Marchessault 1999), but also by the high degree of crystallinity. Insoluble parts may be due to the insoluble crystalline regions of the polymer (Mcchalicher et al. 2009).

PHAs are biodegradable, in that sense they can be assimilated by living organisms and converted into CO_2, H_2O, or methane to generate a new biomass (Lucas et al. 2008). It happens in two steps: first the reduction of the

Fig. 3 **a** General molecular structure of PHAs. **b** Some examples of common monomer units in PHAs. 3HB: 3-hydroxybutyrate, 3HV: 3-hydroxyvalerate, 3HHx: 3-hydroxyhexanoate, 4HB: 4-hydroxybutyrate

Table 4 Typical properties of some PHAs compared to polypropylene (PP)

Parameter	P(3HB)	P(3HB3HV)	P(3HB4HB)	P(3HB3HHx)	PP
Melting temperature (°C)	177	145	150	127	176
Glass transition temperature (T_g, °C)	2	−1	−7	−1	−10
Crystallinity (%)	60	56	45	34	50–70
Tensile strength (MPa)	43	20	26	21	38
Extension to break (%)	5	50	444	400	400

Adapted from Tsuge (2002)

molecular mass of the polymer (by abiotic or biotic reactions) and then the bioassimilation by microorganisms (Lucas et al. 2008; Vroman and Tighzert 2009). PHAs can be degraded by a wide variety of microorganisms, and it has been shown that the biodegradation rate depends strongly on the environment conditions (Mergaert et al. 1992). As a case example, P(3HB) degrades within a few months in anaerobic sewage and within several years in seawater (Madison and Huisman 1999). The decomposition by microorganisms is also highly dependent on the physicochemical properties of the polymer (Jendrossek et al. 1996). The important factors are (i) the stereo configuration and the crystallinity: higher crystallinity ratio or higher crystal perfection decreases the degradability, (ii) the molecular mass: high molecular masses degrade slower than small molecular masses, (iii) the monomeric content: P(3HB) tends to degrade slower than copolymers of 3HV.

Additionally to their biodegradability, PHAs are also biocompatible, which means both polymer and degradation products have no toxic effects on living organisms (Volova et al. 2003; Verlinden et al. 2007).

2.2 PHAs Applications

Due to their biobased origin, their biodegradability, biocompatibility, and versatile properties, PHAs have been

prospected for several applications. Today, two major application sectors are mainly studied for PHAs, the packaging (Bugnicourt et al. 2014; Modi et al. 2011; Ragaert et al. 2019; Hartley Yee and Ray Foster 2014), and the therapeutic sectors (Zhang et al. 2018).

Among the 348 million tonnes of the worldwide plastics production (in 2017), the packaging industry is the main application sector for plastics (Plastics Europe 2018). In Europe, 39.7% (in 2017) of the plastic demand is intended for the packaging sector. Hence, there is a major driving force, related to the transition toward a circular economy, for the development of biopolymers intended for packaging (Ellen MacArthur Foundation 2017). Therefore, PHAs have been of growing concern in the past few years for this application. For the food packaging, it is interesting to look at the barrier properties to gases and moisture. In general, the packaged foods have to be preserved from O_2 and H_2O to avoid microbial growth or deteriorative reactions, that is why the packaging atmosphere is often made out of CO_2 and N_2 (Ragaert et al. 2019). The packaging material has to be the least permeable to these gases. According to their good barrier properties described previously (Part 2.A), PHAs may be suitable for such applications. Indeed, compared to conventional petrobased polymers used in packaging, like PP, PE, and PS, PHAs have lower O_2 permeability, slightly lower CO_2 permeability and higher H_2O permeability (Bugnicourt et al. 2014; Ragaert et al. 2019; Hartley Yee and Ray Foster 2014). It must be mentioned that food packaging often consists of multi layers of different polymer so that combines the barrier properties of each. The degree of heat resistance is also important in food packaging when the food is pasteurized or sterilized. It has been reported that some PHAs have heat resistance similar to PP. Due to their hydrophobicity and film-forming property, PHAs are also suitable for water-resistant surfaces or sealable coating on paperboard packaging (Seoane et al. 2018).

For medical implants, the designed material acts as a functional replacement for the damaged host tissue and must not cause any adverse event. Thanks to their biodegradability and biocompatibility, described above (Part 2.A) PHAs have been studied as medical implant materials. Since

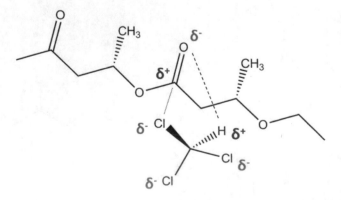

Fig. 4 Interactions between P(3HB) and chloroform

promising in vitro and in vivo studies have shown the potential of PHAs to be used as medical implants (Lukasiewicz et al. 2018; Chen and Wu 2005), various studies are now dealing about the fabrication methods tailoring the scaffold structure and properties (Lim et al. 2017). It has been stated that PHAs can expect to become a family of bioimplant materials with rich applications (Chen 2009). Another medical application is their use for drug delivery systems (Shrivastav et al. 2013; Xiong et al. 2010). Regarding the versatile properties of the existing PHAs, the release rate of the encapsulated drug could be tailored via the use of proper PHAs macromolecules. For example, it has been shown that a higher P(3HB) molecular mass tends to increase the drug release rate (Brophy and Deasy 1986).

In order to extend the field of PHAs applications, some intrinsic problematic of this polyester has to be solved, like the high production cost, its poor thermal stability and its brittleness (Lee 2000). The solutions to their poor thermo-mechanical properties are either their use in polymer blends or their chemical modifications (Li et al. 2016; Ramachandran et al. 2014). Regarding the cost issue, several research focuses have been mentioned in the literature (Chen 2009). Since 40–50% of the PHAs production costs come from the raw materials costs (Shen et al. 2009), one possible solution for a cheaper PHAs production is the use of new and cheap raw materials as growth media (Du et al. 2012; Elain et al. 2016). Molasses, whey, glycerol, lignocellulosic derived materials, fats, oils, and wastewater were successfully used as low-cost raw materials to make PHAs (Du et al. 2012). These PHAs exhibited similar properties to those produced from pure simple sugars. Besides, the high value-added PHAs applications would not suffer from the high production cost (Chen 2009) and the multiple medical applications seem to be an economically practical area.

3 PHAs-Based Membranes

Since the microstructure of the membrane material is one of the parameters tailoring the membrane performances (Mulder 1996), it is important to focus, in a first part, on the structures of PHAs materials and which fabrication process could be used. Then, few examples of concrete applications on PHAs-based separation membranes will be discussed.

3.1 From Dense to Porous Structures

Before being considered as a membrane, the material must be processable in either dense or porous structures. In this case, it may be challenging for several reasons. Firstly, PHAs have a poor solubility behavior and a poor thermal stability, which limit the process. Moreover, it has been reported that the vitrified amorphous parts of PHAs exhibit a progressive crystallization at room temperature (Koning and Lemstra 1993). The latter is the main cause of the PHAs embrittlement. Hence, this slow crystallization must be avoided if a stable structure is desired. This paragraph will deal with the fabrication methods, used in the literature, to get the suitable dense or porous structures.

The blend of plasticizers in the PHA matrix is one easy solution to improve the mechanical properties of the material. Indeed, the IUPAC defined plasticizer as "a substance or material incorporated in a material to increase its flexibility, workability, or distensibility" (Gurgel et al. 2011). The added substance improves the mobility of the macromolecules within the amorphous phase by decreasing the intermolecular forces. Typically, a lower glass transition, lower stiffness, higher elongation at break, and lower elastic modulus could be expected in plasticized samples. Some examples of additives used in PHAs materials are mentioned in Table 5.

Among the additives tested in the literature and reported in Table 5, most of them, except for stearic acid, seem to act as plasticizer. It means that additives molecular chains were uniformly distributed within the polymer. None of them decreased significantly the crystallinity ratio. The water vapor permeability (WVP) was influenced by two phenomena: the additive properties and its hydrophilicity/hydrophobicity (Parra et al. 2006) and the increasing of the free volume in the material (Audic and Lemi 2013). In Table 5, some of the reported additives are biobased and/or biodegradable like soybean oil and broccoli oil (Choi and Park 2004; Audic and Lemi 2013). Indeed, since the main material, PHA, is biobased and biodegradable, the development of plasticizers in accordance with the matrix is relevant. Furthermore, bioplasticizers are also strongly investigated for other common plastics (Gurgel et al. 2011). By adding a plasticizer, except the mechanical properties improvement, a side effect of increased biodegradation rate was sometimes reported (Parra et al. 2006; Yoshie et al. 2000).

Besides plasticizers, other materials were added in the polyester matrix in order to improve the mechanical properties. Boron nitride has been used as nucleating agent in P(3HB) or P(3HB3HV) to improve the crystallization at high temperature and avoid the slow crystallization at room temperature (Puente et al. 2013a, b). Multiwall carbon nanotubes (Yu et al. 2013; Vidhate et al. 2012; Shan et al. 2011; Montanheiro et al. 2015), cellulose nanocrystals (Bhardwaj et al. 2014), cellulose nanofibrils (Jun et al. 2017), clay (Bittmann et al. 2013), and ZnO (Díez-Pascual and Díez-Vicente 2014) were also successfully used as nucleating agents.

Another solution to improve the PHAs properties while reducing the material cost is the blend with other common

Table 5 Some examples of additives tested as plasticizers in PHAs-based films

PHA	Additive		Film fabrication method	Resulting effects	Refs.
	Nature	Concentration (respect to the total solids) (wt%)			
P(3HB3HV) (8 mol% HV)	Polyethylene glycol (PEG) (200–1000–4000 g mol^{-1})	10	Melt blending followed by compression molding	Higher vapor permeability, lower stiffness and resistance to break, lower thermal stability	Requena et al. (2016)
	Lauric acid				
	Stearic acid (SA)			Poor mixing behavior between SA and P(3HB3HV)	
P(3HB) M_w = 380,000 g mol^{-1}	PEG (300 g mol^{-1})	0–40	Casting of P(3HB)/additive solution (in chloroform)	Higher elongation at break, lower T_m, higher water vapor permeability. No major change in crystallinity ratio. Higher biodegradation rate	Parra et al. (2006)
P(3HB)	Dodecanol	0–23	Casting of P(3HB)/additive solution (in chloroform), followed by compression molding	Lower T_g and $T_{cold\ crystallization}$ (T_{CC}) No major effect on the crystallinity ratio, additives migration to the surface, acceleration of film biodegradation	Yoshie et al. (2000)
	Lauric acid				
	Tributyrin				
	Trilaurin				
P(3HB3HV) (6 mol% HV), M_w = 680,000 g mol^{-1}	Soybean oil	20	Casting of P(3HB)/additive solution (in chloroform), followed by compression molding	Higher T_g and T_{cc}. Lower elongation at break	Choi and Park (2004)
	Epoxidized soybean oil			Lower T_g and higher T_{cc}. Higher elongation at break	
	Dibutyl phthalate			Lower T_g and T_{cc}. Higher elongation at break	
	Triethyl citrate				
P(3HB3HV) (8 mol% HV), M_w = 340,000 g mol^{-1}	Epoxidized broccoli oil	0–15	Blend extrusion followed by lamination	Higher elongation at break and lower elastic modulus. Lower T_g. No change in the crystallinity ratio. Changes in the water vapor permeability	Audic et al. (2013)

biopolymers (Ramachandran et al. 2014), such as polylactic acid (PLA), polycaprolacone (PCL), cellulose, starch, or chitosan. The blend of PHA/PLA has been the most investigated since PLA is a biopolymer with a similar structure and similar properties to PHAs (Zhang and Thomas 2011; Loureiro et al. 2015; Szuman et al. 2016). PLA is also one of the most produced biobased/biodegradable polymers and has a competitive price to other polymers (Mirabal et al. 2013). Adding 25 and 50 wt% of PLA in the P(3HB) continuous phase tends to reduce the size of the crystallites and increase the mechanical properties, both elongation and tensile stress at break (Loureiro et al. 2015). Similar observations were described with P(3HB3HV)/PBAT blends (Javadi et al. 2010).

In addition to blends with plasticizers and other polymers, which is not always favored because of poor interfacial interactions, the chemical modification of PHAs is a good way to modulate the material properties. It can be achieved via either graft copolymerization or block copolymerization. These techniques have been previously well summarized (Li et al. 2016). For instance, chemical conjugation with polyethyleneglycols can lead to amphiphilic polymers with suitable properties for drug delivery systems.

Knowing that the PHAs properties may be improved by blending or chemical modification, it is then interesting to focus on the different possible techniques used to process PHAs materials. As mentioned above, PHAs can be processed by thermal techniques to get dense films. Nevertheless, membrane materials are made of both dense and porous structures. Hence, another technique to prepare porous structures must be discussed.

The phase inversion is a well-developed fabrication process and is one of the most commonly used techniques for preparation of polymeric membranes with dense or porous structure (Lalia et al. 2013). This technique involves a demixing mechanism from a homogeneous polymer solution to a solid film. After casting of the polymer solution, the phase inversion can be done in different ways. A first way is by evaporation of solvent, called evaporation induced phase separation (EIPS), a second one is by decreasing the temperature, called temperature-induced phase separation (TIPS), and a third one is by adding a non-solvent to the solution, called non-solvent induced phase separation (NIPS). For the last one, two options are possible, by adding liquid non-solvent, called wet-NIPS or just NIPS, or by adding vaporized non-solvent, called dry-NIPS or vapor-induced phase separation (VIPS). The type of structure, dense or porous, symmetric or asymmetric, obtained by these methods, is highly dependent of the process, its parameters and the chemicals involved (Bouyer et al. 2011). For the case of PHAs, Table 6 summarizes some studies that reported the use of phase inversion to make tailored structures.

Regarding Table 6, PHAs were processed with the three phase inversion techniques (EIPS, NIPS, and TIPS). Moreover, the obtained structures vary from dense symmetric to porous asymmetric. It highlights the potential of PHAs for membrane fabrication. Since the structure formation depends on numerous parameters, from the chemicals interactions to the phase inversion conditions, it is difficult to predict the structure as a function of the process parameters, but some recurrences may be cited.

The EIPS technique tends to form dense structures. However, porous structures could be obtained when salts or other additives, such as PEG, were added. The TIPS technique usually makes porous structures. The NIPS technique mostly leads to porous asymmetric structures. These general observations are similar to what has been reported in the literature for more common membranes made out of petro-based polymers (Lalia et al. 2013; Tae et al. 2016; Broens et al. 1980).

In the case of the NIPS technique with the P(3HB4HB) (13 mol% 4HB)/NMP/water system, the effect of different parameters has been discussed (Marcano et al. 2015, 2017). The discussions mainly refer to the demixing rate. Indeed, a higher demixing rate results in a more porous structure. In that sense, increasing the PHA concentration increases the solution viscosity and so decreases the demixing rate, leading to a denser membrane. Increasing the casting thickness leads to denser membranes. About the influence of the coagulation bath temperature, with an increase of the temperature the viscosity would decrease and the non-solvent and solvent diffusivity would increase, leading to a more porous structure. Nevertheless, for this system, no major changes were observed when the coagulation bath temperature was varied. The influence of the polyvinylpyrrolidone (PVP) addition in the dope solution, its molecular weight and its concentration were studied (Marcano et al. 2017). The authors were able to increase the pore density and

surface porosity by increasing the PVP concentration. It was explained by a decrease of the solution viscosity by adding PVP. However, an additive concentration above 30% had a reversible effect and decreased the surface porosity.

Solvent casting particulate leaching method was used to make highly porous scaffolds. Sodium salt crystals or sugar were used as porogens. The evaporation followed by the porogens leaching leads to highly porous structures with pore sizes around 100 µm. Tan et al. worked on this leaching technique on P(3HB) films and recognized the potential of this material for separation membrane applications (Tan et al. 2016).

At the end, PHAs can be easily processed via phase inversion to form tailored structures. Then, it is of interest to test these structures as potential filtration membranes.

3.2 Separation Purpose

A few examples of concrete separation applications of PHAs-based membranes have been reported in the literature (Mas et al. 1996; Villegas et al. 2011, 2015, 2016; Nicosia et al. 2015; Keawsupsak et al. 2014; Guo et al. 2016).

3.2.1 Pervaporation

Pervaporation is a separation process that finds applications for dehydration of organic solvents, removal of dilute organic molecules or organic–organic mixture separation. In such processes, the solution–diffusion is the most accepted model to describe the mass transport, so that during the separation, the selectivity occurs following unequal affinities between the solutes and the membrane (Jyothi et al. 2019).

Mas et al. were the firsts to consider PHAs-based membranes as candidates for pervaporation applications (Mas et al. 1996). They made dense membranes by EIPS with different P(3HB3HV) having different values of HV content. The membrane performances were analyzed by separation of an ethanol/water mixture. The flux varied from 0.008 to 0.027 kg m^{-2} h^{-1} and the separation factor from 5.0 to 12.6 in favor to water permeation. No major correlation was observed between the HV ratio content and the membrane performances. However, with the increase of the filtration time, the flux tends to increase and the separation factor tends to decrease. It is argued by a progressive deformation of the macromolecular structure accentuated by the plasticizing effect of ethanol molecules.

Villegas et al. evaluated the performances of a PHA-based membrane for pervaporation application on methanol/methyl tertiary butyl ether (MTBE) mixtures (Villegas et al. 2011). They made P(3HB) dense membranes by EIPS. Performances were measured as a function of the feed temperature and feed composition. In the case of 40 mol% methanol/MTBE mixture, the best performances

Table 6 Examples of PHAs materials processed by phase inversion

PHA	Structure preparation Solvent	Polymer concentration	Additive	Phase inversion	Structure morphology	Tested for	Refs.
P(3HB) (Mw = 430,000 g/mol), P(3HB3HV) (9 mol% HV) (596,000 g/mol) and P(3HB3HV) (22 mol% HV) (670,900 g/mol)	Chloroform/tetrahydrofuran (THF) (various ratio)	10 wt%	None	EIPS	Dense	Pervaporation membranes	Mas et al. (1996)
			None	EIPS + VIPS	Porous asymmetric	Microfiltration membranes	
P(3HB) (Mn = 87,000 g/mol), P(3HB3HV) (8 mol% HV) (92,000 g/mol), P(3HB3HV) (14 mol% HV) (56,000 g/mol), and P(3HB3HV) (22 mol% HV) (100,000 g/mol)	Chloroform	1% w/v	None	EIPS	Dense	Pervaporation membranes	Galego et al. (2002)
P(3HB) (Mw = 524,000 g/mol)	Chloroform	6% w/v	None	EIPS	Dense	Pervaporation membranes	Villegas et al. (2011)
P(3HB) (Mw = 3500 g/mol)	Chloroform	0.6 wt%	None	EIPS	Dense	Sorption diffusion of water	Taylor et al. (1998)
P(3HB4HB) (13 mol% 4HB)	N-methly-2-pyrrolidone (NMP)	10, 12 and 17 wt%	PEG 1000 g/mol	VIPS	Porous asymmetric	Antibiofilm materials	Marcano et al. (2015)
P(3HB4HB) (13 mol% 4HB) (Mw = 310,000 g/mol)	NMP	17 wt%	PVP 40,000 g/mol and 360,000 g/mol	VIPS	Porous asymmetric	Antibiofilm materials	Marcano et al. (2017)
P(3HB3HHx) (12 mol% 3HHx) (Mw = 550,000 g/mol)	Dioxane	5 wt%	None	TIPS	Porous symmetric	Biomedical application	Xi et al. (2007)
P(3HB) (Mw = 822,000 g/mol)	Dioxane	1%, 2.3%, 3.3% w/v	None	TIPS	Porous symmetric	Hydrophilic scaffolds	Guzman et al. (2011)
P(3HB3HV) (8 mol% HV)	Chloroform/dichloromethane (1/2 v/v)	4%, 6%, 8% w/v	Sucrose	Evaporation	Porous symmetric	Biomedical application	Torun Kose et al. (2003)
P(3HB3HV) (3 mol% HV) (Mw = 300,000 g/mol)	Chloroform	14% w/v	NaCl	Evaporation	Porous symmetric	Biomedical application	Wang et al. (2013)
P(3HB3HHx) (12 mol% 3HHx)	Chloroform	2% w/v	NaCl	Evaporation	Porous symmetric	Biomedical application	Wang et al. (2005)

(continued)

Table 6 (continued)

PHA	Structure preparation		Additive	Phase inversion	Structure morphology	Tested for	Refs.
	Solvent	Polymer concentration					
P(3HB3HV) (12 mol% HV)	Methylene chloride	5 wt%	Blended with PLA under different ratio	Evaporation	From dense to porous depending of the PLA/PHA ratio	Biomedical application	Santos et al. (2004)
P(3HB)	Chloroform	3 wt%	Glass particles	Evaporation	Porous symmetric	Biomedical application	Misra et al. (2008)
P(3HB) (Mn = 450,000 g/mol)	Chloroform	n.i	Blended with poly (p-dioxane) under different ratio	Evaporation	Porous	Biomedical application	Dias et al. (2008)
P(3HB) (Mw = 430,000 g/mol) (Mn = 290,000 g/mol) (Ip = 1.49)	Chloroform	3% w/v	Blended with PEG under different ratio	Evaporation	n.i	Biomedical application	Cheng et al. (2003)
P(3HB) (Mw = 437,000 g/mol)	Chloroform	0.5%, 1%, 2%, 3% w/v	None	VIPS	Porous, nanofibrous aerogel	Control structure scaffold	Kang et al. (2016)
P(3HB)	Chloroform/methanol	0.045 mol/L	$CuCl_2$ and $NiCl_2$	Evaporation	Porous	Control structure scaffold	Tan et al. (2016)
P(3HB3HV) (3 mol% HV) (Mw = 303,000 g/mol)	Chloroform	5% w/v	None	Evaporation	Dense	Sorption diffusion of water	Follain et al. (2014)
P(3HB3HHx) (17 mol% 3HHx) and P (3HB3HHx) (30 mol% 3HHx)	MIBK	2.5% w/v	None	Evaporation	Porous symmetric	Antibiofilm materials	Kehail and Brigham (2017)
P(3HB) (Mw = 600,000 g/mol) and P(3HB3HV) (12 mol% HV) (Mw = 252,000 g/mol)	Chloroform	2% w/v	Diethylene glycol	Evaporation	Porous	Improved biocompatible material	Chan et al. (2013)
P(3HB)	Chloroform	n.i	Bacterial cellulose	Evaporation	Dense to porous composites	Improved mechanical properties	Barud et al. (2011)

were obtained with a feed temperature of 50 °C. A flux of 0.392 kg m^{-2} h^{-1} and a separation factor of 3.98 in favor to the methanol permeation were measured. By increasing the methanol concentration in the methanol/MTBE feed composition, the separation factor decreases. Besides, it was assumed that MTBE could act as a plasticizer in the membrane and would consequently improve the permeability of both compounds. Later, in another article, the authors went further in the evaluation of the pervaporation performances (Villegas et al. 2016). This time, the authors chemically modified the P(3HB) dense membranes by plasma polymerization with acrylic acid. After membrane modification, due to the additional layer formed and additional flux resistance, the flux was decreased. Nevertheless, the separation factor was significantly improved up to 18.6. The latter is explained by a higher affinity between the plasma polymerized acrylic acid layer and methanol.

Related to pervaporation applications, Galego et al. studied the swelling capacity, vapor permeability and selectivity of PHAs films with different HV content (Galego et al. 2002). The swelling percentage to water–ethanol mixtures and the water vapor permeability of P(3HB) was higher than P(3HB3HV) (22 mol% HV). It is discussed as a consequence of the rougher surface of P(3HB) films and the better ability of P(3HB) to make hydrogen bonding interactions with water molecules.

3.2.2 Air Filtration

Nicosia et al. have made PLA/P(3HB) filters intended for air filtration applications (Nicosia et al. 2015). The filters were made by electrospinning of a PLA/P(3HB) mixture in chloroform. The performances were quantified by measuring the penetration of sodium chloride aerosol particles ranging from 20 to 600 nm. The performances could be improved by superposing several thin layers to reduce the overall packing density of the structure. The results highlight a good potential for air filtration applications. The best collection efficiency was measured at 98.5% for 0.3 μm particles. Additionally, these membranes could be modified by loading of didecyldimethylammonium nitrate in order to improve their antimicrobial activity.

3.2.3 Liquid Filtration

Mas et al. studied PHAs-based membranes intended for microfiltration applications (Mas et al. 1996). The membranes were made with P(3HB3HV) having different HV ratio. For the membrane fabrication, they used the EIPS coupled with the NIPS. For the dope solution, various compositions of chloroform/THF mixtures were used. THF acts as a non-solvent in the dope solution. The non-solvent bath was composed of a mixture of ethanol/water (4/1 v/v). The water permeability and ethanol permeability were measured. The pores size was analyzed by image analysis.

Whether for P(3HB) or for P(3HB3HV) (22 mol% HV), the best permeabilities were obtained for a dope solution containing 8% of THF. The permeability went up to 600 L m^{-2} h^{-1} bar^{-1} but no rejection test was performed. For THF concentrations above 8%, the proper solubility of the polymer was hindered, leading to the densification of the membrane. Among the membranes characterized, the pores size varied from 0.25 to 2 μm.

Keawsupsak et al. have made PLA/P(3HB3HV) blend membranes to test them on water permeability and rejection of bovine serum albumin (Keawsupsak et al. 2014). Membranes were produced by NIPS using NMP as a solvent and water as a non-solvent. At a small amount of P(3HB3HV) added to PLA, the membrane exhibits a better permeability value of 65.2 L m^{-2} h^{-1} bar^{-1}, but it remains quite low. In that case, the BSA rejection was measured at 78.7%. The membrane showed a finger-like structure.

Composite membranes made of P(3HB), calcium alginate and carboxyl multi-walled carbon were made by electrospinning for NF applications (Guo et al. 2016). Their performances were measured by water permeability, separation of emulsified oil–water solutions and rejection of brilliant blue. One membrane demonstrated an excellent antifouling property associated with a pure water flux around 35 L m^{-2} h^{-1} at 2 bar, a 98.2% rejection of brilliant blue and a 98% rejection of oil.

Finally, P(3HB3HV) (3 mol% HV)-based membranes with tailored structures were made in our research group at the ISCR (Institut des Sciences Chimiques de Rennes, France), in the Chemistry and Process Engineering team. The membranes were fabricated via two different techniques, NIPS or EIPS. By NIPS, the NMP was used as solvent and water was used for the non-solvent bath. By EIPS, chloroform was employed as volatile solvent. The dope solutions were casted under controlled conditions. The resulting membranes are described in Table 7.

For each fabrication method, the influence of the dope solution composition was studied by changing the P(3HB3HV) concentration and/or by adding an additive. The membranes properties were highly dependent on the fabrication method. In case of NIPS, the membranes exhibited asymmetric structures. On the other hand, the membranes made by EIPS exhibited symmetric structures.

Figure 5 gives the scanning electron microscopy (SEM) images of the different membranes made by NIPS. The cross sections and surfaces were analyzed.

Asymmetric structures with a finger-like structure at the top and a sponge-like sublayer can be observed for each membrane. Their microstructures properties are detailed in Table 8.

As observed on Figure 5 and confirmed by the Table 8, the increase in P(3HB3HV) concentration tends to decrease the surface porosity but tends to increase the membrane

Table 7 P(3HB3HV)-based membranes made by NIPS and EIPS

Dope solution composition (wt %)	Membrane fabrication technique	Structure	Thickness (μm)	Mechanical integrity
P(3HB3HV)/NMP (15/85)	NIPS—In water bath	Asymmetric	77	Very fragile
P(3HB3HV)/NMP (20/80)	NIPS—In water bath	Asymmetric	113	Very fragile
P(3HB3HV)/NMP (25/75)	NIPS—In water bath	Asymmetric	130	Very fragile
P(3HB3HV)/CHCl$_3$ (10/90)	EIPS	Symmetric	18	Good
P(3HB3HV)/PEG300/CHCl$_3$ (10/5/85)	EIPS	Symmetric	18	Good
P(3HB3HV)/PEG8000/CHCl$_3$ (10/5/85)	EIPS	Symmetric	42	Good

Fig. 5 SEM images of the membranes prepared by NIPS with different P(3HB3HV) concentrations (w/w). Surface on the left and cross section on the right

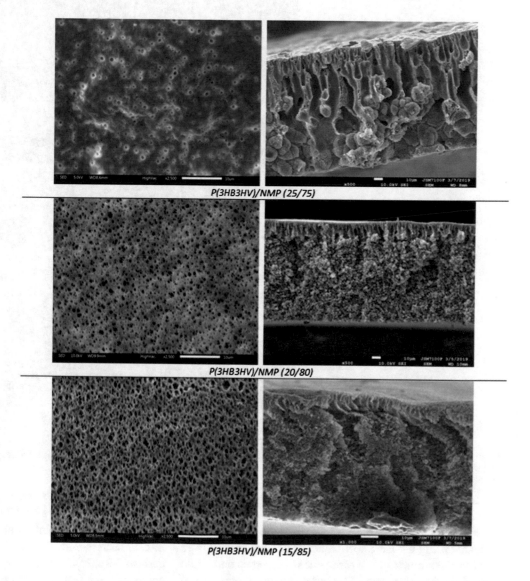

P(3HB3HV)/NMP (25/75)

P(3HB3HV)/NMP (20/80)

P(3HB3HV)/NMP (15/85)

thickness and the finger-like structure section. For filtration applications, such asymmetric finger-like structures would favor high fluxes, owing to the draining effect of the macrovoids, while keeping the sieving rejection thanks to the top surface. Unfortunately, these membranes exhibited poor mechanical properties, hindering their use for filtration tests.

The poor mechanical properties associated with the asymmetric structures and their macrovoids were something already observed in the literature (Tsai et al. 2001). But, if the asymmetric structures may partly cause the poor mechanical properties of these membranes, the nature of the polymer could be involved too. For the membrane

Table 8 Structural properties of the P(3HB3HV) based membranes made by NIPS. Influence of the P(3HB3HV) concentration

Dope solution composition (wt %)	Surface porosity (%)	Structures thicknesses ratiofinger like structure/overall membrane (%)
P(3HB3HV)/NMP (15/85)	2.9	55
P(3HB3HV)/NMP (20/80)	7.7	17
P(3HB3HV)/NMP (25/75)	16.9	10

Fig. 6 SEM images of the membranes prepared by EIPS with different dope solution compositions (w/w). Surface on the left and cross section on the right

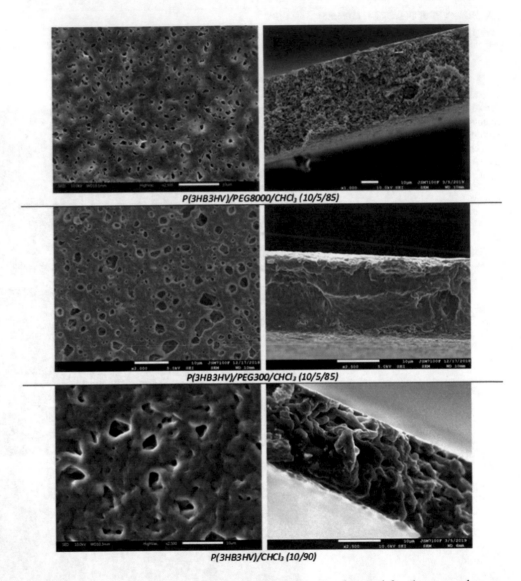

P(3HB3HV)/PEG8000/CHCl₃ (10/5/85)

P(3HB3HV)/PEG300/CHCl₃ (10/5/85)

P(3HB3HV)/CHCl₃ (10/90)

fabrication by NIPS, a PHA with a high HV content should lead to better mechanical properties.

On the other hand, P(3HB3HV) membranes were also fabricated using the EIPS technique. These membranes were found to have much better mechanical properties and could be used for filtration tests. The influence of the presence of additives (PEG of different molecular weights) into the dope solution on P(3HB3HV) membrane properties was thus evaluated. The properties of the membranes made by EIPS are described below:

Figure 6 shows the SEM images of these membranes while Table 9 displays their performances.

Symmetric structures are observed for these membranes. The membrane made with 10 wt% of P(3HB3HV) presents a low water permeability of $4.2 \text{ L m}^{-2} \text{ h}^{-1} \text{ bar}^{-1}$ and a *E. Coli* bacteria rejection of 95.00%. In order to improve the permeability, PEG300 and PEG8000 were added to bring a porogeneous effect. The porogeneous effect depends on the PEG molecular weight.

For the low molecular weight, PEG300, the obtained effect is the opposite of that expected: the final membrane is denser. Porosity decreases to 4%, and it is not permeable to water anymore. This could be explained by a plasticizing

Table 9 Properties of the P(3HB3HV) based membranes made by EIPS. Influence of the dope solution composition

Dope solution composition (weight %)	Overall porosity (%)	Water permeability	Rejection of *E. Coli* bacteria
P(3HB3HV)/CHCl$_3$ (10/90)	9	4.2 L m^{-2} h^{-1} bar^{-1}	95.00%
P(3HB3HV)/PEG300/CHCl$_3$ (10/5/85)	4	Not permeable	/
P(3HB3HV)/PEG8000/CHCl$_3$ (10/5/85)	37	210 L m^{-2} h^{-1} bar^{-1}	99.95%

effect, which predominates and favor the chains rearrangement.

On the contrary, PEG8000 acts as a porogeneous agent, the membrane porosity increases to 37% and membrane performances are greatly improved. The membrane demonstrates a water permeability of 210 L m^{-2} h^{-1} bar^{-1} and a *E. Coli* bacteria rejection of 99.95%. These results are promising within the prospect of making membranes for microfiltration applications.

4 Conclusion

The development of biobased and biodegradable materials for membrane technologies is part of the current challenges of sustainable development. Among the variously considered biopolymers, the PHAs show valuable properties. These versatile biopolyesters present the great interest to cover a wide variety of physicochemical properties tailored by their copolymers composition. Moreover, these properties can be tuned by blend with other biopolymers or by chemical modifications in order to broader their application areas. While the multiple academic studies suggest a great future for high value-added applications, they also strongly focus on alternatives to produce low-cost PHAs. The use of cheap raw materials as growth media seems to be the most viable solution to expend the market. Once this major drawback is solved, PHAs will be able to challenge conventional polymers for common applications. The investigations on PHAs-based membranes show promising results for pressure-driven separation processes, pervaporation, and air filtration. In particular, our research group has demonstrated that the material microstructure can be tailored by the choice of the proper phase inversion parameters, to give similar structures to what is observed with conventional polymeric membranes. The use of tuned PHAs should be considered to improve the mechanical properties of the PHAs-based membranes. Additionally to the mentioned applications, the poor solubility of PHAs in organic solvents is an attractive opportunity for organic solvents filtrations. Moreover, thanks to their biodegradability, PHAs could benefit the area of the single-use filtration membranes. Besides, the use of greener solvents to make membranes is under investigations since it is considered as the main priority to get more sustainable membranes (Marino et al.

2017; Marino et al. 2019; Milescu et al. 2019; Livingston et al. 2014; Tavajohi et al. 2015). Hence, the association of a biobased/biodegradable material with a green solvent must be thought through the future works. Eventually, this review and analysis on PHAs-based membranes aim to help for further developments in this field and sustainable membrane materials.

References

Ahmed, I., Balkhair, K. S., Albeiruttye, M. H., & Shaiban, A. A. J. (2017). Importance and significance of UF/MF membrane systems in desalination water treatment. In *Desalination*. UK: InTechOpen.

Audic, J.-L., Lemiègre L., Corre Y.-M., (2013). Thermal and mechanical properties of a polyhydroxyalkanoate plasticized with biobased epoxidized broccoli oil. *Journal of Applied Polymer Science, 39983*, 1–7. https://doi.org/10.1002/app.39983.

Barud, H. S., Souza, J. L., Santos, D. B., et al. (2011). Bacterial cellulose/poly (3-hydroxybutyrate) composite membranes. *Carbohydrate Polymers, 83*, 1279–1284. https://doi.org/10.1016/j.carbpol.2010.09.049.

Bhardwaj, U., Dhar, P., Kumar, A., & Katiyar, V. (2014). Polyhydroxyalkanoates (PHA)-cellulose based nanobiocomposites for food packaging applications. In *Food additives and packaging* (pp. 275–314).

Bioplastics. European Bioplastics e.V. https://www.european-bioplastics.org/bioplastics/. Accessed July 15, 2019.

Bittmann, B., Bouza, R., Barral, L., et al. (2013). Poly (3-hydroxybutyrate-*co*-3-hydroxyvalerate)/clay nanocomposites for replacement of mineral oil based materials. *Polymer Composites, 34*, 1033–1040. https://doi.org/10.1002/pc.22510.

Bouyer, D., Faur, C., & Pochat, C. (2011) Procédés d'élaboration de membranes par séparation de phases. Tech L'Ingenieur 33.

Broens, L., Altena, P. W., & Smolders, C. A. (1980). Asymmetric membrane structures as a result of phase separation phenomena. *Desalination, 32*, 33–45. https://doi.org/10.1016/S0011-9164(00)86004-X.

Brophy, M. R., & Deasy, P. B. (1986). In vitro and in vivo studies on biodegradable polyester microparticles containing sulphamethizole. *International Journal of Pharmaceutics, 29*, 223–231. https://doi.org/10.1016/0378-5173(86)90119-5.

Bugnicourt, E., Cinelli, P., Lazzeri, A., & Alvarez, V. (2014). Polyhydroxyalkanoate (PHA): Review of synthesis, characteristics, processing and potential applications in packaging. *Express Polymer Letters, 8*, 791–808. https://doi.org/10.3144/expresspolymlett.2014.82.

Cath, T. Y. (2010). Osmotically and thermally driven membrane processes for enhancement of water recovery in desalination processes. *Desalination and Water Treatment, 15*, 279–286. https://doi.org/10.5004/dwt.2010.1760.

Chan, R. T. H., Marc, H., Ahmed, T., et al. (2013). Poly(ethylene glycol)-modulated cellular biocompatibility of polyhydroxyalkanoate films. *Polymer International, 62,* 884–892. https://doi.org/10.1002/pi.4451.

Chaunier, L., Guessasma, S., Belhabib, S., et al. (2018). Material extrusion of plant biopolymers: Opportunities & challenges for 3D printing. *Additive Manufacturing, 21,* 220–233. https://doi.org/10.1016/j.addma.2018.03.016.

Chee, J.-Y., Yoga, S.-S., Lau, N.-S., et al. (2010). Bacterially produced polyhydroxyalkanoate (PHA): Converting renewable resources into bioplastics. In *Current research technology and education topics in applied microbiology and microbial biotechnology* (pp. 1395–1404).

Chen, G.-Q. (2009). A microbial polyhydroxyalkanoates (PHA) based bio- and materials industry. *Chemical Society Reviews, 38,* 2434. https://doi.org/10.1039/b812677c.

Chen, G.-Q., & Wu, Q. (2005). The application of polyhydroxyalkanoates as tissue engineering materials. *Biomaterials, 26,* 6565–6578. https://doi.org/10.1016/j.biomaterials.2005.04.036.

Cheng, G., Cai, Z., & Wang, L. (2003). Biocompatibility and biodegradation of poly(hydroxybutyrate)/poly(ethylene glycol) blend films. *Journal of Materials Science: Materials in Medicine, 14,* 1073–1078. https://doi.org/10.1023/B:JMSM.0000004004.37103.f4.

Choi, J. S., & Park, W. H. (2004). Effect of biodegradable plasticizers on thermal and mechanical properties of poly(3-hydroxybutyrate). *Polymer Testing, 23,* 455–460. https://doi.org/10.1016/j.polymertesting.2003.09.005.

Corre, Y. M., Bruzaud, S., Audic, J. L., & Grohens, Y. (2012). Morphology and functional properties of commercial polyhydroxyalkanoates: A comprehensive and comparative study. *Polymer Testing, 31,* 226–235. https://doi.org/10.1016/j.polymertesting.2011.11.002.

Dawes, E. A., & Senior, P. J. (1973). The role and regulation of energy reserve polymers in micro-organisms. *Advances in Microbial Physiology, 10,* 135–266. https://doi.org/10.1016/S0065-2911(08)60088-0.

de Koning, G. J. M., & Lemstra, P. J. (1993). Crystallization phenomena in bacterial poly[(R)-3-hydroxybutyrate]: 2. embrittlement and rejuvenation. *Polymer (Guildf), 34,* 4089–4094. https://doi.org/10.1016/0032-3861(93)90671-V.

de Paiva, R. G., de Moraes, M. A., de Godoi, F. C., & Beppu, M. M. (2012). Multilayer biopolymer membranes containing copper for antibacterial applications. *Journal of Applied Polymer Science, 126,* E17–E24. https://doi.org/10.1002/app.36666.

Dias, M., Antunes, M. C. M., Santos, A. R., & Felisberti, M. I. (2008). Blends of poly(3-hydroxybutyrate) and poly(p-dioxanone): Miscibility, thermal stability and biocompatibility. *Journal of Materials Science. Materials in Medicine, 19,* 3535–3544. https://doi.org/10.1007/s10856-008-3531-1.

Díez-Pascual, A. M., & Díez-Vicente, A. L. (2014). ZnO-reinforced poly(3-hydroxybutyrate-co-3-hydroxyvalerate) bionanocomposites with antimicrobial function for food packaging. *ACS Applied Materials & Interfaces, 6,* 9822–9834. https://doi.org/10.1021/am502261e.

Doi, Y. (1995). Microbial synthesis, physical properties, and biodegradability of polyhydroxyalkanoates. *Macromolecular Symposium, 98,* 585–599. https://doi.org/10.1002/masy.19950980150.

Du, C., Sabirova, J., Soetaert, W., & Ki Carol Lin, S. (2012). Polyhydroxyalkanoates production from low-cost sustainable raw materials. *Current Chemical Biology, 6,* 14–25. https://doi.org/10.2174/187231312799984394.

Elain, A., Le Grand, A., Corre, Y.-M., et al. (2016). Valorisation of local agro-industrial processing waters as growth media for polyhydroxyalkanoates (PHA) production. *Industrial Crops and Products, 80,* 1–5. https://doi.org/10.1016/j.indcrop.2015.10.052.

Ellen MacArthur Foundation. (2017). The new plastics economy: Rethinking the future of plastics & catalysing action. https://www.ellenmacarthurfoundation.org/publications/the-new-plastics-economy-rethinking-the-future-of-plastics-catalysing-action. Accessed May 14, 2019.

Follain, N., Chappey, C., Dargent, E., et al. (2014). Structure and barrier properties of biodegradable polyhydroxyalkanoate films. *Journal of Physical Chemistry, 118,* 6165–6177.

Galego, N., Miguens, F. C., & Sánchez, R. (2002). Physical and functional characterization of PHA SCL membranes. *Polymer (Guildf), 43,* 3109–3114.

Galiano, F., Briceño, K., Marino, T., et al. (2018). Advances in biopolymer-based membrane preparation and applications. *Journal of Membrane Science, 564,* 562–586. https://doi.org/10.1016/J.MEMSCI.2018.07.059.

Gigault, J., Pedrono, B., Maxit, B., & Terhalle, A. (2016). Marine plastic litter: The unanalyzed nano-fraction. *Environmental Science: Nano, 3,* 346–350. https://doi.org/10.1039/C6EN00008H.

Grassie, N., Murray, E. J., & Holmes, P. A. (1984a). The thermal degradation of poly(-(d)-β-hydroxybutyric acid): Part 1—Identification and quantitative analysis of products. *Polymer Degradation and Stability, 6,* 47–61. https://doi.org/10.1016/0141-3910(84)90016-8.

Grassie, N., Murray, E. J., & Holmes, P. A. (1984b). The thermal degradation of poly(-(d)-β-hydroxybutyric acid): Part 2—Changes in molecular weight. *Polymer Degradation and Stability, 6,* 95–103. https://doi.org/10.1016/0141-3910(84)90075-2.

Grassie, N., Murray, E. J., & Holmes, P. A. (1984c). The thermal degradation of poly(-(d)-β-hydroxybutyric acid): Part 3—The reaction mechanism. *Polymer Degradation and Stability, 6,* 127–134. https://doi.org/10.1016/0141-3910(84)90032-6.

Guo, J., Zhang, Q., Cai, Z., & Zhao, K. (2016). Preparation and dye filtration property of electrospun polyhydroxybutyrate–calcium alginate/carbon nanotubes composite nanofibrous filtration membrane. *Separation and Purification Technology, 161,* 69–79. https://doi.org/10.1016/j.seppur.2016.01.036.

Gurgel, M., Vieira, A., Altenhofen, M., et al. (2011). Natural-based plasticizers and biopolymer films: A review. *European Polymer Journal, 47,* 254–263. https://doi.org/10.1016/j.eurpolymj.2010.12.011.

Guzman, D., Kirsebom, H., Solano, C., et al. (2011). Preparation of hydrophilic poly(3-hydroxybutyrate) macroporous scaffolds through enzyme-mediated modifications. *Journal of Bioactive and Compatable Polymers, 26,* 452–463. https://doi.org/10.1177/0883911511419970.

Hancock, N. T., Black, N. D., & Cath, T. Y. (2012). A comparative life cycle assessment of hybrid osmotic dilution desalination and established seawater desalination and wastewater reclamation processes. *Water Research, 46,* 1145–1154. https://doi.org/10.1016/j.watres.2011.12.004.

Hartley Yee, L., & Ray Foster, L. J. (2014). Polyhydroxyalkanoates as packaging materials: Current applications and future prospects. In *Polyhydroxyalkanoate (PHA) based blends, composites and nanocomposites* (pp. 183–207). London: Royal Society of Chemistry.

Hennessy, J., Livingston, A., & Baker, R. (2017). Membranes from academia to industry. *Nature Materials, 16,* 280–282. https://doi.org/10.1038/nmat4861.

Ioannou-Ttofa, L., Foteinis, S., Chatzisymeon, E., & Fatta-Kassinos, D. (2016). The environmental footprint of a membrane bioreactor treatment process through life cycle analysis. *Science of the Total Environment, 568,* 306–318. https://doi.org/10.1016/j.scitotenv.2016.06.032.

Jacquel, N., Lo, C.-W., Wu, H.-S., et al. (2007). Solubility of polyhydroxyalkanoates by experiment and thermodynamic correlations. *AIChE Journal, 53,* 2704–2714. https://doi.org/10.1002/aic.11274.

Javadi, A., Kramschuster, A.J., Pilla, S., et al. (2010). Processing and characterization of microcellular PHBV/PBAT blends. *Polymer Engineering Science,* 1440–1448. https://doi.org/10.1002/pen.21661.

Jendrossek, D., Schirmer, A., & Schlegel, H. G. (1996). Biodegradation of polyhydroxyalkanoic acids. *Applied Microbiology and Biotechnology, 46,* 451–463. https://doi.org/10.1007/s002530050844.

Jun, D., Guomin, Z., Mingzhu, P., et al. (2017). Crystallization and mechanical properties of reinforced PHBV composites using melt compounding: Effect of CNCs and CNFs. *Carbohydrate Polymers, 168,* 255–262. https://doi.org/10.1016/j.carbpol.2017.03.076.

Jyothi, M. S., Reddy, K. R., Soontarapa, K., et al. (2019). Membranes for dehydration of alcohols via pervaporation. *Journal of Environmental Management, 242,* 415–429. https://doi.org/10.1016/J.JENVMAN.2019.04.043.

Kang, J., Gi, H., Choe, R., & Il, Y. S. (2016). Fabrication and characterization of poly (3-hydroxybutyrate) gels using non-solvent-induced phase separation. *Polymer (Guildf), 104,* 61–71. https://doi.org/10.1016/j.polymer.2016.09.093.

Keawsupsak, K., Jaiyu, A., Pannoi, J., et al. (2014). Poly(lactic acid)/biodegradable polymer blend for the preparation of flat-sheet membrane. *Jurnal Teknologi, 69.* https://doi.org/10.11113/jt.v69.3405.

Kehail, A. A., & Brigham, C. J. (2017). Anti-biofilm activity of solvent-cast and electrospun polyhydroxyalkanoate membranes treated with lysozyme. *Journal of Polymers and the Environment, 26,* 66–72. https://doi.org/10.1007/s10924-016-0921-1.

Ladewig, B., & Al-Shaeli, M. N. Z. (2017). Fundamentals of membrane processes. In *Fundamentals of membrane bioreactors* (p. 150).

Lalia, B. S., Kochkodan, V., Hashaikeh, R., & Hilal, N. (2013). A review on membrane fabrication: Structure, properties and performance relationship. *Desalination, 326,* 77–95. https://doi.org/10.1016/j.desal.2013.06.016.

Lee, S. Y. (2000). Bacterial polyhydroxyalkanoates. *Biotechnology and Bioengineering, 49,* 1–14. https://doi.org/10.1002/(SICI)1097-0290(19960105)49:1%3c1::AID-BIT1%3e3.0.CO;2-P.

Li, Z., Yang, J., & Loh, X. J. (2016). Polyhydroxyalkanoates: Opening doors for a sustainable future. *NPG Asia Materials, 8,* e265. https://doi.org/10.1038/am.2016.48.

Lim, J., You, M., Li, J., & Li, Z. (2017). Emerging bone tissue engineering via polyhydroxyalkanoate (PHA)-based scaffolds. *Materials Science and Engineering C, 79,* 917–929. https://doi.org/10.1016/j.msec.2017.05.132.

Livingston, A. G., Szekely, G., Jimenez-Solomon, M. F., et al. (2014). Sustainability assessment of organic solvent nanofiltration: From fabrication to application. *Green Chemistry, 16,* 4440. https://doi.org/10.1039/c4gc00701h.

Loeb, S. (1981). The Loeb-Sourirajan membrane: How it came about. In *Synthetic membranes* (pp. 1–9).

Loeb, S., & Sourirajan, S. (1963). Sea water demineralization by means of an osmotic membrane. *Saline Water Conversion—II,* 117–132. https://doi.org/10.1021/ba-1963-0038.ch009.

Loureiro, N., Esteves, J., Viana, J., & Ghosh, S. (2015). Mechanical characterization of polyhydroxyalkanoate and poly(lactic acid) blends. *Journal of Thermoplastic Composite Materials, 28,* 195–213. https://doi.org/10.1177/0892705712475020.

Lucas, N., Bienaime, C., Belloy, C., et al. (2008). Polymer biodegradation: Mechanisms and estimation techniques—A review. *Chemosphere, 73,* 429–442. https://doi.org/10.1016/J.CHEMOSPHERE.2008.06.064.

Lukasiewicz, B., Basnett, P., Nigmatullin, R., et al. (2018). Binary polyhydroxyalkanoate systems for soft tissue engineering. *Acta Biomaterialia, 71,* 225–234. https://doi.org/10.1016/J.ACTBIO.2018.02.027.

Macedonio, F., & Drioli, E. (2017). Membrane engineering for green process engineering. *Engineering, 3,* 290–298. https://doi.org/10.1016/J.ENG.2017.03.026.

Madison, L. L., & Huisman, G. W. (1999). Metabolic engineering of poly(3-hydroxyalkanoates): From DNA to plastic. *Microbiology and Molecular Biology Reviews, 63,* 21–53.

Marcano, A., Ba, O., Thebault, P., & Crétois, R. (2015). Elucidation of innovative antibiofilm materials. *Colloids and Surfaces B: Biointerfaces, 136,* 56–63. https://doi.org/10.1016/j.colsurfb.2015.08.007.

Marcano, A., Bou Haidar, N., Marais, S., et al. (2017). Designing biodegradable PHA-based 3D scaffolds with antibiofilm properties for wound dressings: Optimization of the microstructure/nanostructure. *ACS Biomaterials Science and Engineering, 3,* 3654–3661. https://doi.org/10.1021/acsbiomaterials.7b00552.

Marino, T., Galiano, F., Molino, A., & Figoli, A. (2019). New frontiers in sustainable membrane preparation: Cyrene™ as green bioderived solvent. *Journal of Membrane Science, 580,* 224–234. https://doi.org/10.1016/j.memsci.2019.03.034.

Marino, T., Russo, F., Criscuoli, A., & Figoli, A. (2017). TamiSolve ® NxG as novel solvent for polymeric membrane preparation. *Journal of Membrane Science, 542,* 418–429. https://doi.org/10.1016/j.memsci.2017.08.038.

Martins, A. A., Caetano, N. S., & Mata, T. M. (2017). LCA for membrane processes. In *Green chemistry and sustainable technology* (pp. 23–66).

Mas, A., Jaaba, H., Sledz, J., & Schue, F. (1996). Membranes en PHB, P(HB-*co*-9% HV), P(HB-*co*-22% HV) pour la microfiltration ou la pervaporation propriétés filtrantes et état de surface. *European Polymer Journal, 32,* 435–450. https://doi.org/10.1016/0014-3057(96)80013-9.

Mcchalicher, C. W. J., Srienc, F., & Rouse, D. P. (2009). Solubility and degradation of polyhydroxyalkanoate biopolymers in propylene carbonate. *AIChE Journal, 56,* 1616–1625. https://doi.org/10.1002/aic.12087.

Mergaert, J., Anderson, C., Wouters, A., et al. (1992). Biodegradation of polyhydroxyalkanoates. *FEMS Microbiology Letters, 103,* 317–321. https://doi.org/10.1111/j.1574-6968.1992.tb05853.x.

Miguel, O., Fernandez-Berridi, M. J., & Iruin, J. J. (1997). Survey on transport properties of liquids, vapors, and gases in biodegradable poly(3-hydroxybutyrate) (PHB). *Journal of Applied Polymer Science, 64,* 1849–1859. https://doi.org/10.1002/(SICI)1097-4628(19970531)64:9%3c1849::AID-APP22%3e3.0.CO;2-R.

Milescu, R. A., Mcelroy, C. R., Farmer, T. J., et al. (2019). Fabrication of PES/PVP water filtration membranes using cyrene, a safer bio-based polar aprotic solvent. *Advances in Polymer Technology, 2019,* 9692859. https://doi.org/10.1155/2019/9692859.

Mirabal, A. S., Scholz, L., & Carus, M. (2013). Market study on bio-based polymers in the world capacities, production and applications: Status quo and trends towards 2020, Nova-Institute GmbH, http://bio-based.eu/market_study/media/files/13-06-21MSBiopolymersExcerpt.pdf.

Misra, S. K., Mohn, D., Brunner, T. J., et al. (2008). Comparison of nanoscale and microscale bioactive glass on the properties of P (3HB)/bioglass composites. *Biomaterials, 29,* 1750–1761. https://doi.org/10.1016/j.biomaterials.2007.12.040.

Modi, S., Koelling, K., & Vodovotz, Y. (2011). Assessment of PHB with varying hydroxyvalerate content for potential packaging applications. *European Polymer Journal, 47,* 179–186. https://doi.org/10.1016/j.eurpolymj.2010.11.010.

Mohanty, A. K., Misra, M., & Drzal, L. T. (2002). Sustainable bio-composites from renewable resources: Opportunities and challenges in the green materials world. *Journal of Polymers and the Environment, 10*, 19–26. https://doi.org/10.1023/A:1021013921916.

Montanheiro, T. L., do, A., Cristóvan, F. H., Machado, J. P. B., et al. (2015). Effect of MWCNT functionalization on thermal and electrical properties of PHBV/MWCNT nanocomposites. *Journal of Materials Research, 30*, 55–65. https://doi.org/10.1557/jmr.2014.303.

Mulder, M. (1996). Transport in membranes. In *Basic principles of membrane technology* (pp. 210–279). Netherlands: Springer.

Nicosia, A., Gieparda, W., Foksowicz-Flaczyk, J., et al. (2015). Air filtration and antimicrobial capabilities of electrospun PLA/PHB containing ionic liquid. *Separation and Purification Technology, 154*, 154–160. https://doi.org/10.1016/j.seppur.2015.09.037.

Parra, D. F., Fusaro, J., Gaboardi, F., & Rosa, D. S. (2006). Influence of poly (ethylene glycol) on the thermal, mechanical, morphological, physical–chemical and biodegradation properties of poly (3-hydroxybutyrate). *Polymer Degradation and Stability, 91*, 1954–1959. https://doi.org/10.1016/j.polymdegradstab.2006.02.008.

Plastics Europe. (2018). Pastics—The facts 2018. https://www.plasticseurope.org/fr/resources/market-data. Accessed May 14, 2019.

Puente, J. A. S., Esposito, A., Chivrac, F., & Dargent, E. (2013a). Effects of size and specific surface area of boron nitride particles on the crystallization of bacterial poly(3-hydroxybutyrate-*co*-3-hydroxyvalerate). *Macromolecular Symposium, 328*, 8–19. https://doi.org/10.1002/masy.201350601.

Puente, J. A. S., Esposito, A., Chivrac, F., & Dargent, E. (2013b). Effect of boron nitride as a nucleating agent on the crystallization of bacterial poly(3-hydroxybutyrate). *Journal of Applied Polymer Science, 128*, 2586–2594. https://doi.org/10.1002/app.38182.

Ragaert, P., Buntinx, M., Maes, C., et al. (2019). Polyhydroxyalkanoates for food packaging applications. In *Reference module in food science*. Amsterdam: Elsevier.

Ramachandran, H., Kannusamy, S., Huong, K.-H., et al. (2014). Blends of polyhydroxyalkanoates (PHAs). In *Polyhydroxyalkanoate (PHA) based blends, composites and nanocomposites* (pp. 66–97). London: The Royal Society of Chemistry.

Requena, R., Jiménez, A., Vargas, M., & Chiralt, A. (2016). Effect of plasticizers on thermal and physical properties of compression-moulded poly[(3-hydroxybutyrate)-co-(3-hydroxyvalerate)] films. *Polymer Testing, 56*, 45–53. https://doi.org/10.1016/j.polymertesting.2016.09.022.

Santos Jr, A. R., Ferreira, B. M. P., Duek, E. A. R., et al. (2004). Differentiation pattern of vero cells cultured on poly (L-lactic acid)/poly (hydroxybutyrate-co-hydroxyvalerate) blends. *Artificial Organs, 28*, 381–389.

Seoane, I. T., Luzi, F., Puglia, D., et al. (2018). Enhancement of paperboard performance as packaging material by layering with plasticized polyhydroxybutyrate/nanocellulose coatings. *Journal of Applied Polymer Science, 135*, 46872. https://doi.org/10.1002/app.46872.

Shan, G.-F., Gong, X., Chen, W.-P., et al. (2011). Effect of multi-walled carbon nanotubes on crystallization behavior of poly (3-hydroxybutyrate-co-3-hydroxyvalerate). *Colloid and Polymer Science, 289*, 1005–1014. https://doi.org/10.1007/s00396-011-2412-1.

Shen, L., Haufe, J., Patel, M. K. (2009) Product overview and market projection of emerging bio-based plastics PRO-BIP 2009; Report for European Polysaccharide Network of Excellence (EPNOE) and European Bioplastics; European Polysaccharide Network of Excellence (EPNOE) and European Bioplastics: Utrecht, The Netherlands, 2009.

Shrivastav, A., Kim, H.-Y., & Kim, Y.-R. (2013). Advances in the applications of polyhydroxyalkanoate nanoparticles for novel drug delivery system. *BioMed Research International*. https://doi.org/10.1155/2013/581684.

Steinbüchel, A., & Hein, S. (2001). Biochemical and molecular basis of microbial synthesis of polyhydroxyalkanoates in microorganisms. *Advances in Biochemical Engineering, 71*, 81–123. https://doi.org/10.1007/3-540-40021-4_3.

Strathmann, H. (1981). Membrane separation processes. *Journal of Membrane Science, 9*, 121–189. https://doi.org/10.1016/S0376-7388(00)85121-2.

Szuman, K., Krucińska, I., Boguń, M., & Draczyński, Z. (2016). PLA/PHA-biodegradable blends for pneumothermic fabrication of nonwovens. *Autex Research Journal, 16*, 119–127. https://doi.org/10.1515/aut-2015-0047.

Tae, J., Kim, J. F., Hyun, H., et al. (2016). Understanding the non-solvent induced phase separation (NIPS) effect during the fabrication of microporous PVDF membranes via thermally induced phase separation (TIPS). *Journal of Membrane Science, 514*, 250–263. https://doi.org/10.1016/j.memsci.2016.04.069.

Tan, W. L., Yaakob, N. N., Zainal Abidin, A., et al. (2016). Metal chloride induced formation of porous polyhydroxybutyrate (PHB) films : Morphology, thermal properties and crystallinity. In *IOP Conference Series: Materials Science and Engineering* (Vol. 133). https://doi.org/10.1088/1757-899X/133/1/012012.

Tavajohi, N., Cui, Z., Hoon, J., et al. (2015). Microporous poly (vinylidene fluoride) hollow fiber membranes fabricated with PolarClean as water-soluble green diluent and additives. *Journal of Membrane Science, 479*, 204–212. https://doi.org/10.1016/j.memsci.2015.01.031.

Taylor, P., Lordanskii, A. L., Kamaev, P. P., & Zaikov, G. E. (1998). Water sorption and diffusion in poly (3-hydroxybutyrate) films. *International Journal of Polymeric Materials, 41*, 55–63. https://doi.org/10.1080/00914039808034854.

Terada, M., & Marchessault, R. H. (1999). Determination of solubility parameters for poly (3-hydroxyalkanoates). *International Journal of Biological Macromolecules, 25*, 207–215.

The future of plastic (2018). *Nature Communications, 9*, 2157. https://doi.org/10.1038/s41467-018-04565-2.

Torun Kose, G., Kenar, H., Hasırcı, N., & Hasırcı, V. (2003). Macroporous poly (3-hydroxybutyrate-co-3-hydroxyvalerate) matrices for bone tissue engineering. *Biomaterials, 24*, 1949–1958. https://doi.org/10.1016/S0142-9612(02)00613-0.

Tsai, H.-A., Huang, D.-H., Ruaan, R.-C., & Lai, J.-Y. (2001). Mechanical properties of asymmetric polysulfone membranes containing surfactant as additives. *Industrial and Engineering Chemistry Research, 40*, 5917–5922. https://doi.org/10.1021/IE010026E.

Tsuge, T. (2002). Metabolic improvements and use of inexpensive carbon sources in microbial production of polyhydroxyalkanoates. *Journal of Bioscience and Bioengineering, 94*, 579–584. https://doi.org/10.1016/S1389-1723(02)80198-0.

Verlinden, R. A. J., Hill, D. J., Kenward, M. A., et al. (2007). Bacterial synthesis of biodegradable polyhydroxyalkanoates. *Journal of Applied Microbiology, 102*, 1437–1449. https://doi.org/10.1111/j.1365-2672.2007.03335.x.

Vidhate, S., Innocentini-Mei, L., & D'Souza, N. A. (2012). Mechanical and electrical multifunctional poly(3-hydroxybutyrate-co-3-hydroxyvalerate)-multiwall carbon nanotube nanocomposites. *Polymer Engineering & Science, 52*, 1367–1374. https://doi.org/10.1002/pen.23084.

Villegas, M., Castro Vidaurre, E. F., Habert, A. C., & Gottifredi, J. C. (2011). Sorption and pervaporation with poly (3-hydroxybutyrate) membranes: Methanol/methyl tertbutyl ether mixtures. *Journal of*

Membrane Science, 367, 103–109. https://doi.org/10.1016/j.memsci.2010.10.051.

Villegas, M., Castro Vidaurre, E. F., & Gottifredi, J. C. (2015). Sorption and pervaporation of methanol/water mixtures with poly (3-hydroxybutyrate) membranes. *Chemical Engineering Research and Design, 94,* 254–265. https://doi.org/10.1016/j.cherd.2014.07.030.

Villegas, M., Romero, A. I., Parentis, M. L., et al. (2016). Acrylic acid plasma polymerized poly(3-hydroxybutyrate) membranes for methanol/MTBE separation by pervaporation. *Chemical Engineering Research and Design, 109,* 234–248. https://doi.org/10.1016/j.cherd.2016.01.018.

Volova, T., Shishatskaya, E., Sevastianov, V., et al. (2003). Results of biomedical investigations of PHB and PHB/PHV fibers. *Biochemical Engineering Journal, 16,* 125–133. https://doi.org/10.1016/S1369-703X(03)00038-X.

Vroman, I., & Tighzert, L. (2009). Biodegradable polymers. *Materials (Basel), 2,* 307–344. https://doi.org/10.3390/ma2020307.

Wang, N., Zhou, Z., Xia, L., et al. (2013). Fabrication and characterization of bioactive β-Ca_2SiO_4/PHBV composite scaffolds. *Materials Science and Engineering C, 33,* 2294–2301. https://doi.org/10.1016/j.msec.2013.01.059.

Wang, Y., Wu, Q., Chen, J., & Chen, G. (2005). Evaluation of three-dimensional scaffolds made of blends of hydroxyapatite and poly (3-hydroxybutyrate-co-3-hydroxyhexanoate) for bone reconstruction. *Biomaterials, 26,* 899–904. https://doi.org/10.1016/j.biomaterials.2004.03.035.

Xi, J., Li, J., Zhu, L., Gong, Y., Zhao, N., & Zhang, X. (2007). Effects of quenching temperature and time on pore diameter of poly (3-hydroxybutyrate-co-3-hydroxyhexanoate) porous scaffolds and MC3T3-E1 osteoblast response to the scaffolds. *Tsinghua Science and Technology, 12,* 366–371. https://doi.org/10.1016/S1007-0214(07)70055-X.

Xiong, Y.-C., Yao, Y.-C., Zhan, X.-Y., & Chen, G.-Q. (2010). Application of polyhydroxyalkanoates nanoparticles as intracellular sustained drug-release vectors. *Journal of Biomaterials Science, Polymer Edition, 21,* 127–140. https://doi.org/10.1163/156856209X410283.

Yoshie, N., Nakasato, K., Fujiwara, M., et al. (2000). Effect of low molecular weight additives on enzymatic degradation of poly (3-hydroxybutyrate). *Polymer (Guildf), 41,* 3227–3234. https://doi.org/10.1016/S0032-3861(99)00547-9.

Yu, H.-Y., Yao, J.-M., Qin, Z.-Y., et al. (2013). Comparison of covalent and noncovalent interactions of carbon nanotubes on the crystallization behavior and thermal properties of poly (3-hydroxybutyrate-co-3-hydroxyvalerate). *Journal of Applied Polymer Science, 130,* 4299–4307. https://doi.org/10.1002/app.39529.

Zhang, J., Shishatskaya, E. I., Volova, T. G., et al. (2018). Polyhydroxyalkanoates (PHA) for therapeutic applications. *Materials Science and Engineering C, 86,* 144–150. https://doi.org/10.1016/J.MSEC.2017.12.035.

Zhang, M., & Thomas, N. L. (2011). Blending polylactic acid with polyhydroxybutyrate: The effect on thermal, mechanical, and biodegradation properties. *Advances in Polymer Technology, 30,* 67–79. https://doi.org/10.1002/adv.20235.

Zhang, Y., Almodovar-Arbelo, N. E., Weidman, J. L., et al. (2018) Fit-for-purpose block polymer membranes molecularly engineered for water treatment. *npj Clean Water, 1,* 2. https://doi.org/10.1038/s41545-018-0002-1.

Zhu, Y., Romain, C., & Williams, C. K. (2016). Sustainable polymers from renewable resources. *Nature, 540,* 354–362. https://doi.org/10.1038/nature21001.